T0342406

REAL ANALYSIS

Princeton Lectures in Analysis

I Fourier Analysis: An Introduction

II Complex Analysis

III Real Analysis: Measure Theory, Integration, and Hilbert Spaces

IV Functional Analysis: Introduction to Further Topics in Analysis

PRINCETON LECTURES IN ANALYSIS

III

REAL ANALYSIS

MEASURE THEORY, INTEGRATION, AND HILBERT SPACES

Elias M. Stein

&

Rami Shakarchi

PRINCETON UNIVERSITY PRESS
PRINCETON AND OXFORD

Published by Princeton University Press, 41 William Street,
Princeton, New Jersey 08540
In the United Kingdom: Princeton University Press,
3 Market Place, Woodstock, Oxfordshire OX20 1SY

ISBN 0-691-11386-6
LCCN 2004114065

British Library Cataloging-in-Publication Data is available

The publisher would like to acknowledge the authors of this volume for
providing the camera-ready copy from which this book was printed

Printed on acid-free paper. ∞
www.pupress.princeton.edu
Printed in the United States of America

13 15 17 19 20 18 16 14 12

ISBN-13: 978-0-691-11386-9 (cl.)

ISBN-10: 0-691-11386-6 (cl.)

To my grandchildren
CAROLYN, ALISON, JASON

E.M.S.

To my parents
MOHAMED & MIREILLE
AND MY BROTHER
KARIM

R.S.

Foreword

Beginning in the spring of 2000, a series of four one-semester courses were taught at Princeton University whose purpose was to present, in an integrated manner, the core areas of analysis. The objective was to make plain the organic unity that exists between the various parts of the subject, and to illustrate the wide applicability of ideas of analysis to other fields of mathematics and science. The present series of books is an elaboration of the lectures that were given.

While there are a number of excellent texts dealing with individual parts of what we cover, our exposition aims at a different goal: presenting the various sub-areas of analysis not as separate disciplines, but rather as highly interconnected. It is our view that seeing these relations and their resulting synergies will motivate the reader to attain a better understanding of the subject as a whole. With this outcome in mind, we have concentrated on the main ideas and theorems that have shaped the field (sometimes sacrificing a more systematic approach), and we have been sensitive to the historical order in which the logic of the subject developed.

We have organized our exposition into four volumes, each reflecting the material covered in a semester. Their contents may be broadly summarized as follows:

I. Fourier series and integrals.

II. Complex analysis.

III. Measure theory, Lebesgue integration, and Hilbert spaces.

IV. A selection of further topics, including functional analysis, distributions, and elements of probability theory.

However, this listing does not by itself give a complete picture of the many interconnections that are presented, nor of the applications to other branches that are highlighted. To give a few examples: the elements of (finite) Fourier series studied in Book I, which lead to Dirichlet characters, and from there to the infinitude of primes in an arithmetic progression; the X-ray and Radon transforms, which arise in a number of problems in Book I, and reappear in Book III to play an important role in understanding Besicovitch-like sets in two and three dimensions; Fatou's theorem, which guarantees the existence of boundary values of bounded holomorphic functions in the disc, and whose proof relies on ideas developed in each of the first three books; and the theta function, which first occurs in Book I in the solution of the heat equation, and is then used in Book II to find the number of ways an integer can be represented as the sum of two or four squares, and in the analytic continuation of the zeta function.

A few further words about the books and the courses on which they were based. These courses where given at a rather intensive pace, with 48 lecture-hours a semester. The weekly problem sets played an indispensable part, and as a result exercises and problems have a similarly important role in our books. Each chapter has a series of "Exercises" that are tied directly to the text, and while some are easy, others may require more effort. However, the substantial number of hints that are given should enable the reader to attack most exercises. There are also more involved and challenging "Problems"; the ones that are most difficult, or go beyond the scope of the text, are marked with an asterisk.

Despite the substantial connections that exist between the different volumes, enough overlapping material has been provided so that each of the first three books requires only minimal prerequisites: acquaintance with elementary topics in analysis such as limits, series, differentiable functions, and Riemann integration, together with some exposure to linear algebra. This makes these books accessible to students interested in such diverse disciplines as mathematics, physics, engineering, and finance, at both the undergraduate and graduate level.

It is with great pleasure that we express our appreciation to all who have aided in this enterprise. We are particularly grateful to the students who participated in the four courses. Their continuing interest, enthusiasm, and dedication provided the encouragement that made this project possible. We also wish to thank Adrian Banner and José Luis Rodrigo for their special help in running the courses, and their efforts to see that the students got the most from each class. In addition, Adrian Banner also made valuable suggestions that are incorporated in the text.

We wish also to record a note of special thanks for the following individuals: Charles Fefferman, who taught the first week (successfully launching the whole project!); Paul Hagelstein, who in addition to reading part of the manuscript taught several weeks of one of the courses, and has since taken over the teaching of the second round of the series; and Daniel Levine, who gave valuable help in proof-reading. Last but not least, our thanks go to Gerree Pecht, for her consummate skill in typesetting and for the time and energy she spent in the preparation of all aspects of the lectures, such as transparencies, notes, and the manuscript.

We are also happy to acknowledge our indebtedness for the support we received from the 250th Anniversary Fund of Princeton University, and the National Science Foundation's VIGRE program.

Elias M. Stein

Rami Shakarchi

Princeton, New Jersey

August 2002

In this third volume we establish the basic facts concerning measure theory and integration. This allows us to reexamine and develop further several important topics that arose in the previous volumes, as well as to introduce a number of other subjects of substantial interest in analysis. To aid the interested reader, we have starred sections that contain more advanced material. These can be omitted on first reading. We also want to take this opportunity to thank Daniel Levine for his continuing help in proof-reading and the many suggestions he made that are incorporated in the text.

November 2004

Contents

Introduction

I turn away in fright and horror from this lamentable
plague of functions that do not have derivatives.

C. Hermite, 1893

Starting in about 1870 a revolutionary change in the conceptual framework of analysis began to take shape, one that ultimately led to a vast transformation and generalization of the understanding of such basic objects as functions, and such notions as continuity, differentiability, and integrability.

The earlier view that the relevant functions in analysis were given by formulas or other "analytic" expressions, that these functions were by their nature continuous (or nearly so), that by necessity such functions had derivatives for most points, and moreover these were integrable by the accepted methods of integration — all of these ideas began to give way under the weight of various examples and problems that arose in the subject, which could not be ignored and required new concepts to be understood. Parallel with these developments came new insights that were at once both more geometric and more abstract: a clearer understanding of the nature of curves, their rectifiability and their extent; also the beginnings of the theory of sets, starting with subsets of the line, the plane, etc., and the "measure" that could be assigned to each.

That is not to say that there was not considerable resistance to the change of point-of-view that these advances required. Paradoxically, some of the leading mathematicians of the time, those who should have been best able to appreciate the new departures, were among the ones who were most skeptical. That the new ideas ultimately won out can be understood in terms of the many questions that could now be addressed. We shall describe here, somewhat imprecisely, several of the most significant such problems.

1 Fourier series: completion

Whenever f is a (Riemann) integrable function on $[-\pi, \pi]$ we defined in Book I its Fourier series $f \sim \sum a_n e^{inx}$ by

(1)
$$a_n = \frac{1}{2\pi} \int_{-\pi}^{\pi} f(x) e^{-inx} \, dx,$$

and saw then that one had Parseval's identity,

$$\sum_{n=-\infty}^{\infty} |a_n|^2 = \frac{1}{2\pi} \int_{-\pi}^{\pi} |f(x)|^2 \, dx.$$

However, the above relationship between functions and their Fourier coefficients is not completely reciprocal when limited to Riemann integrable functions. Thus if we consider the space $\mathcal{R}$ of such functions with its square norm, and the space $\ell^2(\mathbb{Z})$ with its norm,[1] each element f in $\mathcal{R}$ assigns a corresponding element $\{a_n\}$ in $\ell^2(\mathbb{Z})$, and the two norms are identical. However, it is easy to construct elements in $\ell^2(\mathbb{Z})$ that do not correspond to functions in $\mathcal{R}$. Note also that the space $\ell^2(\mathbb{Z})$ is *complete* in its norm, while $\mathcal{R}$ is not.[2] Thus we are led to two questions:

(i) What are the putative "functions" f that arise when we complete $\mathcal{R}$? In other words: given an arbitrary sequence $\{a_n\} \in \ell^2(\mathbb{Z})$ what is the nature of the (presumed) function f corresponding to these coefficients?

(ii) How do we integrate such functions f (and in particular verify (1))?

2 Limits of continuous functions

Suppose $\{f_n\}$ is a sequence of continuous functions on $[0, 1]$. We assume that $\lim_{n \to \infty} f_n(x) = f(x)$ exists for every x, and inquire as to the nature of the limiting function f.

If we suppose that the convergence is uniform, matters are straightforward and f is then everywhere continuous. However, once we drop the assumption of uniform convergence, things may change radically and the issues that arise can be quite subtle. An example of this is given by the fact that one can construct a sequence of continuous functions $\{f_n\}$ converging everywhere to f so that

---

[1] We use the notation of Chapter 3 in Book I.
[2] See the discussion surrounding Theorem 1.1 in Section 1, Chapter 3 of Book I.

(a) $0 \le f_n(x) \le 1$ for all x.

(b) The sequence $f_n(x)$ is montonically decreasing as $n \to \infty$.

(c) The limiting function f is not Riemann integrable.[3]

However, in view of (a) and (b), the sequence $\int_0^1 f_n(x)\,dx$ converges to a limit. So it is natural to ask: what method of integration can be used to integrate f and obtain that for it

$$\int_0^1 f(x)\,dx = \lim_{n \to \infty} \int_0^1 f_n(x)\,dx\,?$$

It is with Lebesgue integration that we can solve both this problem and the previous one.

3 Length of curves

The study of curves in the plane and the calculation of their lengths are among the first issues dealt with when one learns calculus. Suppose we consider a continuous curve Γ in the plane, given parametrically by $\Gamma = \{(x(t), y(t))\}$, $a \le t \le b$, with x and y continuous functions of t. We define the *length* of Γ in the usual way: as the supremum of the lengths of all polygonal lines joining successively finitely many points of Γ, taken in order of increasing t. We say that Γ is *rectifiable* if its length L is finite. When $x(t)$ and $y(t)$ are continuously differentiable we have the well-known formula,

$$(2) \qquad\qquad L = \int_a^b \left((x'(t))^2 + (y'(t))^2 \right)^{1/2}\,dt.$$

The problems we are led to arise when we consider general curves. More specifically, we can ask:

(i) What are the conditions on the functions $x(t)$ and $y(t)$ that guarantee the rectifiability of Γ?

(ii) When these are satisfied, does the formula (2) hold?

The first question has a complete answer in terms of the notion of functions of "bounded variation." As to the second, it turns out that if x and y are of bounded variation, the integral (2) is always meaningful; however, the equality fails in general, but can be restored under appropriate reparametrization of the curve Γ.

---

[3]The limit f can be highly discontinuous. See, for instance, Exercise 10 in Chapter 1.

There are further issues that arise. Rectifiable curves, because they are endowed with length, are genuinely one-dimensional in nature. Are there (non-rectifiable) curves that are two-dimensional? We shall see that, indeed, there are continuous curves in the plane that fill a square, or more generally have any dimension between 1 and 2, if the notion of fractional dimension is appropriately defined.

4 Differentiation and integration

The so-called "fundamental theorem of the calculus" expresses the fact that differentiation and integration are inverse operations, and this can be stated in two different ways, which we abbreviate as follows:

$$(3) \qquad F(b) - F(a) = \int_a^b F'(x)\,dx,$$

$$(4) \qquad \frac{d}{dx} \int_0^x f(y)\,dy = f(x).$$

For the first assertion, the existence of continuous functions F that are nowhere differentiable, or for which $F'(x)$ exists for every x, but F' is not integrable, leads to the problem of finding a general class of the F for which (3) is valid. As for (4), the question is to formulate properly and establish this assertion for the general class of integrable functions f that arise in the solution of the first two problems considered above. These questions can be answered with the help of certain "covering" arguments, and the notion of absolute continuity.

5 The problem of measure

To put matters clearly, the fundamental issue that must be understood in order to try to answer all the questions raised above is the problem of measure. Stated (imprecisely) in its version in two dimensions, it is the problem of assigning to each subset E of $\mathbb{R}^2$ its two-dimensional measure $m_2(E)$, that is, its "area," extending the standard notion defined for elementary sets. Let us instead state more precisely the analogous problem in one dimension, that of constructing one-dimensional measure $m_1 = m$, which generalizes the notion of length in $\mathbb{R}$.

We are looking for a non-negative function m defined on the family of subsets E of $\mathbb{R}$ that we allow to be extended-valued, that is, to take on the value $+\infty$. We require:

(a) $m(E) = b - a$ if E is the interval $[a, b]$, $a \leq b$, of length $b - a$.

(b) $m(E) = \sum_{n=1}^{\infty} m(E_n)$ whenever $E = \bigcup_{n=1}^{\infty} E_n$ and the sets E_n are disjoint.

Condition (b) is the "countable additivity" of the measure m. It implies the special case:

(b') $m(E_1 \cup E_2) = m(E_1) + m(E_2)$ if E_1 and E_2 are disjoint.

However, to apply the many limiting arguments that arise in the theory the general case (b) is indispensable, and (b') by itself would definitely be inadequate.

To the axioms (a) and (b) one adds the translation-invariance of m, namely

(c) $m(E + h) = m(E)$, for every $h \in \mathbb{R}$.

A basic result of the theory is the existence (and uniqueness) of such a measure, Lebesgue measure, when one limits oneself to a class of reasonable sets, those which are "measurable." This class of sets is closed under countable unions, intersections, and complements, and contains the open sets, the closed sets, and so forth.[4]

It is with the construction of this measure that we begin our study. From it will flow the general theory of integration, and in particular the solutions of the problems discussed above.

A chronology
We conclude this introduction by listing some of the signal events that marked the early development of the subject.

1872 – Weierstrass's construction of a nowhere differentiable function.

1881 – Introduction of functions of bounded variation by Jordan and later (1887) connection with rectifiability.

1883 – Cantor's ternary set.

1890 – Construction of a space-filling curve by Peano.

1898 – Borel's measurable sets.

1902 – Lebesgue's theory of measure and integration.

1905 – Construction of non-measurable sets by Vitali.

1906 – Fatou's application of Lebesgue theory to complex analysis.

---

[4]There is no such measure on the class of all subsets, since there exist non-measurable sets. See the construction of such a set at the end of Section 3, Chapter 1.

1 Measure Theory

> The sets whose measure we can define by virtue of the preceding ideas we will call measurable sets; we do this without intending to imply that it is not possible to assign a measure to other sets.
>
> *E. Borel, 1898*

This chapter is devoted to the construction of Lebesgue measure in $\mathbb{R}^d$ and the study of the resulting class of measurable functions. After some preliminaries we pass to the first important definition, that of exterior measure for any subset E of $\mathbb{R}^d$. This is given in terms of approximations by unions of cubes that cover E. With this notion in hand we can define measurability and thus restrict consideration to those sets that are measurable. We then turn to the fundamental result: the collection of measurable sets is closed under complements and countable unions, and the measure is additive if the subsets in the union are disjoint.

The concept of measurable functions is a natural outgrowth of the idea of measurable sets. It stands in the same relation as the concept of continuous functions does to open (or closed) sets. But it has the important advantage that the class of measurable functions is closed under pointwise limits.

1 Preliminaries

We begin by discussing some elementary concepts which are basic to the theory developed below.

The main idea in calculating the "volume" or "measure" of a subset of $\mathbb{R}^d$ consists of approximating this set by unions of other sets whose geometry is simple and whose volumes are known. It is convenient to speak of "volume" when referring to sets in $\mathbb{R}^d$; but in reality it means "area" in the case $d = 2$ and "length" in the case $d = 1$. In the approach given here we shall use rectangles and cubes as the main building blocks of the theory: in $\mathbb{R}$ we use intervals, while in $\mathbb{R}^d$ we take products of intervals. In all dimensions rectangles are easy to manipulate and have a standard notion of volume that is given by taking the product of the length of all sides.

Next, we prove two simple theorems that highlight the importance of these rectangles in the geometry of open sets: in $\mathbb{R}$ every open set is a countable union of disjoint open intervals, while in $\mathbb{R}^d$, $d \geq 2$, every open set is "almost" the disjoint union of closed cubes, in the sense that only the boundaries of the cubes can overlap. These two theorems motivate the definition of exterior measure given later.

We shall use the following standard notation. A **point** $x \in \mathbb{R}^d$ consists of a d-tuple of real numbers

$$x = (x_1, x_2, \ldots, x_d), \qquad x_i \in \mathbb{R}, \text{ for } i = 1, \ldots, d.$$

Addition of points is componentwise, and so is multiplication by a real scalar. The **norm** of x is denoted by $|x|$ and is defined to be the standard Euclidean norm given by

$$|x| = \left(x_1^2 + \cdots + x_d^2 \right)^{1/2}.$$

The **distance** between two points x and y is then simply $|x - y|$.

The **complement** of a set E in $\mathbb{R}^d$ is denoted by E^c and defined by

$$E^c = \{x \in \mathbb{R}^d : x \notin E\}.$$

If E and F are two subsets of $\mathbb{R}^d$, we denote the complement of F in E by

$$E - F = \{x \in \mathbb{R}^d : x \in E \text{ and } x \notin F\}.$$

The **distance** between two sets E and F is defined by

$$d(E, F) = \inf |x - y|,$$

where the infimum is taken over all $x \in E$ and $y \in F$.

Open, closed, and compact sets

The **open ball** in $\mathbb{R}^d$ centered at x and of radius r is defined by

$$B_r(x) = \{y \in \mathbb{R}^d : |y - x| < r\}.$$

A subset $E \subset \mathbb{R}^d$ is **open** if for every $x \in E$ there exists $r > 0$ with $B_r(x) \subset E$. By definition, a set is **closed** if its complement is open.

We note that any (not necessarily countable) union of open sets is open, while in general the intersection of only finitely many open sets is open. A similar statement holds for the class of closed sets, if one interchanges the roles of unions and intersections.

A set E is **bounded** if it is contained in some ball of finite radius. A bounded set is **compact** if it is also closed. Compact sets enjoy the Heine-Borel covering property:

- Assume E is compact, $E \subset \bigcup_\alpha \mathcal{O}_\alpha$, and each $\mathcal{O}_\alpha$ is open. Then there are finitely many of the open sets, $\mathcal{O}_{\alpha_1}, \mathcal{O}_{\alpha_2}, \ldots, \mathcal{O}_{\alpha_N}$, such that $E \subset \bigcup_{j=1}^{N} \mathcal{O}_{\alpha_j}$.

In words, *any* covering of a compact set by a collection of open sets contains a *finite* subcovering.

A point $x \in \mathbb{R}^d$ is a **limit point** of the set E if for every $r > 0$, the ball $B_r(x)$ contains points of E. This means that there are points in E which are arbitrarily close to x. An **isolated point** of E is a point $x \in E$ such that there exists an $r > 0$ where $B_r(x) \cap E$ is equal to $\{x\}$.

A point $x \in E$ is an **interior point** of E if there exists $r > 0$ such that $B_r(x) \subset E$. The set of all interior points of E is called the **interior** of E. Also, the **closure** $\overline{E}$ of the E consists of the union of E and all its limit points. The **boundary** of a set E, denoted by ∂E, is the set of points which are in the closure of E but not in the interior of E.

Note that the closure of a set is a closed set; every point in E is a limit point of E; and a set is closed if and only if it contains all its limit points. Finally, a closed set E is **perfect** if E does not have any isolated points.

Rectangles and cubes

A (closed) **rectangle** R in $\mathbb{R}^d$ is given by the product of d one-dimensional closed and bounded intervals

$$R = [a_1, b_1] \times [a_2, b_2] \times \cdots \times [a_d, b_d],$$

where $a_j \leq b_j$ are real numbers, $j = 1, 2, \ldots, d$. In other words, we have

$$R = \{(x_1, \ldots, x_d) \in \mathbb{R}^d : a_j \leq x_j \leq b_j \quad \text{for all } j = 1, 2, \ldots, d\}.$$

We remark that in our definition, a rectangle is *closed* and has sides *parallel* to the coordinate axis. In $\mathbb{R}$, the rectangles are precisely the closed and bounded intervals, while in $\mathbb{R}^2$ they are the usual four-sided rectangles. In $\mathbb{R}^3$ they are the closed parallelepipeds.

We say that the lengths of the sides of the rectangle R are $b_1 - a_1, \ldots, b_d - a_d$. The **volume** of the rectangle R is denoted by $|R|$, and

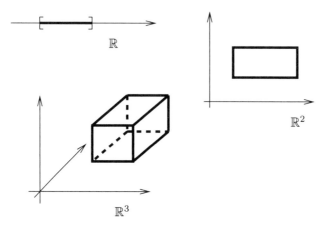

Figure 1. Rectangles in $\mathbb{R}^d$, $d = 1, 2, 3$

is defined to be

$$|R| = (b_1 - a_1) \cdots (b_d - a_d).$$

Of course, when $d = 1$ the "volume" equals length, and when $d = 2$ it equals area.

An open rectangle is the product of open intervals, and the interior of the rectangle R is then

$$(a_1, b_1) \times (a_2, b_2) \times \cdots \times (a_d, b_d).$$

Also, a **cube** is a rectangle for which $b_1 - a_1 = b_2 - a_2 = \cdots = b_d - a_d$. So if $Q \subset \mathbb{R}^d$ is a cube of common side length ℓ, then $|Q| = \ell^d$.

A union of rectangles is said to be **almost disjoint** if the interiors of the rectangles are disjoint.

In this chapter, coverings by rectangles and cubes play a major role, so we isolate here two important lemmas.

Lemma 1.1 *If a rectangle is the almost disjoint union of finitely many other rectangles, say $R = \bigcup_{k=1}^{N} R_k$, then*

$$|R| = \sum_{k=1}^{N} |R_k|.$$

Proof. We consider the grid formed by extending indefinitely the sides of all rectangles $R_1, \ldots, R_N$. This construction yields finitely many rectangles $\tilde{R}_1, \ldots, \tilde{R}_M$, and a partition $J_1, \ldots, J_N$ of the integers between 1 and M, such that the unions

$$R = \bigcup_{j=1}^{M} \tilde{R}_j \quad \text{and} \quad R_k = \bigcup_{j \in J_k} \tilde{R}_j, \quad \text{for } k = 1, \ldots, N$$

are almost disjoint (see the illustration in Figure 2).

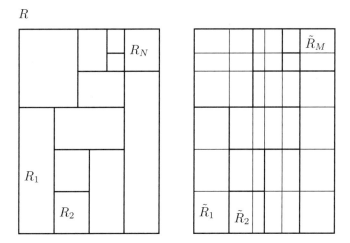

Figure 2. The grid formed by the rectangles R_k

For the rectangle R, for example, we see that $|R| = \sum_{j=1}^{M} |\tilde{R}_j|$, since the grid actually partitions the sides of R and each $\tilde{R}_j$ consists of taking products of the intervals in these partitions. Thus when adding the volumes of the $\tilde{R}_j$ we are summing the corresponding products of lengths of the intervals that arise. Since this also holds for the other rectangles $R_1, \ldots, R_N$, we conclude that

$$|R| = \sum_{j=1}^{M} |\tilde{R}_j| = \sum_{k=1}^{N} \sum_{j \in J_k} |\tilde{R}_j| = \sum_{k=1}^{N} |R_k|.$$

A slight modification of this argument then yields the following:

Lemma 1.2 *If $R, R_1, \ldots, R_N$ are rectangles, and $R \subset \bigcup_{k=1}^{N} R_k$, then*

$$|R| \leq \sum_{k=1}^{N} |R_k|.$$

The main idea consists of taking the grid formed by extending all sides of the rectangles $R, R_1, \ldots, R_N$, and noting that the sets corresponding to the J_k (in the above proof) need not be disjoint any more.

We now proceed to give a description of the structure of open sets in terms of cubes. We begin with the case of $\mathbb{R}$.

Theorem 1.3 *Every open subset $\mathcal{O}$ of $\mathbb{R}$ can be writen uniquely as a countable union of disjoint open intervals.*

Proof. For each $x \in \mathcal{O}$, let I_x denote the largest open interval containing x and contained in $\mathcal{O}$. More precisely, since $\mathcal{O}$ is open, x is contained in some small (non-trivial) interval, and therefore if

$$a_x = \inf\{a < x : (a, x) \subset \mathcal{O}\} \quad \text{and} \quad b_x = \sup\{b > x : (x, b) \subset \mathcal{O}\}$$

we must have $a_x < x < b_x$ (with possibly infinite values for a_x and b_x). If we now let $I_x = (a_x, b_x)$, then by construction we have $x \in I_x$ as well as $I_x \subset \mathcal{O}$. Hence

$$\mathcal{O} = \bigcup_{x \in \mathcal{O}} I_x.$$

Now suppose that two intervals I_x and I_y intersect. Then their union (which is also an open interval) is contained in $\mathcal{O}$ and contains x. Since I_x is maximal, we must have $(I_x \cup I_y) \subset I_x$, and similarly $(I_x \cup I_y) \subset I_y$. This can happen only if $I_x = I_y$; therefore, any two distinct intervals in the collection $\mathcal{I} = \{I_x\}_{x \in \mathcal{O}}$ must be disjoint. The proof will be complete once we have shown that there are only countably many distinct intervals in the collection $\mathcal{I}$. This, however, is easy to see, since every open interval I_x contains a rational number. Since different intervals are disjoint, they must contain distinct rationals, and therefore $\mathcal{I}$ is countable, as desired.

Naturally, if $\mathcal{O}$ is open and $\mathcal{O} = \bigcup_{j=1}^{\infty} I_j$, where the I_j's are disjoint open intervals, the measure of $\mathcal{O}$ ought to be $\sum_{j=1}^{\infty} |I_j|$. Since this representation is unique, we could take this as a definition of measure; we would then note that whenever $\mathcal{O}_1$ and $\mathcal{O}_2$ are open and disjoint, the measure of their union is the sum of their measures. Although this provides a natural notion of measure for an open set, it is not immediately clear how to generalize it to other sets in $\mathbb{R}$. Moreover, a similar approach in higher dimensions already encounters complications even when defining measures of open sets, since in this context the direct analogue of Theorem 1.3 is not valid (see Exercise 12). There is, however, a substitute result.

Theorem 1.4 *Every open subset $\mathcal{O}$ of $\mathbb{R}^d$, $d \geq 1$, can be written as a countable union of almost disjoint closed cubes.*

Proof. We must construct a countable collection $\mathcal{Q}$ of closed cubes whose interiors are disjoint, and so that $\mathcal{O} = \bigcup_{Q \in \mathcal{Q}} Q$.

As a first step, consider the grid in $\mathbb{R}^d$ formed by taking all closed cubes of side length 1 whose vertices have integer coordinates. In other words, we consider the natural grid of lines parallel to the axes, that is, the grid generated by the lattice $\mathbb{Z}^d$. We shall also use the grids formed by cubes of side length 2^{-N} obtained by successively bisecting the original grid.

We either accept or reject cubes in the initial grid as part of $\mathcal{Q}$ according to the following rule: if Q is entirely contained in $\mathcal{O}$ then we accept Q; if Q intersects both $\mathcal{O}$ and $\mathcal{O}^c$ then we tentatively accept it; and if Q is entirely contained in $\mathcal{O}^c$ then we reject it.

As a second step, we bisect the tentatively accepted cubes into 2^d cubes with side length $1/2$. We then repeat our procedure, by accepting the smaller cubes if they are completely contained in $\mathcal{O}$, tentatively accepting them if they intersect both $\mathcal{O}$ and $\mathcal{O}^c$, and rejecting them if they are contained in $\mathcal{O}^c$. Figure 3 illustrates these steps for an open set in $\mathbb{R}^2$.

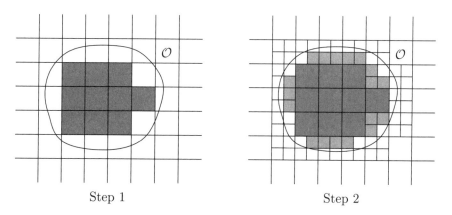

Step 1   Step 2

Figure 3. Decomposition of $\mathcal{O}$ into almost disjoint cubes

This procedure is then repeated indefinitely, and (by construction) the resulting collection $\mathcal{Q}$ of all accepted cubes is countable and consists of almost disjoint cubes. To see why their union is all of $\mathcal{O}$, we note that given $x \in \mathcal{O}$ there exists a cube of side length 2^{-N} (obtained from successive bisections of the original grid) that contains x and that is entirely contained in $\mathcal{O}$. Either this cube has been accepted, or it is contained in a cube that has been previously accepted. This shows that the union of all cubes in $\mathcal{Q}$ covers $\mathcal{O}$.

Once again, if $\mathcal{O} = \bigcup_{j=1}^{\infty} R_j$ where the rectangles R_j are almost disjoint, it is reasonable to assign to $\mathcal{O}$ the measure $\sum_{j=1}^{\infty} |R_j|$. This is natural since the volume of the boundary of each rectangle should be 0, and the overlap of the rectangles should not contribute to the volume of $\mathcal{O}$. We note, however, that the above decomposition into cubes is not unique, and it is not immediate that the sum is independent of this decomposition. So in $\mathbb{R}^d$, with $d \geq 2$, the notion of volume or area, even for open sets, is more subtle.

The general theory developed in the next section actually yields a notion of volume that is consistent with the decompositions of open sets of the previous two theorems, and applies to all dimensions. Before we come to that, we discuss an important example in $\mathbb{R}$.

The Cantor set

The Cantor set plays a prominent role in set theory and in analysis in general. It and its variants provide a rich source of enlightening examples.

We begin with the closed unit interval $C_0 = [0, 1]$ and let C_1 denote the set obtained from deleting the middle third open interval from $[0, 1]$, that is,

$$C_1 = [0, 1/3] \cup [2/3, 1].$$

Next, we repeat this procedure for each sub-interval of C_1; that is, we delete the middle third open interval. At the second stage we get

$$C_2 = [0, 1/9] \cup [2/9, 1/3] \cup [2/3, 7/9] \cup [8/9, 1].$$

We repeat this process for each sub-interval of C_2, and so on (Figure 4).

This procedure yields a sequence C_k, $k = 0, 1, 2, \ldots$ of compact sets with

$$C_0 \supset C_1 \supset C_2 \supset \cdots \supset C_k \supset C_{k+1} \supset \cdots.$$

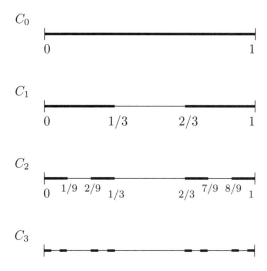

C_0

0   1

C_1

0   1/3   2/3   1

C_2

0 1/9 2/9 1/3   2/3 7/9 8/9 1

C_3

Figure 4. Construction of the Cantor set

The **Cantor set** $\mathcal{C}$ is by definition the intersection of all C_k's:

$$\mathcal{C} = \bigcap_{k=0}^{\infty} C_k.$$

The set $\mathcal{C}$ is not empty, since all end-points of the intervals in C_k (all k) belong to $\mathcal{C}$.

Despite its simple construction, the Cantor set enjoys many interesting topological and analytical properties. For instance, $\mathcal{C}$ is closed and bounded, hence compact. Also, $\mathcal{C}$ is totally disconnected: given any $x, y \in \mathcal{C}$ there exists $z \notin \mathcal{C}$ that lies between x and y. Finally, $\mathcal{C}$ is perfect: it has no isolated points (Exercise 1).

Next, we turn our attention to the question of determining the "size" of $\mathcal{C}$. This is a delicate problem, one that may be approached from different angles depending on the notion of size we adopt. For instance, in terms of cardinality the Cantor set is rather large: it is not countable. Since it can be mapped to the interval $[0, 1]$, the Cantor set has the cardinality of the continuum (Exercise 2).

However, from the point of view of "length" the size of $\mathcal{C}$ is small. Roughly speaking, the Cantor set has length zero, and this follows from the following intuitive argument: the set $\mathcal{C}$ is covered by sets C_k whose lengths go to zero. Indeed, C_k is a disjoint union of 2^k intervals of length

3^{-k}, making the total length of C_k equal to $(2/3)^k$. But $\mathcal{C} \subset C_k$ for all k, and $(2/3)^k \to 0$ as k tends to infinity. We shall define a notion of measure and make this argument precise in the next section.

2 The exterior measure

The notion of exterior measure is the first of two important concepts needed to develop a theory of measure. We begin with the definition and basic properties of exterior measure. Loosely speaking, the exterior measure m_* assigns to *any* subset of $\mathbb{R}^d$ a first notion of size; various examples show that this notion coincides with our earlier intuition. However, the exterior measure lacks the desirable property of additivity when taking the union of disjoint sets. We remedy this problem in the next section, where we discuss in detail the other key concept of measure theory, the notion of measurable sets.

The exterior measure, as the name indicates, attempts to describe the volume of a set E by approximating it from the outside. The set E is covered by cubes, and if the covering gets finer, with fewer cubes overlapping, the volume of E should be close to the sum of the volumes of the cubes.

The precise definition is as follows: if E is *any* subset of $\mathbb{R}^d$, the **exterior measure**[1] of E is

$$(1) \qquad\qquad m_*(E) = \inf \sum_{j=1}^{\infty} |Q_j|,$$

where the infimum is taken over all countable coverings $E \subset \bigcup_{j=1}^{\infty} Q_j$ by closed cubes. The exterior measure is always non-negative but could be infinite, so that in general we have $0 \le m_*(E) \le \infty$, and therefore takes values in the extended positive numbers.

We make some preliminary remarks about the definition of the exterior measure given by (1).

(i) It is important to note that it would not suffice to allow *finite* sums in the definition of $m_*(E)$. The quantity that would be obtained if one considered only coverings of E by finite unions of cubes is in general larger than $m_*(E)$. (See Exercise 14.)

(ii) One can, however, replace the coverings by cubes, with coverings by rectangles; or with coverings by balls. That the former alternative

---

[1] Some authors use the term **outer measure** instead of exterior measure.

yields the same exterior measure is quite direct. (See Exercise 15.) The
equivalence with the latter is more subtle. (See Exercise 26 in Chapter 3.)

We begin our investigation of this new notion by providing examples
of sets whose exterior measures can be calculated, and we check that
the latter matches our intuitive idea of volume (length in one dimension,
area in two dimensions, etc.)

EXAMPLE 1. The exterior measure of a point is zero. This is clear once
we observe that a point is a cube with volume zero, and which covers
itself. Of course the exterior measure of the empty set is also zero.

EXAMPLE 2. The exterior measure of a closed cube is equal to its volume.
Indeed, suppose Q is a closed cube in $\mathbb{R}^d$. Since Q covers itself, we must
have $m_*(Q) \leq |Q|$. Therefore, it suffices to prove the reverse inequality.
We consider an arbitrary covering $Q \subset \bigcup_{j=1}^{\infty} Q_j$ by cubes, and note
that it suffices to prove that

$$(2) \qquad |Q| \leq \sum_{j=1}^{\infty} |Q_j|.$$

For a fixed $\epsilon > 0$ we choose for each j an open cube S_j which contains Q_j,
and such that $|S_j| \leq (1 + \epsilon)|Q_j|$. From the open covering $\bigcup_{j=1}^{\infty} S_j$ of the
compact set Q, we may select a finite subcovering which, after possibly
renumbering the rectangles, we may write as $Q \subset \bigcup_{j=1}^{N} S_j$. Taking the
closure of the cubes S_j, we may apply Lemma 1.2 to conclude that $|Q| \leq \sum_{j=1}^{N} |S_j|$. Consequently,

$$|Q| \leq (1 + \epsilon) \sum_{j=1}^{N} |Q_j| \leq (1 + \epsilon) \sum_{j=1}^{\infty} |Q_j|.$$

Since ϵ is arbitrary, we find that the inequality (2) holds; thus $|Q| \leq m_*(Q)$, as desired.

EXAMPLE 3. If Q is an open cube, the result $m_*(Q) = |Q|$ still holds.
Since Q is covered by its closure $\overline{Q}$, and $|\overline{Q}| = |Q|$, we immediately see
that $m_*(Q) \leq |Q|$. To prove the reverse inequality, we note that if Q_0 is
a closed cube contained in Q, then $m_*(Q_0) \leq m_*(Q)$, since any covering
of Q by a countable number of closed cubes is also a covering of Q_0 (see
Observation 1 below). Hence $|Q_0| \leq m_*(Q)$, and since we can choose Q_0
with a volume as close as we wish to $|Q|$, we must have $|Q| \leq m_*(Q)$.

EXAMPLE 4. The exterior measure of a rectangle R is equal to its volume. Indeed, arguing as in Example 2, we see that $|R| \leq m_*(R)$. To obtain the reverse inequality, consider a grid in $\mathbb{R}^d$ formed by cubes of side length $1/k$. Then, if $\mathcal{Q}$ consists of the (finite) collection of all cubes entirely contained in R, and $\mathcal{Q}'$ the (finite) collection of all cubes that intersect the complement of R, we first note that $R \subset \bigcup_{Q \in (\mathcal{Q} \cup \mathcal{Q}')} Q$. Also, a simple argument yields

$$\sum_{Q \in \mathcal{Q}} |Q| \leq |R|.$$

Moreover, there are $O(k^{d-1})$ cubes[2] in $\mathcal{Q}'$, and these cubes have volume k^{-d}, so that $\sum_{Q \in \mathcal{Q}'} |Q| = O(1/k)$. Hence

$$\sum_{Q \in (\mathcal{Q} \cup \mathcal{Q}')} |Q| \leq |R| + O(1/k),$$

and letting k tend to infinity yields $m_*(R) \leq |R|$, as desired.

EXAMPLE 5. The exterior measure of $\mathbb{R}^d$ is infinite. This follows from the fact that any covering of $\mathbb{R}^d$ is also a covering of any cube $Q \subset \mathbb{R}^d$, hence $|Q| \leq m_*(\mathbb{R}^d)$. Since Q can have arbitrarily large volume, we must have $m_*(\mathbb{R}^d) = \infty$.

EXAMPLE 6. The Cantor set $\mathcal{C}$ has exterior measure 0. From the construction of $\mathcal{C}$, we know that $\mathcal{C} \subset C_k$, where each C_k is a disjoint union of 2^k closed intervals, each of length 3^{-k}. Consequently, $m_*(\mathcal{C}) \leq (2/3)^k$ for all k, hence $m_*(\mathcal{C}) = 0$.

Properties of the exterior measure

The previous examples and comments provide some intuition underlying the definition of exterior measure. Here, we turn to the further study of m_* and prove five properties of exterior measure that are needed in what follows.

First, we record the following remark that is immediate from the definition of m_*:

---

[2]We remind the reader of the notation $f(x) = O(g(x))$, which means that $|f(x)| \leq C|g(x)|$ for some constant C and all x in a given range. In this particular example, there are fewer than Ck^{d-1} cubes in question, as $k \to \infty$.

- For every $\epsilon > 0$, there exists a covering $E \subset \bigcup_{j=1}^{\infty} Q_j$ with

$$\sum_{j=1}^{\infty} m_*(Q_j) \leq m_*(E) + \epsilon.$$

The relevant properties of exterior measure are listed in a series of observations.

Observation 1 (Monotonicity) *If $E_1 \subset E_2$, then $m_*(E_1) \leq m_*(E_2)$.*

This follows once we observe that any covering of E_2 by a countable collection of cubes is also a covering of E_1.

In particular, monotonicity implies that every bounded subset of $\mathbb{R}^d$ has finite exterior measure.

Observation 2 (Countable sub-additivity) *If $E = \bigcup_{j=1}^{\infty} E_j$, then $m_*(E) \leq \sum_{j=1}^{\infty} m_*(E_j)$.*

First, we may assume that each $m_*(E_j) < \infty$, for otherwise the inequality clearly holds. For any $\epsilon > 0$, the definition of the exterior measure yields for each j a covering $E_j \subset \bigcup_{k=1}^{\infty} Q_{k,j}$ by closed cubes with

$$\sum_{k=1}^{\infty} |Q_{k,j}| \leq m_*(E_j) + \frac{\epsilon}{2^j}.$$

Then, $E \subset \bigcup_{j,k=1}^{\infty} Q_{k,j}$ is a covering of E by closed cubes, and therefore

$$\begin{aligned}
m_*(E) \leq \sum_{j,k} |Q_{k,j}| &= \sum_{j=1}^{\infty} \sum_{k=1}^{\infty} |Q_{k,j}| \\
&\leq \sum_{j=1}^{\infty} \left(m_*(E_j) + \frac{\epsilon}{2^j} \right) \\
&= \sum_{j=1}^{\infty} m_*(E_j) + \epsilon.
\end{aligned}$$

Since this holds true for every $\epsilon > 0$, the second observation is proved.

Observation 3 *If $E \subset \mathbb{R}^d$, then $m_*(E) = \inf m_*(\mathcal{O})$, where the infimum is taken over all open sets $\mathcal{O}$ containing E.*

By monotonicity, it is clear that the inequality $m_*(E) \leq \inf m_*(\mathcal{O})$ holds. For the reverse inequality, let $\epsilon > 0$ and choose cubes Q_j such that $E \subset \bigcup_{j=1}^{\infty} Q_j$, with

$$\sum_{j=1}^{\infty} |Q_j| \leq m_*(E) + \frac{\epsilon}{2}.$$

Let Q_j^0 denote an open cube containing Q_j, and such that $|Q_j^0| \leq |Q_j| + \epsilon/2^{j+1}$. Then $\mathcal{O} = \bigcup_{j=1}^{\infty} Q_j^0$ is open, and by Observation 2

$$m_*(\mathcal{O}) \leq \sum_{j=1}^{\infty} m_*(Q_j^0) = \sum_{j=1}^{\infty} |Q_j^0|$$

$$\leq \sum_{j=1}^{\infty} \left(|Q_j| + \frac{\epsilon}{2^{j+1}} \right)$$

$$\leq \sum_{j=1}^{\infty} |Q_j| + \frac{\epsilon}{2}$$

$$\leq m_*(E) + \epsilon.$$

Hence $\inf m_*(\mathcal{O}) \leq m_*(E)$, as was to be shown.

Observation 4 *If $E = E_1 \cup E_2$, and $d(E_1, E_2) > 0$, then*

$$m_*(E) = m_*(E_1) + m_*(E_2).$$

By Observation 2, we already know that $m_*(E) \leq m_*(E_1) + m_*(E_2)$, so it suffices to prove the reverse inequality. To this end, we first select δ such that $d(E_1, E_2) > \delta > 0$. Next, we choose a covering $E \subset \bigcup_{j=1}^{\infty} Q_j$ by closed cubes, with $\sum_{j=1}^{\infty} |Q_j| \leq m_*(E) + \epsilon$. We may, after subdividing the cubes Q_j, assume that each Q_j has a diameter less than δ. In this case, each Q_j can intersect at most one of the two sets E_1 or E_2. If we denote by J_1 and J_2 the sets of those indices j for which Q_j intersects E_1 and E_2, respectively, then $J_1 \cap J_2$ is empty, and we have

$$E_1 \subset \bigcup_{j \in J_1}^{\infty} Q_j \quad \text{as well as} \quad E_2 \subset \bigcup_{j \in J_2}^{\infty} Q_j.$$

Therefore,

$$m_*(E_1) + m_*(E_2) \le \sum_{j \in J_1} |Q_j| + \sum_{j \in J_2} |Q_j|$$

$$\le \sum_{j=1}^{\infty} |Q_j|$$

$$\le m_*(E) + \epsilon.$$

Since ϵ is arbitrary, the proof of Observation 4 is complete.

Observation 5 *If a set E is the countable union of almost disjoint cubes $E = \bigcup_{j=1}^{\infty} Q_j$, then*

$$m_*(E) = \sum_{j=1}^{\infty} |Q_j|.$$

Let $\tilde{Q}_j$ denote a cube strictly contained in Q_j such that $|Q_j| \le |\tilde{Q}_j| + \epsilon/2^j$, where ϵ is arbitrary but fixed. Then, for every N, the cubes $\tilde{Q}_1, \tilde{Q}_2, \ldots, \tilde{Q}_N$ are disjoint, hence at a finite distance from one another, and repeated applications of Observation 4 imply

$$m_* \left(\bigcup_{j=1}^{N} \tilde{Q}_j \right) = \sum_{j=1}^{N} |\tilde{Q}_j| \ge \sum_{j=1}^{N} \left(|Q_j| - \epsilon/2^j \right).$$

Since $\bigcup_{j=1}^{N} \tilde{Q}_j \subset E$, we conclude that for every integer N,

$$m_*(E) \ge \sum_{j=1}^{N} |Q_j| - \epsilon.$$

In the limit as N tends to infinity we deduce $\sum_{j=1}^{\infty} |Q_j| \le m_*(E) + \epsilon$ for every $\epsilon > 0$, hence $\sum_{j=1}^{\infty} |Q_j| \le m_*(E)$. Therefore, combined with Observation 2, our result proves that we have equality.

This last property shows that if a set can be decomposed into almost disjoint cubes, its exterior measure equals the sum of the volumes of the cubes. In particular, by Theorem 1.4 we see that the exterior measure of an open set equals the sum of the volumes of the cubes in a decomposition, and this coincides with our initial guess. Moreover, this also yields a proof that the sum is independent of the decomposition.

One can see from this that the volumes of simple sets that are calculated by elementary calculus agree with their exterior measure. This assertion can be proved most easily once we have developed the requisite tools in integration theory. (See Chapter 2.) In particular, we can then verify that the exterior measure of a ball (either open or closed) equals its volume.

Despite observations 4 and 5, one *cannot* conclude in general that if $E_1 \cup E_2$ is a disjoint union of subsets of $\mathbb{R}^d$, then

$$(3) \qquad m_*(E_1 \cup E_2) = m_*(E_1) + m_*(E_2).$$

In fact (3) holds when the sets in question are not highly irregular or "pathological" but are measurable in the sense described below.

3 Measurable sets and the Lebesgue measure

The notion of measurability isolates a collection of subsets in $\mathbb{R}^d$ for which the exterior measure satisfies all our desired properties, including additivity (and in fact countable additivity) for disjoint unions of sets.

There are a number of different ways of defining measurability, but these all turn out to be equivalent. Probably the simplest and most intuitive is the following: A subset E of $\mathbb{R}^d$ is **Lebesgue measurable**, or simply **measurable**, if for any $\epsilon > 0$ there exists an open set $\mathcal{O}$ with $E \subset \mathcal{O}$ and

$$m_*(\mathcal{O} - E) \leq \epsilon.$$

This should be compared to Observation 3, which holds for *all* sets E.

If E is measurable, we define its **Lebesgue measure** (or **measure**) $m(E)$ by

$$m(E) = m_*(E).$$

Clearly, the Lebesgue measure inherits all the features contained in Observations 1 - 5 of the exterior measure.

Immediately from the definition, we find:

Property 1 *Every open set in $\mathbb{R}^d$ is measurable.*

Our immediate goal now is to gather various further properties of measurable sets. In particular, we shall prove that the collection of measurable sets behave well under the various operations of set theory: countable unions, countable intersections, and complements.

Property 2 *If $m_*(E) = 0$, then E is measurable. In particular, if F is a subset of a set of exterior measure 0, then F is measurable.*

By Observation 3 of the exterior measure, for every $\epsilon > 0$ there exists an open set $\mathcal{O}$ with $E \subset \mathcal{O}$ and $m_*(\mathcal{O}) \leq \epsilon$. Since $(\mathcal{O} - E) \subset \mathcal{O}$, monotonicity implies $m_*(\mathcal{O} - E) \leq \epsilon$, as desired.

As a consequence of this property, we deduce that the Cantor set $\mathcal{C}$ in Example 6 is measurable and has measure 0.

Property 3 *A countable union of measurable sets is measurable.*

Suppose $E = \bigcup_{j=1}^{\infty} E_j$, where each E_j is measurable. Given $\epsilon > 0$, we may choose for each j an open set $\mathcal{O}_j$ with $E_j \subset \mathcal{O}_j$ and $m_*(\mathcal{O}_j - E_j) \leq \epsilon/2^j$. Then the union $\mathcal{O} = \bigcup_{j=1}^{\infty} \mathcal{O}_j$ is open, $E \subset \mathcal{O}$, and $(\mathcal{O} - E) \subset \bigcup_{j=1}^{\infty}(\mathcal{O}_j - E_j)$, so monotonicity and sub-additivity of the exterior measure imply

$$m_*(\mathcal{O} - E) \leq \sum_{j=1}^{\infty} m_*(\mathcal{O}_j - E_j) \leq \epsilon.$$

Property 4 *Closed sets are measurable.*

First, we observe that it suffices to prove that compact sets are measurable. Indeed, any closed set F can be written as the union of compact sets, say $F = \bigcup_{k=1}^{\infty} F \cap B_k$, where B_k denotes the closed ball of radius k centered at the origin; then Property 3 applies.

So, suppose F is compact (so that in particular $m_*(F) < \infty$), and let $\epsilon > 0$. By Observation 3 we can select an open set $\mathcal{O}$ with $F \subset \mathcal{O}$ and $m_*(\mathcal{O}) \leq m_*(F) + \epsilon$. Since F is closed, the difference $\mathcal{O} - F$ is open, and by Theorem 1.4 we may write this difference as a countable union of almost disjoint cubes

$$\mathcal{O} - F = \bigcup_{j=1}^{\infty} Q_j.$$

For a fixed N, the finite union $K = \bigcup_{j=1}^{N} Q_j$ is compact; therefore $d(K, F) > 0$ (we isolate this little fact in a lemma below). Since $(K \cup F) \subset \mathcal{O}$, Observations 1, 4, and 5 of the exterior measure imply

$$m_*(\mathcal{O}) \geq m_*(F) + m_*(K)$$

$$= m_*(F) + \sum_{j=1}^{N} m_*(Q_j).$$

Hence $\sum_{j=1}^{N} m_*(Q_j) \leq m_*(\mathcal{O}) - m_*(F) \leq \epsilon$, and this also holds in the limit as N tends to infinity. Invoking the sub-additivity property of the exterior measure finally yields

$$m_*(\mathcal{O} - F) \leq \sum_{j=1}^{\infty} m_*(Q_j) \leq \epsilon,$$

as desired.

We digress briefly to complete the above argument by proving the following.

Lemma 3.1 *If F is closed, K is compact, and these sets are disjoint, then $d(F, K) > 0$.*

Proof. Since F is closed, for each point $x \in K$, there exists $\delta_x > 0$ so that $d(x, F) > 3\delta_x$. Since $\bigcup_{x \in K} B_{2\delta_x}(x)$ covers K, and K is compact, we may find a subcover, which we denote by $\bigcup_{j=1}^{N} B_{2\delta_j}(x_j)$. If we let $\delta = \min(\delta_1, \ldots, \delta_N)$, then we must have $d(K, F) \geq \delta > 0$. Indeed, if $x \in K$ and $y \in F$, then for some j we have $|x_j - x| \leq 2\delta_j$, and by construction $|y - x_j| \geq 3\delta_j$. Therefore

$$|y - x| \geq |y - x_j| - |x_j - x| \geq 3\delta_j - 2\delta_j \geq \delta,$$

and the lemma is proved.

Property 5 *The complement of a measurable set is measurable.*

If E is measurable, then for every positive integer n we may choose an open set $\mathcal{O}_n$ with $E \subset \mathcal{O}_n$ and $m_*(\mathcal{O}_n - E) \leq 1/n$. The complement $\mathcal{O}_n^c$ is closed, hence measurable, which implies that the union $S = \bigcup_{n=1}^{\infty} \mathcal{O}_n^c$ is also measurable by Property 3. Now we simply note that $S \subset E^c$, and

$$(E^c - S) \subset (\mathcal{O}_n - E),$$

such that $m_*(E^c - S) \leq 1/n$ for all n. Therefore, $m_*(E^c - S) = 0$, and $E^c - S$ is measurable by Property 2. Therefore E^c is measurable since it is the union of two measurable sets, namely S and $(E^c - S)$.

Property 6 *A countable intersection of measurable sets is measurable.*

This follows from Properties 3 and 5, since

$$\bigcap_{j=1}^{\infty} E_j = \left(\bigcup_{j=1}^{\infty} E_j^c \right)^c.$$

In conclusion, we find that the family of measurable sets is closed under the familiar operations of set theory. We point out that we have shown more than simply closure with respect to finite unions and intersections: we have proved that the collection of measurable sets is closed under *countable* unions and intersections. This passage from finite operations to infinite ones is crucial in the context of analysis. We emphasize, however, that the operations of *uncountable* unions or intersections are not permissible when dealing with measurable sets!

Theorem 3.2 *If $E_1, E_2, \ldots,$ are disjoint measurable sets, and $E = \bigcup_{j=1}^{\infty} E_j$, then*

$$m(E) = \sum_{j=1}^{\infty} m(E_j).$$

Proof. First, we assume further that each E_j is bounded. Then, for each j, by applying the definition of measurability to E_j^c, we can choose a closed subset F_j of E_j with $m_*(E_j - F_j) \leq \epsilon/2^j$. For each fixed N, the sets $F_1, \ldots, F_N$ are compact and disjoint, so that $m\left(\bigcup_{j=1}^{N} F_j\right) = \sum_{j=1}^{N} m(F_j)$. Since $\bigcup_{j=1}^{N} F_j \subset E$, we must have

$$m(E) \geq \sum_{j=1}^{N} m(F_j) \geq \sum_{j=1}^{N} m(E_j) - \epsilon.$$

Letting N tend to infinity, since ϵ was arbitrary we find that

$$m(E) \geq \sum_{j=1}^{\infty} m(E_j).$$

Since the reverse inequality always holds (by sub-additivity in Observation 2), this concludes the proof when each E_j is bounded.

In the general case, we select any sequence of cubes $\{Q_k\}_{k=1}^{\infty}$ that increases to $\mathbb{R}^d$, in the sense that $Q_k \subset Q_{k+1}$ for all $k \geq 1$ and $\bigcup_{k=1}^{\infty} Q_k = \mathbb{R}^d$. We then let $S_1 = Q_1$ and $S_k = Q_k - Q_{k-1}$ for $k \geq 2$. If we define measurable sets by $E_{j,k} = E_j \cap S_k$, then

$$E = \bigcup_{j,k} E_{j,k}.$$

The union above is disjoint and every $E_{j,k}$ is bounded. Moreover $E_j = \bigcup_{k=1}^{\infty} E_{j,k}$, and this union is also disjoint. Putting these facts together, and using what has already been proved, we obtain

$$m(E) = \sum_{j,k} m(E_{j,k}) = \sum_j \sum_k m(E_{j,k}) = \sum_j m(E_j),$$

as claimed.

With this, the countable additivity of the Lebesgue measure on measurable sets has been established. This result provides the necessary connection between the following:

- our primitive notion of volume given by the exterior measure,

- the more refined idea of measurable sets, and

- the countably infinite operations allowed on these sets.

We make two definitions to state succinctly some further consequences. If $E_1, E_2, \ldots$ is a countable collection of subsets of $\mathbb{R}^d$ that increases to E in the sense that $E_k \subset E_{k+1}$ for all k, and $E = \bigcup_{k=1}^{\infty} E_k$, then we write $E_k \nearrow E$.

Similarly, if $E_1, E_2, \ldots$ decreases to E in the sense that $E_k \supset E_{k+1}$ for all k, and $E = \bigcap_{k=1}^{\infty} E_k$, we write $E_k \searrow E$.

Corollary 3.3 *Suppose $E_1, E_2, \ldots$ are measurable subsets of $\mathbb{R}^d$.*

(i) *If $E_k \nearrow E$, then $m(E) = \lim_{N \to \infty} m(E_N)$.*

(ii) *If $E_k \searrow E$ and $m(E_k) < \infty$ for some k, then*

$$m(E) = \lim_{N \to \infty} m(E_N).$$

Proof. For the first part, let $G_1 = E_1$, $G_2 = E_2 - E_1$, and in general $G_k = E_k - E_{k-1}$ for $k \geq 2$. By their construction, the sets G_k are measurable, disjoint, and $E = \bigcup_{k=1}^{\infty} G_k$. Hence

$$m(E) = \sum_{k=1}^{\infty} m(G_k) = \lim_{N \to \infty} \sum_{k=1}^{N} m(G_k) = \lim_{N \to \infty} m\left(\bigcup_{k=1}^{N} G_k \right),$$

and since $\bigcup_{k=1}^{N} G_k = E_N$ we get the desired limit.

For the second part, we may clearly assume that $m(E_1) < \infty$. Let $G_k = E_k - E_{k+1}$ for each k, so that

$$E_1 = E \cup \bigcup_{k=1}^{\infty} G_k$$

is a disjoint union of measurable sets. As a result, we find that

$$m(E_1) = m(E) + \lim_{N \to \infty} \sum_{k=1}^{N-1} (m(E_k) - m(E_{k+1}))$$
$$= m(E) + m(E_1) - \lim_{N \to \infty} m(E_N).$$

Hence, since $m(E_1) < \infty$, we see that $m(E) = \lim_{N \to \infty} m(E_N)$, and the proof is complete.

The reader should note that the second conclusion may fail without the assumption that $m(E_k) < \infty$ for some k. This is shown by the simple example when $E_n = (n, \infty) \subset \mathbb{R}$, for all n.

What follows provides an important geometric and analytic insight into the nature of measurable sets, in terms of their relation to open and closed sets. Its thrust is that, in effect, an arbitrary measurable set can be well approximated by the open sets that contain it, and alternatively, by the closed sets it contains.

Theorem 3.4 *Suppose E is a measurable subset of $\mathbb{R}^d$. Then, for every $\epsilon > 0$:*

(i) *There exists an open set $\mathcal{O}$ with $E \subset \mathcal{O}$ and $m(\mathcal{O} - E) \leq \epsilon$.*

(ii) *There exists a closed set F with $F \subset E$ and $m(E - F) \leq \epsilon$.*

(iii) *If $m(E)$ is finite, there exists a compact set K with $K \subset E$ and $m(E - K) \leq \epsilon$.*

(iv) *If $m(E)$ is finite, there exists a finite union $F = \bigcup_{j=1}^{N} Q_j$ of closed cubes such that*

$$m(E \triangle F) \leq \epsilon.$$

The notation $E \triangle F$ stands for the **symmetric difference** between the sets E and F, defined by $E \triangle F = (E - F) \cup (F - E)$, which consists of those points that belong to only one of the two sets E or F.

Proof. Part (i) is just the definition of measurability. For the second part, we know that E^c is measurable, so there exists an open set $\mathcal{O}$ with $E^c \subset \mathcal{O}$ and $m(\mathcal{O} - E^c) \leq \epsilon$. If we let $F = \mathcal{O}^c$, then F is closed, $F \subset E$, and $E - F = \mathcal{O} - E^c$. Hence $m(E - F) \leq \epsilon$ as desired.

For (iii), we first pick a closed set F so that $F \subset E$ and $m(E - F) \leq \epsilon/2$. For each n, we let B_n denote the ball centered at the origin of radius

n, and define compact sets $K_n = F \cap B_n$. Then $E - K_n$ is a sequence of measurable sets that decreases to $E - F$, and since $m(E) < \infty$, we conclude that for all large n one has $m(E - K_n) \leq \epsilon$.

For the last part, choose a family of closed cubes $\{Q_j\}_{j=1}^{\infty}$ so that

$$E \subset \bigcup_{j=1}^{\infty} Q_j \quad \text{and} \quad \sum_{j=1}^{\infty} |Q_j| \leq m(E) + \epsilon/2.$$

Since $m(E) < \infty$, the series converges and there exists $N > 0$ such that $\sum_{j=N+1}^{\infty} |Q_j| < \epsilon/2$. If $F = \bigcup_{j=1}^{N} Q_j$, then

$$m(E \triangle F) = m(E - F) + m(F - E)$$

$$\leq m \left(\bigcup_{j=N+1}^{\infty} Q_j \right) + m \left(\bigcup_{j=1}^{\infty} Q_j - E \right)$$

$$\leq \sum_{j=N+1}^{\infty} |Q_j| + \sum_{j=1}^{\infty} |Q_j| - m(E)$$

$$\leq \epsilon.$$

Invariance properties of Lebesgue measure

A crucial property of Lebesgue measure in $\mathbb{R}^d$ is its translation-invariance, which can be stated as follows: if E is a measurable set and $h \in \mathbb{R}^d$, then the set $E_h = E + h = \{x + h : x \in E\}$ is also measurable, and $m(E + h) = m(E)$. With the observation that this holds for the special case when E is a cube, one passes to the exterior measure of arbitrary sets E, and sees from the definition of m_* given in Section 2 that $m_*(E_h) = m_*(E)$. To prove the measurability of E_h under the assumption that E is measurable, we note that if $\mathcal{O}$ is open, $\mathcal{O} \supset E$, and $m_*(\mathcal{O} - E) < \epsilon$, then $\mathcal{O}_h$ is open, $\mathcal{O}_h \supset E_h$, and $m_*(\mathcal{O}_h - E_h) < \epsilon$.

In the same way one can prove the relative dilation-invariance of Lebesgue measure. Suppose $\delta > 0$, and denote by δE the set $\{\delta x : x \in E\}$. We can then assert that δE is measurable whenever E is, and $m(\delta E) = \delta^d m(E)$. One can also easily see that Lebesgue measure is reflection-invariant. That is, whenever E is measurable, so is $-E = \{-x : x \in E\}$ and $m(-E) = m(E)$.

Other invariance properties of Lebesgue measure are in Exercise 7 and 8, and Problem 4 of Chapter 2.

σ-algebras and Borel sets

A **σ-algebra** of sets is a collection of subsets of $\mathbb{R}^d$ that is closed under countable unions, countable intersections, and complements.

The collection of all subsets of $\mathbb{R}^d$ is of course a σ-algebra. A more interesting and relevant example consists of all measurable sets in $\mathbb{R}^d$, which we have just shown also forms a σ-algebra.

Another σ-algebra, which plays a vital role in analysis, is the **Borel σ-algebra** in $\mathbb{R}^d$, denoted by $\mathcal{B}_{\mathbb{R}^d}$, which by definition is the smallest σ-algebra that contains all open sets. Elements of this σ-algebra are called **Borel sets**.

The definition of the Borel σ-algebra will be meaningful once we have defined the term "smallest," and shown that such a σ-algebra exists and is unique. The term "smallest" means that if $\mathcal{S}$ is any σ-algebra that contains all open sets in $\mathbb{R}^d$, then necessarily $\mathcal{B}_{\mathbb{R}^d} \subset \mathcal{S}$. Since we observe that any intersection (not necessarily countable) of σ-algebras is again a σ-algebra, we may define $\mathcal{B}_{\mathbb{R}^d}$ as the intersection of all σ-algebras that contain the open sets. This shows the existence and uniqueness of the Borel σ-algebra.

Since open sets are measurable, we conclude that the Borel σ-algebra is contained in the σ-algebra of measurable sets. Naturally, we may ask if this inclusion is strict: do there exist Lebesgue measurable sets which are not Borel sets? The answer is "yes." (See Exercise 35.)

From the point of view of the Borel sets, the Lebesgue sets arise as the **completion** of the σ-algebra of Borel sets, that is, by adjoining all subsets of Borel sets of measure zero. This is an immediate consequence of Corollary 3.5 below.

Starting with the open and closed sets, which are the simplest Borel sets, one could try to list the Borel sets in order of their complexity. Next in order would come countable intersections of open sets; such sets are called G_δ sets. Alternatively, one could consider their complements, the countable union of closed sets, called the F_σ sets.[3]

Corollary 3.5 *A subset E of $\mathbb{R}^d$ is measurable*

  (i) *if and only if E differs from a G_δ by a set of measure zero,*

  (ii) *if and only if E differs from an F_σ by a set of measure zero.*

Proof. Clearly E is measurable whenever it satisfies either (i) or (ii), since the F_σ, G_δ, and sets of measure zero are measurable.

---

[3]The terminology G_δ comes from German "Gebiete" and "Durschnitt"; F_σ comes from French "fermé" and "somme."

Conversely, if E is measurable, then for each integer $n \geq 1$ we may select an open set $\mathcal{O}_n$ that contains E, and such that $m(\mathcal{O}_n - E) \leq 1/n$. Then $S = \bigcap_{n=1}^{\infty} \mathcal{O}_n$ is a G_δ that contains E, and $(S - E) \subset (\mathcal{O}_n - E)$ for all n. Therefore $m(S - E) \leq 1/n$ for all n; hence $S - E$ has exterior measure zero, and is therefore measurable.

For the second implication, we simply apply part (ii) of Theorem 3.4 with $\epsilon = 1/n$, and take the union of the resulting closed sets.

Construction of a non-measurable set

Are all subsets of $\mathbb{R}^d$ measurable? In this section, we answer this question when $d = 1$ by constructing a subset of $\mathbb{R}$ which is *not* measurable.[4] This justifies the conclusion that a satisfactory theory of measure cannot encompass *all* subsets of $\mathbb{R}$.

The construction of a non-measurable set $\mathcal{N}$ uses the axiom of choice, and rests on a simple equivalence relation among real numbers in $[0, 1]$.

We write $x \sim y$ whenever $x - y$ is rational, and note that this is an equivalence relation since the following properties hold:

- $x \sim x$ for every $x \in [0, 1]$

- if $x \sim y$, then $y \sim x$

- if $x \sim y$ and $y \sim z$, then $x \sim z$.

Two equivalence classes either are disjoint or coincide, and $[0, 1]$ is the disjoint union of all equivalence classes, which we write as

$$[0, 1] = \bigcup_{\alpha} \mathcal{E}_\alpha.$$

Now we construct the set $\mathcal{N}$ by choosing exactly one element x_α from each $\mathcal{E}_\alpha$, and setting $\mathcal{N} = \{x_\alpha\}$. This (seemingly obvious) step requires further comment, which we postpone until after the proof of the following theorem.

Theorem 3.6 *The set $\mathcal{N}$ is not measurable.*

The proof is by contradiction, so we assume that $\mathcal{N}$ is measurable. Let $\{r_k\}_{k=1}^{\infty}$ be an enumeration of all the rationals in $[-1, 1]$, and consider the translates

$$\mathcal{N}_k = \mathcal{N} + r_k.$$

---

[4]The existence of such a set in $\mathbb{R}$ implies the existence of corresponding non-measurable subsets of $\mathbb{R}^d$ for each d, as a consequence of Proposition 3.4 in the next chapter.

We claim that the sets $\mathcal{N}_k$ are disjoint, and

$$(4) \qquad\qquad [0,1] \subset \bigcup_{k=1}^{\infty} \mathcal{N}_k \subset [-1,2].$$

To see why these sets are disjoint, suppose that the intersection $\mathcal{N}_k \cap \mathcal{N}_{k'}$ is non-empty. Then there exist rationals $r_k \neq r'_k$ and α and β with $x_\alpha + r_k = x_\beta + r_{k'}$; hence

$$x_\alpha - x_\beta = r_{k'} - r_k.$$

Consequently $\alpha \neq \beta$ and $x_\alpha - x_\beta$ is rational; hence $x_\alpha \sim x_\beta$, which contradicts the fact that $\mathcal{N}$ contains only *one* representative of each equivalence class.

The second inclusion is straightforward since each $\mathcal{N}_k$ is contained in $[-1,2]$ by construction. Finally, if $x \in [0,1]$, then $x \sim x_\alpha$ for some α, and therefore $x - x_\alpha = r_k$ for some k. Hence $x \in \mathcal{N}_k$, and the first inclusion holds.

Now we may conclude the proof of the theorem. If $\mathcal{N}$ were measurable, then so would be $\mathcal{N}_k$ for all k, and since the union $\bigcup_{k=1}^{\infty} \mathcal{N}_k$ is disjoint, the inclusions in (4) yield

$$1 \leq \sum_{k=1}^{\infty} m(\mathcal{N}_k) \leq 3.$$

Since $\mathcal{N}_k$ is a translate of $\mathcal{N}$, we must have $m(\mathcal{N}_k) = m(\mathcal{N})$ for all k. Consequently,

$$1 \leq \sum_{k=1}^{\infty} m(\mathcal{N}) \leq 3.$$

This is the desired contradiction, since neither $m(\mathcal{N}) = 0$ nor $m(\mathcal{N}) > 0$ is possible.

Axiom of choice

That the construction of the set $\mathcal{N}$ is possible is based on the following general proposition.

- Suppose E is a set and $\{E_\alpha\}$ is a collection of non-empty subsets of E. (The indexing set of α's is not assumed to be countable.) Then there is a function $\alpha \mapsto x_\alpha$ (a "choice function") such that $x_\alpha \in E_\alpha$, for all α.

In this general form this assertion is known as the **axiom of choice**. This axiom occurs (at least implicitly) in many proofs in mathematics, but because of its seeming intuitive self-evidence, its significance was not at first understood. The initial realization of the importance of this axiom was in its use to prove a famous assertion of Cantor, the **well-ordering principle**. This proposition (sometimes referred to as "transfinite induction") can be formulated as follows.

A set E is **linearly ordered** if there is a binary relation $\leq$ such that:

(a) $x \leq x$ for all $x \in E$.

(b) If $x, y \in E$ are distinct, then either $x \leq y$ or $y \leq x$ (but not both).

(c) If $x \leq y$ and $y \leq z$, then $x \leq z$.

We say that a set E can be **well-ordered** if it can be linearly ordered in such a way that *every* non-empty subset $A \subset E$ has a smallest element in that ordering (that is, an element $x_0 \in A$ such that $x_0 \leq x$ for any other $x \in A$).

A simple example of a well-ordered set is $\mathbb{Z}^+$, the positive integers with their usual ordering. The fact that $\mathbb{Z}^+$ is well-ordered is an essential part of the usual (finite) induction principle. More generally, the well-ordering principle states:

- Any set E can be well-ordered.

It is in fact nearly obvious that the well-ordering principle implies the axiom of choice: if we well-order E, we can choose x_α to be the smallest element in E_α, and in this way we have constructed the required choice function. It is also true, but not as easy to show, that the axiom of choice implies the well-ordering principle. (See Problem 6 for another equivalent formulation of the Axiom of Choice.)

We shall follow the common practice of assuming the axiom of choice (and hence the validity of the well-ordering principle).[5] However, we should point out that while the axiom of choice seems self-evident the well-ordering principle leads quickly to some baffling conclusions: one only needs to spend a little time trying to imagine what a well-ordering of the reals might look like!

---

[5]It can be proved that in an appropriate formulation of the axioms of set theory, the axiom of choice is independent of the other axioms; thus we are free to accept its validity.

4 Measurable functions

With the notion of measurable sets in hand, we now turn our attention to the objects that lie at the heart of integration theory: measurable functions.

The starting point is the notion of a **characteristic function** of a set E, which is defined by

$$\chi_E(x) = \begin{cases} 1 & \text{if } x \in E, \\ 0 & \text{if } x \notin E. \end{cases}$$

The next step is to pass to the functions that are the building blocks of integration theory. For the Riemann integral it is in effect the class of **step functions**, with each given as a finite sum

$$(5) \qquad\qquad f = \sum_{k=1}^{N} a_k \chi_{R_k},$$

where each R_k is a rectangle, and the a_k are constants.

However, for the Lebesgue integral we need a more general notion, as we shall see in the next chapter. A **simple function** is a finite sum

$$(6) \qquad\qquad f = \sum_{k=1}^{N} a_k \chi_{E_k}$$

where each E_k is a measurable set of finite measure, and the a_k are constants.

4.1 Definition and basic properties

We begin by considering only real-valued functions f on $\mathbb{R}^d$, which we allow to take on the infinite values $+\infty$ and $-\infty$, so that $f(x)$ belongs to the extended real numbers

$$-\infty \leq f(x) \leq \infty.$$

We shall say that f is **finite-valued** if $-\infty < f(x) < \infty$ for all x. In the theory that follows, and the many applications of it, we shall almost always find ourselves in situations where a function takes on infinite values on at most a set of measure zero.

A function f defined on a measurable subset E of $\mathbb{R}^d$ is **measurable**, if for all $a \in \mathbb{R}$, the set

$$f^{-1}([-\infty, a)) = \{x \in E : f(x) < a\}$$

is measurable. To simplify our notation, we shall often denote the set $\{x \in E : f(x) < a\}$ simply by $\{f < a\}$ whenever no confusion is possible.

First, we note that there are many equivalent definitions of measurable functions. For example, we may require instead that the inverse image of closed intervals be measurable. Indeed, to prove that f is measurable if and only if $\{x : f(x) \leq a\} = \{f \leq a\}$ is measurable for every a, we note that in one direction, one has

$$\{f \leq a\} = \bigcap_{k=1}^{\infty} \{f < a + 1/k\},$$

and recall that the countable intersection of measurable sets is measurable. For the other direction, we observe that

$$\{f < a\} = \bigcup_{k=1}^{\infty} \{f \leq a - 1/k\}.$$

Similarly, f is measurable if and only if $\{f \geq a\}$ (or $\{f > a\}$) is measurable for every a. In the first case this is immediate from our definition and the fact that $\{f \geq a\}$ is the complement of $\{f < a\}$, and in the second case this follows from what we have just proved and the fact that $\{f \leq a\} = \{f > a\}^c$. A simple consequence is that $-f$ is measurable whenever f is measurable.

In the same way, one can show that if f is finite-valued, then it is measurable if and only if the sets $\{a < f < b\}$ are measurable for every $a, b \in \mathbb{R}$. Similar conclusions hold for whichever combination of strict or weak inequalities one chooses. For example, if f is finite-valued, then it is measurable if and only if $\{a \leq f < b\}$ for all $a, b \in \mathbb{R}$. By the same arguments one sees the following:

Property 1 *The finite-valued function f is measurable if and only if $f^{-1}(\mathcal{O})$ is measurable for every open set $\mathcal{O}$, and if and only if $f^{-1}(F)$ is measurable for every closed set F.*

Note that this property also applies to extended-valued functions, if we make the additional hypothesis that both $f^{-1}(\infty)$ and $f^{-1}(-\infty)$ are measurable sets.

Property 2 *If f is continuous on $\mathbb{R}^d$, then f is measurable. If f is measurable and finite-valued, and Φ is continuous, then $\Phi \circ f$ is measurable.*

In fact, Φ is continuous, so $\Phi^{-1}((-\infty, a))$ is an open set $\mathcal{O}$, and hence $(\Phi \circ f)^{-1}((-\infty, a)) = f^{-1}(\mathcal{O})$ is measurable.

It should be noted, however, that in general it is not true that $f \circ \Phi$ is measurable whenever f is measurable and Φ is continuous. See Exercise 35.

Property 3 *Suppose $\{f_n\}_{n=1}^{\infty}$ is a sequence of measurable functions. Then*

$$\sup_n f_n(x), \quad \inf_n f_n(x), \quad \limsup_{n \to \infty}, f_n(x) \quad and \quad \liminf_{n \to \infty} f_n(x)$$

are measurable.

Proving that $\sup_n f_n$ is measurable requires noting that $\{\sup_n f_n > a\} = \bigcup_n \{f_n > a\}$. This also yields the result for $\inf_n f_n(x)$, since this quantity equals $-\sup_n(-f_n(x))$.

The result for the limsup and liminf also follows from the two observations

$$\limsup_{n \to \infty} f_n(x) = \inf_k \{\sup_{n \geq k} f_n\} \quad and \quad \liminf_{n \to \infty} f_n(x) = \sup_k \{\inf_{n \geq k} f_n\}.$$

Property 4 *If $\{f_n\}_{n=1}^{\infty}$ is a collection of measurable functions, and*

$$\lim_{n \to \infty} f_n(x) = f(x),$$

then f is measurable.

Since $f(x) = \limsup_{n \to \infty} f_n(x) = \liminf_{n \to \infty} f_n(x)$, this property is a consequence of property 3.

Property 5 *If f and g are measurable, then*

(i) *The integer powers f^k, $k \geq 1$ are measurable.*

(ii) *$f + g$ and fg are measurable if both f and g are finite-valued.*

For (i) we simply note that if k is odd, then $\{f^k > a\} = \{f > a^{1/k}\}$, and if k is even and $a \geq 0$, then $\{f^k > a\} = \{f > a^{1/k}\} \cup \{f < -a^{1/k}\}$.

For (ii), we first see that $f + g$ is measurable because

$$\{f + g > a\} = \bigcup_{r \in \mathbb{Q}} \{f > a - r\} \cap \{g > r\},$$

with $\mathbb{Q}$ denoting the rationals.

Finally, fg is measurable because of the previous results and the fact that

$$fg = \frac{1}{4}[(f + g)^2 - (f - g)^2].$$

We shall say that two functions f and g defined on a set E are equal **almost everywhere**, and write

$$f(x) = g(x) \quad \text{a.e. } x \in E,$$

if the set $\{x \in E : f(x) \neq g(x)\}$ has measure zero. We sometimes abbreviate this by saying that $f = g$ a.e. More generally, a property or statement is said to hold almost everywhere (a.e.) if it is true except on a set of measure zero.

One sees easily that if f is measurable and $f = g$ a.e., then g is measurable. This follows at once from the fact that $\{f < a\}$ and $\{g < a\}$ differ by a set of measure zero. Moreover, all the properties above can be relaxed to conditions holding almost everywhere. For instance, if $\{f_n\}_{n=1}^{\infty}$ is a collection of measurable functions, and

$$\lim_{n \to \infty} f_n(x) = f(x) \quad \text{a.e.,}$$

then f is measurable.

Note that if f and g are defined almost everywhere on a measurable subset $E \subset \mathbb{R}^d$, then the functions $f + g$ and fg can only be defined on the intersection of the domains of f and g. Since the union of two sets of measure zero has again measure zero, $f + g$ is defined almost everywhere on E. We summarize this discussion as follows.

Property 6 *Suppose f is measurable, and $f(x) = g(x)$ for a.e. x. Then g is measurable.*

In this light, Property 5 (ii) also holds when f and g are finite-valued almost everywhere.

4.2 Approximation by simple functions or step functions

The theorems in this section are all of the same nature and provide further insight in the structure of measurable functions. We begin by approximating pointwise, non-negative measurable functions by simple functions.

Theorem 4.1 *Suppose f is a non-negative measurable function on $\mathbb{R}^d$. Then there exists an increasing sequence of non-negative simple functions $\{\varphi_k\}_{k=1}^{\infty}$ that converges pointwise to f, namely,*

$$\varphi_k(x) \le \varphi_{k+1}(x) \quad and \quad \lim_{k\to\infty} \varphi_k(x) = f(x), \text{ for all } x.$$

Proof. We begin first with a truncation. For $N \ge 1$, let Q_N denote the cube centered at the origin and of side length N. Then we define

$$F_N(x) = \begin{cases} f(x) & \text{if } x \in Q_N \text{ and } f(x) \le N, \\ N & \text{if } x \in Q_N \text{ and } f(x) > N, \\ 0 & \text{otherwise.} \end{cases}$$

Then, $F_N(x) \to f(x)$ as N tends to infinity for all x. Now, we partition the range of F_N, namely $[0, N]$, as follows. For fixed $N, M \ge 1$, we define

$$E_{\ell,M} = \left\{ x \in Q_N : \frac{\ell}{M} < F_N(x) \le \frac{\ell+1}{M} \right\}, \quad \text{for } 0 \le \ell < NM.$$

Then we may form

$$F_{N,M}(x) = \sum_{\ell} \frac{\ell}{M} \, \chi_{E_{\ell,M}}(x).$$

Each $F_{N,M}$ is a simple function that satisfies $0 \le F_N(x) - F_{N,M}(x) \le 1/M$ for all x. If we now choose $N = M = 2^k$ with $k \ge 1$ integral, and let $\varphi_k = F_{2^k, 2^k}$, then we see that $0 \le F_M(x) - \varphi_k(x) \le 1/2^k$ for all x, $\{\varphi_k\}$ is increasing, and this sequence satisfies all the desired properties.

Note that the result holds for non-negative functions that are extended-valued, if the limit $+\infty$ is allowed. We now drop the assumption that f is non-negative, and also allow the extended limit $-\infty$.

Theorem 4.2 *Suppose f is measurable on $\mathbb{R}^d$. Then there exists a sequence of simple functions $\{\varphi_k\}_{k=1}^{\infty}$ that satisfies*

$$|\varphi_k(x)| \le |\varphi_{k+1}(x)| \quad and \quad \lim_{k\to\infty} \varphi_k(x) = f(x), \text{ for all } x.$$

In particular, we have $|\varphi_k(x)| \le |f(x)|$ for all x and k.

Proof. We use the following decomposition of the function f: $f(x) = f^+(x) - f^-(x)$, where

$$f^+(x) = \max(f(x), 0) \quad and \quad f^-(x) = \max(-f(x), 0).$$

Since both f^+ and f^- are non-negative, the previous theorem yields two increasing sequences of non-negative simple functions $\{\varphi_k^{(1)}(x)\}_{k=1}^\infty$ and $\{\varphi_k^{(2)}(x)\}_{k=1}^\infty$ which converge pointwise to f^+ and f^-, respectively. Then, if we let

$$\varphi_k(x) = \varphi_k^{(1)}(x) - \varphi_k^{(2)}(x),$$

we see that $\varphi_k(x)$ converges to $f(x)$ for all x. Finally, the sequence $\{|\varphi_k|\}$ is increasing because the definition of f^+, f^- and the properties of $\varphi_k^{(1)}$ and $\varphi_k^{(2)}$ imply that

$$|\varphi_k(x)| = \varphi_k^{(1)}(x) + \varphi_k^{(2)}(x).$$

We may now go one step further, and approximate by step functions. Here, in general, the convergence may hold only almost everywhere.

Theorem 4.3 *Suppose f is measurable on $\mathbb{R}^d$. Then there exists a sequence of step functions $\{\psi_k\}_{k=1}^\infty$ that converges pointwise to $f(x)$ for almost every x.*

Proof. By the previous theorem, there are simple functions $\{\varphi_k\}$ so that $\lim_{k\to\infty} \varphi_k(x) = f(x)$ for all x. To approximate each φ_k by a step function, we recall part (iv) of Theorem 3.4, which states that if E is a measurable set of finite measure, then for every ϵ there exist cubes $Q_1, \ldots, Q_N$ such that $m(E \triangle \bigcup_{j=1}^N Q_j) \leq \epsilon$. By considering the grid formed by extending the sides of these cubes, we see that there exist almost disjoint rectangles $\tilde{R}_1, \ldots, \tilde{R}_M$ such that $\bigcup_{j=1}^N Q_j = \bigcup_{j=1}^M \tilde{R}_j$. By taking closed rectangles R_j contained in $\tilde{R}_j$, and slightly smaller in size, we find a collection of disjoint closed rectangles that satisfy $m(E \triangle \bigcup_{j=1}^M R_j) \leq 2\epsilon$. It follows from this observation and the definition of a simple function that for each k, there exists a step function ψ_k and a measurable set F_k so that $m(F_k) < 2^{-k}$ and $\varphi_k(x) = \psi_k(x)$ for all $x \notin F_k$.

If we define $F = \bigcap_{\ell=1}^\infty \bigcup_{k>\ell}^\infty F_k$, then $m(F) = 0$, because $m(\bigcup_{k>\ell}^\infty F_k) \leq \sum_{k>\ell} m(F_k) \leq 2^{-\ell}$. For $x \notin F$, there exists k_0 so that $x \in \bigcap_{k>k_0} F_k^c$, so for all $k > k_0$ one has

$$|f(x) - \psi_k(x)| \leq |f(x) - \varphi_k(x)| + |\varphi_k(x) - \psi_k(x)| = |f(x) - \varphi_k(x)|,$$

and since $\lim_{k\to\infty} \varphi_k(x) = f(x)$ we conclude that $\lim_{k\to\infty} \psi_k(x) = f(x)$ for all $x \notin F$, as desired.

4.3 Littlewood's three principles

Although the notions of measurable sets and measurable functions represent new tools, we should not overlook their relation to the older concepts they replaced. Littlewood aptly summarized these connections in the form of three principles that provide a useful intuitive guide in the initial study of the theory.

(i) Every set is nearly a finite union of intervals.

(ii) Every function is nearly continuous.

(iii) Every convergent sequence is nearly uniformly convergent.

The sets and functions referred to above are of course assumed to be measurable. The catch is in the word "nearly," which has to be understood appropriately in each context. A precise version of the first principle appears in part (iv) of Theorem 3.4. An exact formulation of the third principle is given in the following important result.

Theorem 4.4 (Egorov) *Suppose $\{f_k\}_{k=1}^{\infty}$ is a sequence of measurable functions defined on a measurable set E with $m(E) < \infty$, and assume that $f_k \to f$ a.e on E. Given $\epsilon > 0$, we can find a closed set $A_\epsilon \subset E$ such that $m(E - A_\epsilon) \le \epsilon$ and $f_k \to f$ uniformly on A_ϵ.*

Proof. We may assume without loss of generality that $f_k(x) \to f(x)$ for every $x \in E$. For each pair of non-negative integers n and k, let

$$E_k^n = \{x \in E : |f_j(x) - f(x)| < 1/n, \text{ for all } j > k\}.$$

Now fix n and note that $E_k^n \subset E_{k+1}^n$, and $E_k^n \nearrow E$ as k tends to infinity. By Corollary 3.3, we find that there exists k_n such that $m(E - E_{k_n}^n) < 1/2^n$. By construction, we then have

$$|f_j(x) - f(x)| < 1/n \quad \text{whenever } j > k_n \text{ and } x \in E_{k_n}^n.$$

We choose N so that $\sum_{n=N}^{\infty} 2^{-n} < \epsilon/2$, and let

$$\tilde{A}_\epsilon = \bigcap_{n \ge N} E_{k_n}^n.$$

We first observe that

$$m(E - \tilde{A}_\epsilon) \le \sum_{n=N}^{\infty} m(E - E_{k_n}^n) < \epsilon/2.$$

Next, if $\delta > 0$, we choose $n \geq N$ such that $1/n < \delta$, and note that $x \in \tilde{A}_\epsilon$ implies $x \in E^n_{k_n}$. We see therefore that $|f_j(x) - f(x)| < \delta$ whenever $j > k_n$. Hence f_k converges uniformly to f on $\tilde{A}_\epsilon$.

Finally, using Theorem 3.4 choose a closed subset $A_\epsilon \subset \tilde{A}_\epsilon$ with $m(\tilde{A}_\epsilon - A_\epsilon) < \epsilon/2$. As a result, we have $m(E - A_\epsilon) < \epsilon$ and the theorem is proved.

The next theorem attests to the validity of the second of Littlewood's principle.

Theorem 4.5 (Lusin) *Suppose f is measurable and finite valued on E with E of finite measure. Then for every $\epsilon > 0$ there exists a closed set F_ϵ, with*

$$F_\epsilon \subset E, \quad \text{and} \quad m(E - F_\epsilon) \leq \epsilon$$

and such that $f|_{F_\epsilon}$ is continuous.

By $f|_{F_\epsilon}$ we mean the restriction of f to the set F_ϵ. The conclusion of the theorem states that if f is viewed as a function defined only on F_ϵ, then f is continuous. However, the theorem does *not* make the stronger assertion that the function f defined on E is continuous at the points of F_ϵ.

Proof. Let f_n be a sequence of step functions so that $f_n \to f$ a.e. Then we may find sets E_n so that $m(E_n) < 1/2^n$ and f_n is continuous outside E_n. By Egorov's theorem, we may find a set $A_{\epsilon/3}$ on which $f_n \to f$ uniformly and $m(E - A_{\epsilon/3}) \leq \epsilon/3$. Then we consider

$$F' = A_{\epsilon/3} - \bigcup_{n \geq N} E_n$$

for N so large that $\sum_{n \geq N} 1/2^n < \epsilon/3$. Now for every $n \geq N$ the function f_n is continuous on F'; thus f (being the uniform limit of $\{f_n\}$) is also continuous on F'. To finish the proof, we merely need to approximate the set F' by a closed set $F_\epsilon \subset F'$ such that $m(F' - F_\epsilon) < \epsilon/3$.

5* The Brunn-Minkowski inequality

Since addition and multiplication by scalars are basic features of vector spaces, it is not surprising that properties of these operations arise in a fundamental way in the theory of Lebesgue measure on $\mathbb{R}^d$. We have already discussed in this connection the translation-invariance and relative dilation-invariance of Lebesgue measure. Here we come to the study of the sum of two measurable sets A and B, defined as

$$A + B = \{x \in \mathbb{R}^d : \; x = x' + x'' \;\; \text{with } x' \in A \text{ and } x'' \in B\}.$$

This notion is of importance in a number of questions, in particular in the theory of convex sets; we shall apply it to the isoperimetric problem in Chapter 3.

In this regard the first (admittedly vague) question we can pose is whether one can establish any general estimate for the measure of $A + B$ in terms of the measures of A and B (assuming that these three sets are measurable). We can see easily that it is not possible to obtain an upper bound for $m(A + B)$ in terms of $m(A)$ and $m(B)$. Indeed, simple examples show that we may have $m(A) = m(B) = 0$ while $m(A + B) > 0$. (See Exercise 20.)

In the converse direction one might ask for a general estimate of the form

$$m(A + B)^\alpha \geq c_\alpha \left(m(A)^\alpha + m(B)^\alpha\right),$$

where α is a positive number and the constant c_α is independent of A and B. Clearly, the best one can hope for is $c_\alpha = 1$. The role of the exponent α can be understood by considering **convex sets**. Such sets A are defined by the property that whenever x and y are in A then the line segment joining them, $\{xt + y(1 - t) : \; 0 \leq t \leq 1\}$, also belongs to A. If we recall the definition $\lambda A = \{\lambda x, \; x \in A\}$ for $\lambda > 0$, we note that whenever A is convex, then $A + \lambda A = (1 + \lambda)A$. However, $m((1 + \lambda)A) = (1 + \lambda)^d m(A)$, and thus the presumed inequality can hold only if $(1 + \lambda)^{d\alpha} \geq 1 + \lambda^{d\alpha}$, for all $\lambda > 0$. Now

(7) $$(a + b)^\gamma \geq a^\gamma + b^\gamma \quad \text{if } \gamma \geq 1 \text{ and } a, b \geq 0,$$

while the reverse inequality holds if $0 \leq \gamma \leq 1$. (See Exercise 38.) This yields $\alpha \geq 1/d$. Moreover, (7) shows that the inequality with the exponent $1/d$ implies the corresponding inequality with $\alpha \geq 1/d$, and so we are naturally led to the inequality

(8) $$m(A + B)^{1/d} \geq m(A)^{1/d} + m(B)^{1/d}.$$

Before proceeding with the proof of (8), we need to mention a technical impediment that arises. While we may assume that A and B are measurable, it does not necessarily follow that then $A + B$ is measurable. (See Exercise 13 in the next chapter.) However it is easily seen that this difficulty does not occur when, for example, A and B are closed sets, or when one of them is open. (See Exercise 19.)

With the above considerations in mind we can state the main result.

Theorem 5.1 *Suppose A and B are measurable sets in $\mathbb{R}^d$ and their sum $A + B$ is also measurable. Then the inequality (8) holds.*

Let us first check (8) when A and B are rectangles with side lengths $\{a_j\}_{j=1}^d$ and $\{b_j\}_{j=1}^d$, respectively. Then (8) becomes

$$(9) \qquad \left(\prod_{j=1}^d (a_j + b_j)\right)^{1/d} \geq \left(\prod_{j=1}^d a_j\right)^{1/d} + \left(\prod_{j=1}^d b_j\right)^{1/d},$$

which by homogeneity we can reduce to the special case where $a_j + b_j = 1$ for each j. In fact, notice that if we replace a_j, b_j by $\lambda_j a_j, \lambda_j b_j$, with $\lambda_j > 0$, then both sides of (9) are multiplied by $(\lambda_1 \lambda_2 \cdots \lambda_d)^{1/d}$. We then need only choose $\lambda_j = (a_j + b_j)^{-1}$. With this reduction, the inequality (9) is an immediate consequence of the arithmetic-geometric inequality (Exercise 39)

$$\frac{1}{d} \sum_{j=1}^d x_j \geq \left(\prod_{j=1}^d x_j\right)^{1/d}, \qquad \text{for all } x_j \geq 0:$$

we add the two inequalities that result when we set $x_j = a_j$ and $x_j = b_j$, respectively.

We next turn to the case when each A and B are the union of finitely many rectangles whose interiors are disjoint. We shall prove (8) in this case by induction on the total number of rectangles in A and B. We denote this number by n. Here it is important to note that the desired inequality is unchanged when we translate A and B independently. In fact, replacing A by $A + h$ and B by $B + h'$ replaces $A + B$ by $A + B + h + h'$, and thus the corresponding measures remain the same. We now choose a pair of disjoint rectangles R_1 and R_2 in the collection making up A, and we note that they can be separated by a coordinate hyperplane. Thus we may assume that for some j, after translation by an appropriate h, R_1 lies in $A_- = A \cap \{x_j \leq 0\}$, and R_2 in $A_+ = A \cap \{0 \leq x_j\}$. Observe also that both A_+ and A_- contain at least one less rectangle than A does, and $A = A_- \cup A_+$.

We next translate B so that $B_- = B \cap \{x_j \leq 0\}$ and $B_+ = B \cap \{x_j \geq 0\}$ satisfy

$$\frac{m(B_\pm)}{m(B)} = \frac{m(A_\pm)}{m(A)}.$$

However, $A + B \supset (A_+ + B_+) \cup (A_- + B_-)$, and the union on the right is essentially disjoint, since the two parts lie in different half-spaces. Moreover, the total number of rectangles in either A_+ and B_+, or A_- and B_- is also less than n. Thus the induction hypothesis applies and

$$m(A + B) \geq m(A_+ + B_+) + m(A_- + B_-)$$
$$\geq \left(m(A_+)^{1/d} + m(B_+)^{1/d} \right)^d + \left(m(A_-)^{1/d} + m(B_-)^{1/d} \right)^d$$
$$= m(A_+) \left[1 + \left(\frac{m(B)}{m(A)} \right)^{1/d} \right]^d + m(A_-) \left[1 + \left(\frac{m(B)}{m(A)} \right)^{1/d} \right]^d$$
$$= \left(m(A)^{1/d} + m(B)^{1/d} \right)^d,$$

which gives the desired inequality (8) when A and B are both finite unions of rectangles with disjoint interiors.

Next, this quickly implies the result when A and B are open sets of finite measure. Indeed, by Theorem 1.4, for any $\epsilon > 0$ we can find unions of almost disjoint rectangles A_ϵ and B_ϵ, such that $A_\epsilon \subset A$, $B_\epsilon \subset B$, with $m(A) \leq m(A_\epsilon) + \epsilon$ and $m(B) \leq m(B_\epsilon) + \epsilon$. Since $A + B \supset A_\epsilon + B_\epsilon$, the inequality (8) for A_ϵ and B_ϵ and a passage to a limit gives the desired result. From this, we can pass to the case where A and B are arbitrary compact sets, by noting first that $A + B$ is then compact, and that if we define $A^\epsilon = \{x : d(x, A) < \epsilon\}$, then A^ϵ are open, and $A^\epsilon \searrow A$ as $\epsilon \to 0$. With similar definitions for B^ϵ and $(A + B)^\epsilon$, we observe also that $A + B \subset A^\epsilon + B^\epsilon \subset (A + B)^{2\epsilon}$. Hence, letting $\epsilon \to 0$, we see that (8) for A^ϵ and B^ϵ implies the desired result for A and B. The general case, in which we assume that A, B, and $A + B$ are measurable, then follows by approximating A and B from inside by compact sets, as in (iii) of Theorem 3.4.

6 Exercises

1. Prove that the Cantor set $\mathcal{C}$ constructed in the text is totally disconnected and perfect. In other words, given two distinct points $x, y \in \mathcal{C}$, there is a point $z \notin \mathcal{C}$ that lies between x and y, and yet $\mathcal{C}$ has no isolated points.

[Hint: If $x, y \in \mathcal{C}$ and $|x - y| > 1/3^k$, then x and y belong to two different intervals in C_k. Also, given any $x \in \mathcal{C}$ there is an end-point y_k of some interval in C_k that satisfies $x \neq y_k$ and $|x - y_k| \leq 1/3^k$.]

2. The Cantor set $\mathcal{C}$ can also be described in terms of ternary expansions.

(a) Every number in $[0, 1]$ has a ternary expansion

$$x = \sum_{k=1}^{\infty} a_k 3^{-k}, \qquad \text{where } a_k = 0, 1, \text{ or } 2.$$

Note that this decomposition is not unique since, for example, $1/3 = \sum_{k=2}^{\infty} 2/3^k$.
Prove that $x \in \mathcal{C}$ if and only if x has a representation as above where every a_k is either 0 or 2.

(b) The **Cantor-Lebesgue function** is defined on $\mathcal{C}$ by

$$F(x) = \sum_{k=1}^{\infty} \frac{b_k}{2^k} \qquad \text{if } x = \sum_{k=1}^{\infty} a_k 3^{-k}, \text{ where } b_k = a_k/2.$$

In this definition, we choose the expansion of x in which $a_k = 0$ or 2.

Show that F is well defined and continuous on $\mathcal{C}$, and moreover $F(0) = 0$ as well as $F(1) = 1$.

(c) Prove that $F : \mathcal{C} \to [0, 1]$ is surjective, that is, for every $y \in [0, 1]$ there exists $x \in \mathcal{C}$ such that $F(x) = y$.

(d) One can also extend F to be a continuous function on $[0, 1]$ as follows. Note that if (a, b) is an open interval of the complement of $\mathcal{C}$, then $F(a) = F(b)$. Hence we may define F to have the constant value $F(a)$ in that interval.

A geometrical construction of F is described in Chapter 3.

3. Cantor sets of constant dissection. Consider the unit interval $[0, 1]$, and let ξ be a fixed real number with $0 < \xi < 1$ (the case $\xi = 1/3$ corresponds to the Cantor set $\mathcal{C}$ in the text).

In stage 1 of the construction, remove the centrally situated open interval in $[0, 1]$ of length ξ. In stage 2, remove two central intervals each of relative length ξ, one in each of the remaining intervals after stage 1, and so on.

Let $\mathcal{C}_\xi$ denote the set which remains after applying the above procedure indefinitely.[6]

(a) Prove that the complement of $\mathcal{C}_\xi$ in $[0, 1]$ is the union of open intervals of total length equal to 1.

(b) Show directly that $m_*(\mathcal{C}_\xi) = 0$.

[Hint: After the kth stage, show that the remaining set has total length $= (1 - \xi)^k$.]

4. Cantor-like sets. Construct a closed set $\hat{\mathcal{C}}$ so that at the kth stage of the construction one removes 2^{k-1} centrally situated open intervals each of length ℓ_k, with

$$\ell_1 + 2\ell_2 + \cdots + 2^{k-1}\ell_k < 1.$$

---

[6]The set we call $\mathcal{C}_\xi$ is sometimes denoted by $\mathcal{C}_{\frac{1-\xi}{2}}$.

(a) If ℓ_j are chosen small enough, then $\sum_{k=1}^{\infty} 2^{k-1}\ell_k < 1$. In this case, show that $m(\hat{\mathcal{C}}) > 0$, and in fact, $m(\hat{\mathcal{C}}) = 1 - \sum_{k=1}^{\infty} 2^{k-1}\ell_k$.

(b) Show that if $x \in \hat{\mathcal{C}}$, then there exists a sequence of points $\{x_n\}_{n=1}^{\infty}$ such that $x_n \notin \hat{\mathcal{C}}$, yet $x_n \to x$ and $x_n \in I_n$, where I_n is a sub-interval in the complement of $\hat{\mathcal{C}}$ with $|I_n| \to 0$.

(c) Prove as a consequence that $\hat{\mathcal{C}}$ is perfect, and contains no open interval.

(d) Show also that $\hat{\mathcal{C}}$ is uncountable.

5. Suppose E is a given set, and $\mathcal{O}_n$ is the open set:

$$\mathcal{O}_n = \{x : \ d(x, E) < 1/n\}.$$

Show:

(a) If E is compact, then $m(E) = \lim_{n\to\infty} m(\mathcal{O}_n)$.

(b) However, the conclusion in (a) may be false for E closed and unbounded; or E open and bounded.

6. Using translations and dilations, prove the following: Let B be a ball in $\mathbb{R}^d$ of radius r. Then $m(B) = v_d r^d$, where $v_d = m(B_1)$, and B_1 is the unit ball, $B_1 = \{x \in \mathbb{R}^d : \ |x| < 1\}$.

A calculation of the constant v_d is postponed until Exercise 14 in the next chapter.

7. If $\delta = (\delta_1, \dots, \delta_d)$ is a d-tuple of positive numbers $\delta_i > 0$, and E is a subset of $\mathbb{R}^d$, we define δE by

$$\delta E = \{(\delta_1 x_1, \dots, \delta_d x_d) : \ \text{where } (x_1, \dots, x_d) \in E\}.$$

Prove that δE is measurable whenever E is measurable, and

$$m(\delta E) = \delta_1 \cdots \delta_d m(E).$$

8. Suppose L is a linear transformation of $\mathbb{R}^d$. Show that if E is a measurable subset of $\mathbb{R}^d$, then so is $L(E)$, by proceeding as follows:

(a) Note that if E is compact, so is $L(E)$. Hence if E is an F_σ set, so is $L(E)$.

(b) Because L automatically satisfies the inequality

$$|L(x) - L(x')| \leq M|x - x'|$$

for some M, we can see that L maps any cube of side length ℓ into a cube of side length $c_d M \ell$, with $c_d = 2\sqrt{d}$. Now if $m(E) = 0$, there is a collection of cubes $\{Q_j\}$ such that $E \subset \bigcup_j Q_j$, and $\sum_j m(Q_j) < \epsilon$. Thus $m_*(L(E)) \leq c'\epsilon$, and hence $m(L(E)) = 0$. Finally, use Corollary 3.5.

One can show that $m(L(E)) = |\det L| \, m(E)$; see Problem 4 in the next chapter.

9. Give an example of an open set $\mathcal{O}$ with the following property: the boundary of the closure of $\mathcal{O}$ has positive Lebesgue measure.

[Hint: Consider the set obtained by taking the union of open intervals which are deleted at the odd steps in the construction of a Cantor-like set.]

10. This exercise provides a construction of a decreasing sequence of positive continuous functions on the interval $[0, 1]$, whose pointwise limit is *not* Riemann integrable.

Let $\hat{\mathcal{C}}$ denote a Cantor-like set obtained from the construction detailed in Exercise 4, so that in particular $m(\hat{\mathcal{C}}) > 0$. Let F_1 denote a piecewise-linear and continuous function on $[0, 1]$, with $F_1 = 1$ in the complement of the first interval removed in the construction of $\hat{\mathcal{C}}$, $F_1 = 0$ at the center of this interval, and $0 \leq F_1(x) \leq 1$ for all x. Similarly, construct $F_2 = 1$ in the complement of the intervals in stage two of the construction of $\hat{\mathcal{C}}$, with $F_2 = 0$ at the center of these intervals, and $0 \leq F_2 \leq 1$. Continuing this way, let $f_n = F_1 \cdot F_2 \cdots F_n$ (see Figure 5).

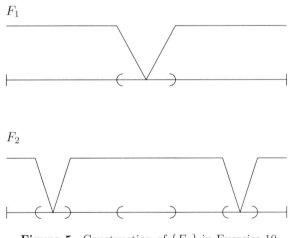

Figure 5. Construction of $\{F_n\}$ in Exercise 10

Prove the following:

(a) For all $n \geq 1$ and all $x \in [0, 1]$, one has $0 \leq f_n(x) \leq 1$ and $f_n(x) \geq f_{n+1}(x)$. Therefore, $f_n(x)$ converges to a limit as $n \to \infty$ which we denote by $f(x)$.

(b) The function f is discontinuous at every point of $\hat{\mathcal{C}}$.

   [Hint: Note that $f(x) = 1$ if $x \in \hat{\mathcal{C}}$, and find a sequence of points $\{x_n\}$ so that $x_n \to x$ and $f(x_n) = 0$.]

Now $\int f_n(x) \, dx$ is decreasing, hence $\int f_n$ converges. However, a bounded function is Riemann integrable if and only if its set of discontinuities has measure zero.

(The proof of this fact, which is given in the Appendix of Book I, is outlined in Problem 4.) Since f is discontinuous on a set of positive measure, we find that f is not Riemann integrable.

11. Let A be the subset of $[0, 1]$ which consists of all numbers which do not have the digit 4 appearing in their decimal expansion. Find $m(A)$.

12. Theorem 1.3 states that every open set in $\mathbb{R}$ is the disjoint union of open intervals. The analogue in $\mathbb{R}^d$, $d \geq 2$, is generally false. Prove the following:

(a) An open disc in $\mathbb{R}^2$ is not the disjoint union of open rectangles.

[Hint: What happens to the boundary of any of these rectangles?]

(b) An open connected set Ω is the disjoint union of open rectangles if and only if Ω is itself an open rectangle.

13. The following deals with G_δ and F_σ sets.

(a) Show that a closed set is a G_δ and an open set an F_σ.

[Hint: If F is closed, consider $\mathcal{O}_n = \{x : d(x, F) < 1/n\}$.]

(b) Give an example of an F_σ which is not a G_δ.

[Hint: This is more difficult; let F be a denumerable set that is dense.]

(c) Give an example of a Borel set which is not a G_δ nor an F_σ.

14. The purpose of this exercise is to show that covering by a *finite* number of intervals will not suffice in the definition of the outer measure m_*.

The **outer Jordan content** $J_*(E)$ of a set E in $\mathbb{R}$ is defined by

$$J_*(E) = \inf \sum_{j=1}^{N} |I_j|,$$

where the inf is taken over every *finite* covering $E \subset \bigcup_{j=1}^{N} I_j$, by intervals I_j.

(a) Prove that $J_*(E) = J_*(\overline{E})$ for every set E (here $\overline{E}$ denotes the closure of E).

(b) Exhibit a countable subset $E \subset [0, 1]$ such that $J_*(E) = 1$ while $m_*(E) = 0$.

15. At the start of the theory, one might define the outer measure by taking coverings by rectangles instead of cubes. More precisely, we define

$$m_*^{\mathcal{R}}(E) = \inf \sum_{j=1}^{\infty} |R_j|,$$

where the inf is now taken over all countable coverings $E \subset \bigcup_{j=1}^{\infty} R_j$ by (closed) rectangles.

Show that this approach gives rise to the same theory of measure developed in the text, by proving that $m_*(E) = m_*^{\mathcal{R}}(E)$ for every subset E of $\mathbb{R}^d$.

[Hint: Use Lemma 1.1.]

16. The Borel-Cantelli lemma. Suppose $\{E_k\}_{k=1}^{\infty}$ is a countable family of measuable subsets of $\mathbb{R}^d$ and that

$$\sum_{k=1}^{\infty} m(E_k) < \infty.$$

Let

$$E = \{x \in \mathbb{R}^d : x \in E_k, \text{ for infinitely many } k\}$$
$$= \limsup_{k \to \infty} (E_k).$$

(a) Show that E is measurable.

(b) Prove $m(E) = 0$.

[Hint: Write $E = \bigcap_{n=1}^{\infty} \bigcup_{k \geq n} E_k$.]

17. Let $\{f_n\}$ be a sequence of measurable functions on $[0, 1]$ with $|f_n(x)| < \infty$ for a.e x. Show that there exists a sequence c_n of positive real numbers such that

$$\frac{f_n(x)}{c_n} \to 0 \quad \text{a.e. } x$$

[Hint: Pick c_n such that $m(\{x : |f_n(x)/c_n| > 1/n\}) < 2^{-n}$, and apply the Borel-Cantelli lemma.]

18. Prove the following assertion: Every measurable function is the limit a.e. of a sequence of continuous functions.

19. Here are some observations regarding the set operation $A + B$.

(a) Show that if either A and B is open, then $A + B$ is open.

(b) Show that if A and B are closed, then $A + B$ is measurable.

(c) Show, however, that $A + B$ might not be closed even though A and B are closed.

[Hint: For (b) show that $A + B$ is an F_σ set.]

20. Show that there exist closed sets A and B with $m(A) = m(B) = 0$, but $m(A + B) > 0$:

(a) In $\mathbb{R}$, let $A = \mathcal{C}$ (the Cantor set), $B = \mathcal{C}/2$. Note that $A + B \supset [0, 1]$.

(b) In $\mathbb{R}^2$, observe that if $A = I \times \{0\}$ and $B = \{0\} \times I$ (where $I = [0, 1]$), then $A + B = I \times I$.

21. Prove that there is a continuous function that maps a Lebesgue measurable set to a non-measurable set.

[Hint: Consider a non-measurable subset of $[0, 1]$, and its inverse image in $\mathcal{C}$ by the function F in Exercise 2.]

22. Let $\chi_{[0,1]}$ be the characteristic function of $[0, 1]$. Show that there is no everywhere continuous function f on $\mathbb{R}$ such that

$$f(x) = \chi_{[0,1]}(x) \quad \text{almost everywhere.}$$

23. Suppose $f(x, y)$ is a function on $\mathbb{R}^2$ that is separately continuous: for each fixed variable, f is continuous in the other variable. Prove that f is measurable on $\mathbb{R}^2$.

[Hint: Approximate f in the variable x by piecewise-linear functions f_n so that $f_n \to f$ pointwise.]

24. Does there exist an enumeration $\{r_n\}_{n=1}^{\infty}$ of the rationals, such that the complement of

$$\bigcup_{n=1}^{\infty} \left(r_n - \frac{1}{n}, r_n + \frac{1}{n} \right)$$

in $\mathbb{R}$ is non-empty?

[Hint: Find an enumeration where the only rationals outside of a fixed bounded interval take the form r_n, with $n = m^2$ for some integer m.]

25. An alternative definition of measurability is as follows: E is measurable if for every $\epsilon > 0$ there is a *closed* set F contained in E with $m_*(E - F) < \epsilon$. Show that this definition is equivalent with the one given in the text.

26. Suppose $A \subset E \subset B$, where A and B are measurable sets of finite measure. Prove that if $m(A) = m(B)$, then E is measurable.

27. Suppose E_1 and E_2 are a pair of compact sets in $\mathbb{R}^d$ with $E_1 \subset E_2$, and let $a = m(E_1)$ and $b = m(E_2)$. Prove that for any c with $a < c < b$, there is a compact set E with $E_1 \subset E \subset E_2$ and $m(E) = c$.

[Hint: As an example, if $d = 1$ and E is a measurable subset of $[0, 1]$, consider $m(E \cap [0, t])$ as a function of t.]

28. Let E be a subset of $\mathbb{R}$ with $m_*(E) > 0$. Prove that for each $0 < \alpha < 1$, there exists an open interval I so that

$$m_*(E \cap I) \geq \alpha\, m_*(I).$$

Loosely speaking, this estimate shows that E contains almost a whole interval.

[Hint: Choose an open set $\mathcal{O}$ that contains E, and such that $m_*(E) \geq \alpha\, m_*(\mathcal{O})$. Write $\mathcal{O}$ as the countable union of disjoint open intervals, and show that one of these intervals must satisfy the desired property.]

29. Suppose E is a measurable subset of $\mathbb{R}$ with $m(E) > 0$. Prove that the **difference set** of E, which is defined by

$$\{z \in \mathbb{R} : z = x - y \text{ for some } x, y \in E\},$$

contains an open interval centered at the origin.

If E contains an interval, then the conclusion is straightforward. In general, one may rely on Exercise 28.

[Hint: Indeed, by Exercise 28, there exists an open interval I so that $m(E \cap I) \geq (9/10)\, m(I)$. If we denote $E \cap I$ by E_0, and suppose that the difference set of E_0 does not contain an open interval around the origin, then for arbitrarily small a the sets E_0, and $E_0 + a$ are disjoint. From the fact that $(E_0 \cup (E_0 + a)) \subset (I \cup (I + a))$ we get a contradiction, since the left-hand side has measure $2m(E_0)$, while the right-hand side has measure only slightly larger than $m(I)$.]

A more general formulation of this result is as follows.

30. If E and F are measurable, and $m(E) > 0$, $m(F) > 0$, prove that

$$E + F = \{x + y : x \in E,\ x \in F\}$$

contains an interval.

31. The result in Exercise 29 provides an alternate proof of the non-measurability of the set $\mathcal{N}$ studied in the text. In fact, we may also prove the non-measurability of a set in $\mathbb{R}$ that is very closely related to the set $\mathcal{N}$.

Given two real numbers x and y, we shall write as before that $x \sim y$ whenever the difference $x - y$ is rational. Let $\mathcal{N}^*$ denote a set that consists of one element in each equivalence class of $\sim$. Prove that $\mathcal{N}^*$ is non-measurable by using the result in Exercise 29.

[Hint: If $\mathcal{N}^*$ is measurable, then so are its translates $\mathcal{N}_n^* = \mathcal{N}^* + r_n$, where $\{r_n\}_{n=1}^{\infty}$ is an enumeration of $\mathbb{Q}$. How does this imply that $m(\mathcal{N}^*) > 0$? Can the difference set of $\mathcal{N}^*$ contain an open interval centered at the origin?]

32. Let $\mathcal{N}$ denote the non-measurable subset of $I = [0, 1]$ constructed at the end of Section 3.

(a) Prove that if E is a measurable subset of $\mathcal{N}$, then $m(E) = 0$.

(b) If G is a subset of $\mathbb{R}$ with $m_*(G) > 0$, prove that a subset of G is non-measurable.

[Hint: For (a) use the translates of E by the rationals.]

33. Let $\mathcal{N}$ denote the non-measurable set constructed in the text. Recall from the exercise above that measurable subsets of $\mathcal{N}$ have measure zero.

Show that the set $\mathcal{N}^c = I - \mathcal{N}$ satisfies $m_*(\mathcal{N}^c) = 1$, and conclude that if $E_1 = \mathcal{N}$ and $E_2 = \mathcal{N}^c$, then

$$m_*(E_1) + m_*(E_2) \neq m_*(E_1 \cup E_2),$$

although E_1 and E_2 are disjoint.

[Hint: To prove that $m_*(\mathcal{N}^c) = 1$, argue by contradiction and pick a measurable set U such that $U \subset I$, $\mathcal{N}^c \subset U$ and $m_*(U) < 1 - \epsilon$.]

34. Let $\mathcal{C}_1$ and $\mathcal{C}_2$ be any two Cantor sets (constructed in Exercise 3). Show that there exists a function $F : [0,1] \to [0,1]$ with the following properties:

(i) F is continuous and bijective,

(ii) F is monotonically increasing,

(iii) F maps $\mathcal{C}_1$ surjectively onto $\mathcal{C}_2$.

[Hint: Copy the construction of the standard Cantor-Lebesgue function.]

35. Give an example of a measurable function f and a continuous function Φ so that $f \circ \Phi$ is non-measurable.

[Hint: Let $\Phi : \mathcal{C}_1 \to \mathcal{C}_2$ as in Exercise 34, with $m(\mathcal{C}_1) > 0$ and $m(\mathcal{C}_2) = 0$. Let $N \subset \mathcal{C}_1$ be non-measurable, and take $f = \chi_{\Phi(N)}$.]

Use the construction in the hint to show that there exists a Lebesgue measurable set that is not a Borel set.

36. This exercise provides an example of a measurable function f on $[0,1]$ such that every function g equivalent to f (in the sense that f and g differ only on a set of measure zero) is discontinuous at *every* point.

(a) Construct a measurable set $E \subset [0,1]$ such that for any non-empty open sub-interval I in $[0,1]$, both sets $E \cap I$ and $E^c \cap I$ have positive measure.

(b) Show that $f = \chi_E$ has the property that whenever $g(x) = f(x)$ a.e x, then g must be discontinuous at every point in $[0,1]$.

[Hint: For the first part, consider a Cantor-like set of positive measure, and add in each of the intervals that are omitted in the first step of its construction, another Cantor-like set. Continue this procedure indefinitely.]

37. Suppose Γ is a curve $y = f(x)$ in $\mathbb{R}^2$, where f is continuous. Show that $m(\Gamma) = 0$.

[Hint: Cover Γ by rectangles, using the uniform continuity of f.]

38. Prove that $(a + b)^\gamma \geq a^\gamma + b^\gamma$ whenever $\gamma \geq 1$ and $a, b \geq 0$. Also, show that the reverse inequality holds when $0 \leq \gamma \leq 1$.

[Hint: Integrate the inequality between $(a + t)^{\gamma-1}$ and $t^{\gamma-1}$ from 0 to b.]

39. Establish the inequality

$$(10) \qquad \frac{x_1 + \cdots + x_d}{d} \geq (x_1 \cdots x_d)^{1/d} \qquad \text{for all } x_j \geq 0, \ j = 1, \ldots, d$$

by using backward induction as follows:

(a) The inequality is true whenever d is a power of 2 ($d = 2^k$, $k \geq 1$).

(b) If (10) holds for some integer $d \geq 2$, then it must hold for $d - 1$, that is, one has $(y_1 + \cdots + y_{d-1})/(d - 1) \geq (y_1 \cdots y_{d-1})^{1/(d-1)}$ for all $y_j \geq 0$, with $j = 1, \ldots, d - 1$.

[Hint: For (a), if $k \geq 2$, write $(x_1 + \cdots + x_{2^k})/2^k$ as $(A + B)/2$, where $A = (x_1 + \cdots + x_{2^{k-1}})/2^{k-1}$, and apply the inequality when $d = 2$. For (b), apply the inequality to $x_1 = y_1, \ldots, x_{d-1} = y_{d-1}$ and $x_d = (y_1 + \cdots + y_{d-1})/(d - 1)$.]

7 Problems

1. Given an irrational x, one can show (using the pigeon-hole principle, for example) that there exists infinitely many fractions p/q, with relatively prime integers p and q such that

$$\left| x - \frac{p}{q} \right| \leq \frac{1}{q^2}.$$

However, prove that the set of those $x \in \mathbb{R}$ such that there exist infinitely many fractions p/q, with relatively prime integers p and q such that

$$\left| x - \frac{p}{q} \right| \leq \frac{1}{q^3} \qquad (\text{or} \leq 1/q^{2+\epsilon}),$$

is a set of measure zero.

[Hint: Use the Borel-Cantelli lemma.]

2. Any open set Ω can be written as the union of closed cubes, so that $\Omega = \bigcup Q_j$ with the following properties

(i) The Q_j's have disjoint interiors.

(ii) $d(Q_j, \Omega^c) \approx$ side length of Q_j. This means that there are positive constants c and C so that $c \leq d(Q_j, \Omega^c)/\ell(Q_j) \leq C$, where $\ell(Q_j)$ denotes the side length of Q_j.

3. Find an example of a measurable subset C of $[0, 1]$ such that $m(C) = 0$, yet the difference set of C contains a non-trivial interval centered at the origin. Compare with the result in Exercise 29.

[Hint: Pick the Cantor set $C = \mathcal{C}$. For a fixed $a \in [-1, 1]$, consider the line $y = x + a$ in the plane, and copy the construction of the Cantor set, but in the cube $Q = [0, 1] \times [0, 1]$. First, remove all but four closed cubes of side length $1/3$, one at each corner of Q; then, repeat this procedure in each of the remaining cubes (see Figure 6). The resulting set is sometimes called a Cantor dust. Use the property of nested compact sets to show that the line intersects this Cantor dust.]

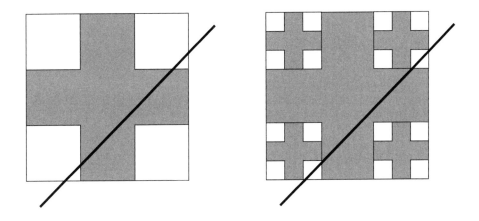

Figure 6. Construction of the Cantor dust

4. Complete the following outline to prove that a bounded function on an interval $[a, b]$ is Riemann integrable if and only if its set of discontinuities has measure zero. This argument is given in detail in the appendix to Book I.

Let f be a *bounded* function on a compact interval J, and let $I(c, r)$ denote the open interval centered at c of radius $r > 0$. Let $\operatorname{osc}(f, c, r) = \sup |f(x) - f(y)|$, where the supremum is taken over all $x, y \in J \cap I(c, r)$, and define the oscillation of f at c by $\operatorname{osc}(f, c) = \lim_{r \to 0} \operatorname{osc}(f, c, r)$. Clearly, f is continuous at $c \in J$ if and only if $\operatorname{osc}(f, c) = 0$.

Prove the following assertions:

(a) For every $\epsilon > 0$, the set of points c in J such that $\operatorname{osc}(f, c) \geq \epsilon$ is compact.

(b) If the set of discontinuities of f has measure 0, then f is Riemann integrable.

[Hint: Given $\epsilon > 0$ let $A_\epsilon = \{c \in J : \operatorname{osc}(f, c) \geq \epsilon\}$. Cover A_ϵ by a finite number of open intervals whose total length is $\leq \epsilon$. Select an appropriate partition of J and estimate the difference between the upper and lower sums of f over this partition.]

(c) Conversely, if f is Riemann integrable on J, then its set of discontinuities has measure 0.

[Hint: The set of discontinuities of f is contained in $\bigcup_n A_{1/n}$. Choose a partition P such that $U(f, P) - L(f, P) < \epsilon/n$. Show that the total length of the intervals in P whose interior intersect $A_{1/n}$ is $\leq \epsilon$.]

5. Suppose E is measurable with $m(E) < \infty$, and

$$E = E_1 \cup E_2, \qquad E_1 \cap E_2 = \emptyset.$$

If $m(E) = m_*(E_1) + m_*(E_2)$, then E_1 and E_2 are measurable.

In particular, if $E \subset Q$, where Q is a finite cube, then E is measurable if and only if $m(Q) = m_*(E) + m_*(Q - E)$.

6.* The fact that the axiom of choice and the well-ordering principle are equivalent is a consequence of the following considerations.

One begins by defining a partial ordering on a set E to be a binary relation $\leq$ on the set E that satisfies:

(i) $x \leq x$ for all $x \in E$.

(ii) If $x \leq y$ and $y \leq x$, then $x = y$.

(iii) If $x \leq y$ and $y \leq z$, then $x \leq z$.

If in addition $x \leq y$ or $y \leq x$ whenever $x, y \in E$, then $\leq$ is a linear ordering of E.

The axiom of choice and the well-ordering principle are then logically equivalent to the Hausdorff maximal principle:

> *Every non-empty partially ordered set has a (non-empty) maximal linearly ordered subset.*

In other words, if E is partially ordered by $\leq$, then E contains a non-empty subset F which is linearly ordered by $\leq$ and such that if F is contained in a set G also linearly ordered by $\leq$, then $F = G$.

An application of the Hausdorff maximal principle to the collection of all well-orderings of subsets of E implies the well-ordering principle for E. However, the proof that the axiom of choice implies the Hausdorff maximal principle is more complicated.

7.* Consider the curve $\Gamma = \{y = f(x)\}$ in $\mathbb{R}^2$, $0 \leq x \leq 1$. Assume that f is twice continuously differentiable in $0 \leq x \leq 1$. Then show that $m(\Gamma + \Gamma) > 0$ if and only if $\Gamma + \Gamma$ contains an open set, if and only if f is not linear.

8.* Suppose A and B are open sets of finite positive measure. Then we have equality in the Brunn-Minkowski inequality (8) if and only if A and B are convex and similar, that is, there are a $\delta > 0$ and an $h \in \mathbb{R}^d$ such that

$$A = \delta B + h.$$

2 Integration Theory

...amongst the many definitions that have been successively proposed for the integral of real-valued functions of a real variable, I have retained only those which, in my opinion, are indispensable to understand the transformations undergone by the problem of integration, and to capture the relationship between the notion of area, so simple in appearance, and certain more complicated analytical definitions of the integral.

One might ask if there is sufficient interest to occupy oneself with such complications, and if it is not better to restrict oneself to the study of functions that necessitate only simple definitions.... As we shall see in this course, we would then have to renounce the possibility of resolving many problems posed long ago, and which have simple statements. It is to solve these problems, and not for love of complications, that I have introduced in this book a definition of the integral more general than that of Riemann.

H. Lebesgue, 1903

1 The Lebesgue integral: basic properties and convergence theorems

The general notion of the Lebesgue integral on $\mathbb{R}^d$ will be defined in a step-by-step fashion, proceeding successively to increasingly larger families of functions. At each stage we shall see that the integral satisfies elementary properties such as linearity and monotonicity, and we prove appropriate convergence theorems that amount to interchanging the integral with limits. At the end of the process we shall have achieved a general theory of integration that will be decisive in the study of further problems.

We proceed in four stages, by progressively integrating:

1. Simple functions

2. Bounded functions supported on a set of finite measure

3. Non-negative functions

4. Integrable functions (the general case).

We emphasize from the onset that all functions are assumed to be measurable. At the beginning we also consider only finite-valued functions which take on real values. Later we shall also consider extended-valued functions, and also complex-valued functions.

Stage one: simple functions

Recall from the previous chapter that a simple function φ is a finite sum

$$(1) \qquad\qquad \varphi(x) = \sum_{k=1}^{N} a_k \chi_{E_k}(x),$$

where the E_k are measurable sets of finite measure and the a_k are constants. A complication that arises from this definition is that a simple function can be written in a multitude of ways as such finite linear combinations; for example, $0 = \chi_E - \chi_E$ for any measurable set E of finite measure. Fortunately, there is an unambiguous choice for the representation of a simple function, which is natural and useful in applications.

The **canonical form** of φ is the unique decomposition as in (1), where the numbers a_k are distinct and non-zero, and the sets E_k are disjoint.

Finding the canonical form of φ is straightforward: since φ can take only finitely many distinct and non-zero values, say $c_1, \ldots, c_M$, we may set $F_k = \{x : \varphi(x) = c_k\}$, and note that the sets F_k are disjoint. Therefore $\varphi = \sum_{k=1}^{M} c_k \chi_{F_k}$ is the desired canonical form of φ.

If φ is a simple function with canonical form $\varphi(x) = \sum_{k=1}^{M} c_k \chi_{F_k}(x)$, then we *define* the **Lebesgue integral** of φ by

$$\int_{\mathbb{R}^d} \varphi(x)\, dx = \sum_{k=1}^{M} c_k m(F_k).$$

If E is a measurable subset of $\mathbb{R}^d$ with finite measure, then $\varphi(x)\chi_E(x)$ is also a simple function, and we *define*

$$\int_E \varphi(x)\, dx = \int \varphi(x)\chi_E(x)\, dx.$$

To emphasize the choice of the Lebesgue measure m in the definition of the integral, one sometimes writes

$$\int_{\mathbb{R}^d} \varphi(x)\, dm(x)$$

for the Lebesgue integral of φ. In fact, as a matter of convenience, we shall often write $\int \varphi(x)\,dx$ or simply $\int \varphi$ for the integral of φ over $\mathbb{R}^d$.

Proposition 1.1 *The integral of simple functions defined above satisfies the following properties:*

(i) *Independence of the representation. If $\varphi = \sum_{k=1}^{N} a_k \chi_{E_k}$ is any representation of φ, then*

$$\int \varphi = \sum_{k=1}^{N} a_k m(E_k).$$

(ii) *Linearity. If φ and ψ are simple, and $a, b \in \mathbb{R}$, then*

$$\int (a\varphi + b\psi) = a \int \varphi + b \int \psi.$$

(iii) *Additivity. If E and F are disjoint subsets of $\mathbb{R}^d$ with finite measure, then*

$$\int_{E \cup F} \varphi = \int_{E} \varphi + \int_{F} \varphi.$$

(iv) *Monotonicity. If $\varphi \leq \psi$ are simple, then*

$$\int \varphi \leq \int \psi.$$

(v) *Triangle inequality. If φ is a simple function, then so is $|\varphi|$, and*

$$\left| \int \varphi \right| \leq \int |\varphi|.$$

Proof.   The only conclusion that is a little tricky is the first, which asserts that the integral of a simple function can be calculated by using any of its decompositions as a linear combination of characteristic functions.

Suppose that $\varphi = \sum_{k=1}^{N} a_k \chi_{E_k}$, where we assume that the sets E_k are disjoint, but we do not suppose that the numbers a_k are distinct and non-zero. For each distinct non-zero value a among the $\{a_k\}$ we define $E'_a = \bigcup E_k$, where the union is taken over those indices k such that $a_k = a$. Note then that the sets E'_a are disjoint, and $m(E'_a) = \sum m(E_k)$, where the sum is taken over the same set of k's. Then clearly $\varphi = \sum a \chi_{E'_a}$, where the sum is over the distinct non-zero values of $\{a_k\}$. Thus

$$\int \varphi = \sum am(E'_a) = \sum_{k=1}^{N} a_k m(E_k).$$

Next, suppose $\varphi = \sum_{k=1}^{N} a_k \chi_{E_k}$, where we no longer assume that the E_k are disjoint. Then we can "refine" the decomposition $\bigcup_{k=1}^{N} E_k$ by finding sets $E_1^*, E_2^*, \ldots, E_n^*$ with the property that $\bigcup_{k=1}^{N} E_k = \bigcup_{j=1}^{n} E_j^*$; the sets E_j^* ($j = 1, \ldots, n$) are mutually disjoint; and for each k, $E_k = \bigcup E_j^*$, where the union is taken over those E_j^* that are contained in E_k. (A proof of this elementary fact can be found in Exercise 1.) For each j, let now $a_j^* = \sum a_k$, with the summation taken over all k such that E_k contains E_j^*. Then clearly $\varphi = \sum_{j=1}^{n} a_j^* \chi_{E_j^*}$. However, this is a decomposition already dealt with above because the E_j^* are disjoint. Thus

$$\int \varphi = \sum a_j^* m(E_j^*) = \sum \sum_{E_k \supset E_j^*} a_k m(E_j^*) = \sum a_k m(E_k),$$

and conclusion (i) is established.

Conclusion (ii) follows by using any representation of φ and ψ, and the obvious linearity of (i).

For the additivity over sets, one must note that if E and F are disjoint, then

$$\chi_{E \cup F} = \chi_E + \chi_F,$$

and we may use the linearity of the integral to see that $\int_{E \cup F} \varphi = \int_E \varphi + \int_F \varphi$.

If $\eta \geq 0$ is a simple function, then its canonical form is everywhere non-negative, and therefore $\int \eta \geq 0$ by the definition of the integral. Applying this argument to $\psi - \varphi$ gives the desired monotonicity property.

Finally, for the triangle inequality, it suffices to write φ in its canonical form $\varphi = \sum_{k=1}^{N} a_k \chi_{E_k}$ and observe that

$$|\varphi| = \sum_{k=1}^{N} |a_k| \chi_{E_k}(x).$$

Therefore, by the triangle inequality applied to the definition of the integral, one sees that

$$\left| \int \varphi \right| = \left| \sum_{k=1}^{N} a_k m(E_k) \right| \leq \sum_{k=1}^{N} |a_k| m(E_k) = \int |\varphi|.$$

Incidentally, it is worthwhile to point out the following easy fact: whenever f and g are a pair of simple functions that agree almost everywhere, then $\int f = \int g$. The identity of the integrals of two functions that agree almost everywhere will continue to hold for the successive definitions of the integral that follow.

Stage two: bounded functions supported on a set of finite measure

The **support** of a measurable function f is defined to be the set of all points where f does not vanish,

$$\mathrm{supp}(f) = \{x : f(x) \neq 0\}.$$

We shall also say that f is **supported** on a set E, if $f(x) = 0$ whenever $x \notin E$.

Since f is measurable, so is the set $\mathrm{supp}(f)$. We shall next be interested in those bounded measurable functions that have $m(\mathrm{supp}(f)) < \infty$.

An important result in the previous chapter (Theorem 4.2) states the following: if f is a function bounded by M and supported on a set E, then there exists a sequence $\{\varphi_n\}$ of simple functions, with each φ_n bounded by M and supported on E, and such that

$$\varphi_n(x) \to f(x) \quad \text{for all } x.$$

The key lemma that follows allows us to define the integral for the class of bounded functions supported on sets of finite measure.

Lemma 1.2 *Let f be a bounded function supported on a set E of finite measure. If $\{\varphi_n\}_{n=1}^{\infty}$ is any sequence of simple functions bounded by M, supported on E, and with $\varphi_n(x) \to f(x)$ for a.e. x, then:*

(i) *The limit $\displaystyle\lim_{n\to\infty} \int \varphi_n$ exists.*

(ii) *If $f = 0$ a.e., then the limit $\displaystyle\lim_{n\to\infty} \int \varphi_n$ equals 0.*

Proof. The assertions of the lemma would be nearly obvious if we had that φ_n converges to f uniformly on E. Instead, we recall one of Littlewood's principles, which states that the convergence of a sequence of measurable functions is "nearly" uniform. The precise statement lying behind this principle is Egorov's theorem, which we proved in Chapter 1, and which we apply here.

Since the measure of E is finite, given $\epsilon > 0$ Egorov's theorem guarantees the existence of a (closed) measurable subset A_ϵ of E such that $m(E - A_\epsilon) \leq \epsilon$, and $\varphi_n \to f$ uniformly on A_ϵ. Therefore, setting $I_n = \int \varphi_n$ we have that

$$
\begin{aligned}
|I_n - I_m| &\leq \int_E |\varphi_n(x) - \varphi_m(x)|\, dx \\
&= \int_{A_\epsilon} |\varphi_n(x) - \varphi_m(x)|\, dx + \int_{E - A_\epsilon} |\varphi_n(x) - \varphi_m(x)|\, dx \\
&\leq \int_{A_\epsilon} |\varphi_n(x) - \varphi_m(x)|\, dx + 2M\, m(E - A_\epsilon) \\
&\leq \int_{A_\epsilon} |\varphi_n(x) - \varphi_m(x)|\, dx + 2M\epsilon.
\end{aligned}
$$

By the uniform convergence, one has, for all $x \in A_\epsilon$ and all large n and m, the estimate $|\varphi_n(x) - \varphi_m(x)| < \epsilon$, so we deduce that

$$
|I_n - I_m| \leq m(E)\epsilon + 2M\epsilon \quad \text{for all large } n \text{ and } m.
$$

Since ϵ is arbitrary and $m(E) < \infty$, this proves that $\{I_n\}$ is a Cauchy sequence and hence converges, as desired.

For the second part, we note that if $f = 0$, we may repeat the argument above to find that $|I_n| \leq m(E)\epsilon + M\epsilon$, which yields $\lim_{n\to\infty} I_n = 0$, as was to be shown.

Using Lemma 1.2 we can now turn to the integration of bounded functions that are supported on sets of finite measure. For such a function f we *define* its **Lebesgue integral** by

$$
\int f(x)\, dx = \lim_{n\to\infty} \int \varphi_n(x)\, dx,
$$

where $\{\varphi_n\}$ is any sequence of simple functions satisfying: $|\varphi_n| \leq M$, each φ_n is supported on the support of f, and $\varphi_n(x) \to f(x)$ for a.e. x as n tends to infinity. By the previous lemma, we know that this limit exists.

Next, we must first show that $\int f$ is independent of the limiting sequence $\{\varphi_n\}$ used, in order for the integral to be well-defined. Therefore, suppose that $\{\psi_n\}$ is another sequence of simple functions that is bounded by M, supported on $\text{supp}(f)$, and such that $\psi_n(x) \to f(x)$ for a.e. x as n tends to infinity. Then, if $\eta_n = \varphi_n - \psi_n$, the sequence $\{\eta_n\}$ consists of simple functions bounded by $2M$, supported on a set of finite measure, and such that $\eta_n \to 0$ a.e. as n tends to infinity. We may therefore conclude, by the second part of the lemma, that $\int \eta_n \to 0$ as n tends to infinity. Consequently, the two limits

$$\lim_{n \to \infty} \int \varphi_n(x)\, dx \quad \text{and} \quad \lim_{n \to \infty} \int \psi_n(x)\, dx$$

(which exist by the lemma) are indeed equal.

If E is a subset of $\mathbb{R}^d$ with finite measure, and f is bounded with $m(\operatorname{supp}(f)) < \infty$, then it is natural to define

$$\int_E f(x)\, dx = \int f(x)\chi_E(x)\, dx.$$

Clearly, if f is itself simple, then $\int f$ as defined above coincides with the integral of simple functions studied earlier. This extension of the definition of integration also satisfies all the basic properties of the integral of simple functions.

Proposition 1.3 *Suppose f and g are bounded functions supported on sets of finite measure. Then the following properties hold.*

(i) *Linearity. If $a, b \in \mathbb{R}$, then*

$$\int (af + bg) = a \int f + b \int g.$$

(ii) *Additivity. If E and F are disjoint subsets of $\mathbb{R}^d$, then*

$$\int_{E \cup F} f = \int_E f + \int_F f.$$

(iii) *Monotonicity. If $f \leq g$, then*

$$\int f \leq \int g.$$

(iv) *Triangle inequality. $|f|$ is also bounded, supported on a set of finite measure, and*

$$\left| \int f \right| \leq \int |f|.$$

All these properties follow by using approximations by simple functions, and the properties of the integral of simple functions given in Proposition 1.1.

We are now in a position to prove the first important convergence theorem.

Theorem 1.4 (Bounded convergence theorem) *Suppose that* $\{f_n\}$ *is a sequence of measurable functions that are all bounded by* M, *are supported on a set* E *of finite measure, and* $f_n(x) \to f(x)$ *a.e.* x *as* $n \to \infty$. *Then* f *is measurable, bounded, supported on* E *for a.e.* x, *and*

$$\int |f_n - f| \to 0 \qquad as \ n \to \infty.$$

Consequently,

$$\int f_n \to \int f \qquad as \ n \to \infty.$$

Proof. From the assumptions one sees at once that f is bounded by M almost everywhere and vanishes outside E, except possibly on a set of measure zero. Clearly, the triangle inequality for the integral implies that it suffices to prove that $\int |f_n - f| \to 0$ as n tends to infinity.

The proof is a reprise of the argument in Lemma 1.2. Given $\epsilon > 0$, we may find, by Egorov's theorem, a measurable subset A_ϵ of E such that $m(E - A_\epsilon) \leq \epsilon$ and $f_n \to f$ uniformly on A_ϵ. Then, we know that for all sufficiently large n we have $|f_n(x) - f(x)| \leq \epsilon$ for all $x \in A_\epsilon$. Putting these facts together yields

$$\int |f_n(x) - f(x)|\,dx \leq \int_{A_\epsilon} |f_n(x) - f(x)|\,dx + \int_{E-A_\epsilon} |f_n(x) - f(x)|\,dx$$
$$\leq \epsilon m(E) + 2M\,m(E - A_\epsilon)$$

for all large n. Since ϵ is arbitrary, the proof of the theorem is complete.

We note that the above convergence theorem is a statement about the interchange of an integral and a limit, since its conclusion simply says

$$\lim_{n\to\infty} \int f_n = \int \lim_{n\to\infty} f_n.$$

A useful observation that we can make at this point is the following: if $f \geq 0$ is bounded and supported on a set of finite measure E and $\int f = 0$, then $f = 0$ almost everywhere. Indeed, if for each integer $k \geq 1$ we set $E_k = \{x \in E : f(x) \geq 1/k\}$, then the fact that $k^{-1}\chi_{E_k}(x) \leq f(x)$ implies

$$k^{-1}m(E_k) \leq \int f,$$

by monotonicity of the integral. Thus $m(E_k) = 0$ for all k, and since $\{x : f(x) > 0\} = \bigcup_{k=1}^{\infty} E_k$, we see that $f = 0$ almost everywhere.

Return to Riemann integrable functions

We shall now show that Riemann integrable functions are also Lebesgue integrable. When we combine this with the bounded convergence theorem we have just proved, we see that Lebesgue integration resolves the second problem in the Introduction.

Theorem 1.5 *Suppose f is Riemann integrable on the closed interval $[a, b]$. Then f is measurable, and*

$$\int_{[a,b]}^{\mathcal{R}} f(x)\, dx = \int_{[a,b]}^{\mathcal{L}} f(x)\, dx,$$

where the integral on the left-hand side is the standard Riemann integral, and that on the right-hand side is the Lebesgue integral.

Proof. By definition, a Riemann integrable function is bounded, say $|f(x)| \leq M$, so we need to prove that f is measurable, and then establish the equality of integrals.

Again, by definition of Riemann integrability,[1] we may construct two sequences of step functions $\{\varphi_k\}$ and $\{\psi_k\}$ that satisfy the following properties: $|\varphi_k(x)| \leq M$ and $|\psi_k(x)| \leq M$ for all $x \in [a, b]$ and $k \geq 1$,

$$\varphi_1(x) \leq \varphi_2(x) \leq \cdots \leq f \leq \cdots \leq \psi_2(x) \leq \psi_1(x),$$

and

$$(2) \qquad \lim_{k \to \infty} \int_{[a,b]}^{\mathcal{R}} \varphi_k(x)\, dx = \lim_{k \to \infty} \int_{[a,b]}^{\mathcal{R}} \psi_k(x)\, dx = \int_{[a,b]}^{\mathcal{R}} f(x)\, dx.$$

Several observations are in order. First, it follows immediately from their definition that for step functions the Riemann and Lebesgue integrals agree; therefore

(3)

$$\int_{[a,b]}^{\mathcal{R}} \varphi_k(x)\, dx = \int_{[a,b]}^{\mathcal{L}} \varphi_k(x)\, dx \quad \text{and} \quad \int_{[a,b]}^{\mathcal{R}} \psi_k(x)\, dx = \int_{[a,b]}^{\mathcal{L}} \psi_k(x)\, dx$$

---

[1]See also Section 1 of the Appendix in Book I.

for all $k \geq 1$. Next, if we let

$$\tilde{\varphi}(x) = \lim_{k \to \infty} \varphi_k(x) \quad \text{and} \quad \tilde{\psi}(x) = \lim_{k \to \infty} \psi_k(x),$$

we have $\tilde{\varphi} \leq f \leq \tilde{\psi}$. Moreover, both $\tilde{\varphi}$ and $\tilde{\psi}$ are measurable (being the limit of step functions), and the bounded convergence theorem yields

$$\lim_{k \to \infty} \int_{[a,b]}^{\mathcal{L}} \varphi_k(x) \, dx = \int_{[a,b]}^{\mathcal{L}} \tilde{\varphi}(x) \, dx$$

and

$$\lim_{k \to \infty} \int_{[a,b]}^{\mathcal{L}} \psi_k(x) \, dx = \int_{[a,b]}^{\mathcal{L}} \tilde{\psi}(x) \, dx.$$

This together with (2) and (3) yields

$$\int_{[a,b]}^{\mathcal{L}} (\tilde{\psi}(x) - \tilde{\varphi}(x)) \, dx = 0,$$

and since $\psi_k - \varphi_k \geq 0$, we must have $\tilde{\psi} - \tilde{\varphi} \geq 0$. By the observation following the proof of the bounded convergence theorem, we conclude that $\tilde{\psi} - \tilde{\varphi} = 0$ a.e., and therefore $\tilde{\varphi} = \tilde{\psi} = f$ a.e., which proves that f is measurable. Finally, since $\varphi_k \to f$ almost everywhere, we have (by definition)

$$\lim_{k \to \infty} \int_{[a,b]}^{\mathcal{L}} \varphi_k(x) \, dx = \int_{[a,b]}^{\mathcal{L}} f(x) \, dx,$$

and by (2) and (3) we see that $\int_{[a,b]}^{\mathcal{R}} f(x) \, dx = \int_{[a,b]}^{\mathcal{L}} f(x) \, dx$, as desired.

Stage three: non-negative functions

We proceed with the integrals of functions that are measurable and non-negative but not necessarily bounded. It will be important to allow these functions to be extended-valued, that is, these functions may take on the value $+\infty$ (on a measurable set). We recall in this connection the convention that one defines the supremum of a set of positive numbers to be $+\infty$ if the set is unbounded.

In the case of such a function f we *define* its (extended) **Lebesgue integral** by

$$\int f(x) \, dx = \sup_g \int g(x) \, dx,$$

where the supremum is taken over all measurable functions g such that $0 \leq g \leq f$, and where g is bounded and supported on a set of finite measure.

With the above definition of the integral, there are only two possible cases; the supremum is either finite, or infinite. In the first case, when $\int f(x)\,dx < \infty$, we shall say that f is **Lebesgue integrable** or simply **integrable**.

Clearly, if E is any measurable subset of $\mathbb{R}^d$, and $f \geq 0$, then $f\chi_E$ is also positive, and we define

$$\int_E f(x)\,dx = \int f(x)\chi_E(x)\,dx.$$

Simple examples of functions on $\mathbb{R}^d$ that are integrable (or non-integrable) are given by

$$f_a(x) = \begin{cases} |x|^{-a} & \text{if } |x| \leq 1, \\ 0 & \text{if } |x| > 1. \end{cases}$$

$$F_a(x) = \frac{1}{1 + |x|^a}, \qquad \text{all } x \in \mathbb{R}^d.$$

Then f_a is integrable exactly when $a < d$, while F_a is integrable exactly when $a > d$. See the discussion following Corollary 1.10 and also Exercise 10.

Proposition 1.6 *The integral of non-negative measurable functions enjoys the following properties:*

(i) *Linearity. If $f, g \geq 0$, and a, b are positive real numbers, then*

$$\int (af + bg) = a \int f + b \int g.$$

(ii) *Additivity. If E and F are disjoint subsets of $\mathbb{R}^d$, and $f \geq 0$, then*

$$\int_{E \cup F} f = \int_E f + \int_F f.$$

(iii) *Monotonicity. If $0 \leq f \leq g$, then*

$$\int f \leq \int g.$$

(iv) *If g is integrable and $0 \le f \le g$, then f is integrable.*

(v) *If f is integrable, then $f(x) < \infty$ for almost every x.*

(vi) *If $\int f = 0$, then $f(x) = 0$ for almost every x.*

Proof.   Of the first four assertions, only (i) is not an immediate consequence of the definitions, and to prove it we argue as follows. We take $a = b = 1$ and note that if $\varphi \le f$ and $\psi \le g$, where both φ and ψ are bounded and supported on sets of finite measure, then $\varphi + \psi \le f + g$, and $\varphi + \psi$ is also bounded and supported on a set of finite measure. Consequently

$$\int f + \int g \le \int (f + g).$$

To prove the reverse inequality, suppose η is bounded and supported on a set of finite measure, and $\eta \le f + g$. If we define $\eta_1(x) = \min(f(x), \eta(x))$ and $\eta_2 = \eta - \eta_1$, we note that

$$\eta_1 \le f \quad \text{and} \quad \eta_2 \le g.$$

Moreover both η_1, η_2 are bounded and supported on sets of finite measure. Hence

$$\int \eta = \int (\eta_1 + \eta_2) = \int \eta_1 + \int \eta_2 \le \int f + \int g.$$

Taking the supremum over η yields the required inequality.

To prove the conclusion (v) we argue as follows. Suppose $E_k = \{x : f(x) \ge k\}$, and $E_\infty = \{x : f(x) = \infty\}$. Then

$$\int f \ge \int \chi_{E_k} f \ge k m(E_k),$$

hence $m(E_k) \to 0$ as $k \to \infty$. Since $E_k \searrow E_\infty$, Corollary 3.3 in the previous chapter implies that $m(E_\infty) = 0$.

The proof of (vi) is the same as the observation following Theorem 1.4.

We now turn our attention to some important convergence theorems for the class of non-negative measurable functions. To motivate the results that follow, we ask the following question: Suppose $f_n \ge 0$ and $f_n(x) \to f(x)$ for almost every x. Is it true that $\int f_n \, dx \to \int f \, dx$? Unfortunately, the example that follows provides a negative answer to this, and shows that we must change our formulation of the question to obtain a positive convergence result.

Let

$$f_n(x) = \begin{cases} n & \text{if } 0 < x < 1/n, \\ 0 & \text{otherwise.} \end{cases}$$

Then $f_n(x) \to 0$ for all x, yet $\int f_n(x)\,dx = 1$ for all n. In this particular example, the limit of the integrals is greater than the integral of the limit function. This turns out to be the case in general, as we shall see now.

Lemma 1.7 (Fatou) *Suppose $\{f_n\}$ is a sequence of measurable functions with $f_n \geq 0$. If $\lim_{n\to\infty} f_n(x) = f(x)$ for a.e. x, then*

$$\int f \leq \liminf_{n\to\infty} \int f_n.$$

Proof. Suppose $0 \leq g \leq f$, where g is bounded and supported on a set E of finite measure. If we set $g_n(x) = \min(g(x), f_n(x))$, then g_n is measurable, supported on E, and $g_n(x) \to g(x)$ a.e., so by the bounded convergence theorem

$$\int g_n \to \int g.$$

By construction, we also have $g_n \leq f_n$, so that $\int g_n \leq \int f_n$, and therefore

$$\int g \leq \liminf_{n\to\infty} \int f_n.$$

Taking the supremum over all g yields the desired inequality.

In particular, we do not exclude the cases $\int f = \infty$, or $\liminf_{n\to\infty} f_n = \infty$.

We can now immediately deduce the following series of corollaries.

Corollary 1.8 *Suppose f is a non-negative measurable function, and $\{f_n\}$ a sequence of non-negative measurable functions with $f_n(x) \leq f(x)$ and $f_n(x) \to f(x)$ for almost every x. Then*

$$\lim_{n\to\infty} \int f_n = \int f.$$

Proof. Since $f_n(x) \leq f(x)$ a.e x, we necessarily have $\int f_n \leq \int f$ for all n; hence

$$\limsup_{n\to\infty} \int f_n \leq \int f.$$

This inequality combined with Fatou's lemma proves the desired limit.

In particular, we can now obtain a basic convergence theorem for the class of non-negative measurable functions. Its statement requires the following notation.

In analogy with the symbols $\nearrow$ and $\searrow$ used to describe increasing and decreasing sequences of sets, we shall write

$$f_n \nearrow f$$

whenever $\{f_n\}_{n=1}^\infty$ is a sequence of measurable functions that satisfies

$$f_n(x) \leq f_{n+1}(x) \quad \text{a.e } x, \text{ all } n \geq 1 \quad \text{and} \quad \lim_{n\to\infty} f_n(x) = f(x) \quad \text{a.e } x.$$

Similarly, we write $f_n \searrow f$ whenever

$$f_n(x) \geq f_{n+1}(x) \quad \text{a.e } x, \text{ all } n \geq 1 \quad \text{and} \quad \lim_{n\to\infty} f_n(x) = f(x) \quad \text{a.e } x.$$

Corollary 1.9 (Monotone convergence theorem) *Suppose $\{f_n\}$ is a sequence of non-negative measurable functions with $f_n \nearrow f$. Then*

$$\lim_{n\to\infty} \int f_n = \int f.$$

The monotone convergence theorem has the following useful consequence:

Corollary 1.10 *Consider a series $\sum_{k=1}^\infty a_k(x)$, where $a_k(x) \geq 0$ is measurable for every $k \geq 1$. Then*

$$\int \sum_{k=1}^\infty a_k(x)\,dx = \sum_{k=1}^\infty \int a_k(x)\,dx.$$

If $\sum_{k=1}^\infty \int a_k(x)\,dx$ is finite, then the series $\sum_{k=1}^\infty a_k(x)$ converges for a.e. x.

Proof. Let $f_n(x) = \sum_{k=1}^n a_k(x)$ and $f(x) = \sum_{k=1}^\infty a_k(x)$. The functions f_n are measurable, $f_n(x) \leq f_{n+1}(x)$, and $f_n(x) \to f(x)$ as n tends to infinity. Since

$$\int f_n = \sum_{k=1}^n \int a_k(x)\,dx,$$

the monotone convergence theorem implies

$$\sum_{k=1}^{\infty} \int a_k(x)\, dx = \int \sum_{k=1}^{\infty} a_k(x)\, dx.$$

If $\sum \int a_k < \infty$, then the above implies that $\sum_{k=1}^{\infty} a_k(x)$ is integrable, and by our earlier observation, we conclude that $\sum_{k=1}^{\infty} a_k(x)$ is finite almost everywhere.

We give two nice illustrations of this last corollary.

The first consists of another proof of the Borel-Cantelli lemma (see Exercise 16, Chapter 1), which says that if $E_1, E_2, \ldots$ is a collection of measurable subsets with $\sum m(E_k) < \infty$, then the set of points that belong to infinitely many sets E_k has measure zero. To prove this fact, we let

$$a_k(x) = \chi_{E_k}(x),$$

and note that a point x belongs to infinitely many sets E_k if and only if $\sum_{k=1}^{\infty} a_k(x) = \infty$. Our assumption on $\sum m(E_k)$ says precisely that $\sum_{k=1}^{\infty} \int a_k(x)\, dx < \infty$, and the corollary implies that $\sum_{k=1}^{\infty} a_k(x)$ is finite except possibly on a set of measure zero, and thus the Borel-Cantelli lemma is proved.

The second illustration will be useful in our discussion of approximations to the identity in Chapter 3. Consider the function

$$f(x) = \begin{cases} \frac{1}{|x|^{d+1}} & \text{if } x \neq 0, \\ 0 & \text{otherwise.} \end{cases}$$

We prove that f is integrable outside any ball, $|x| \geq \epsilon$, and moreover

$$\int_{|x| \geq \epsilon} f(x)\, dx \leq \frac{C}{\epsilon}, \quad \text{for some constant } C > 0.$$

Indeed, if we let $A_k = \{x \in \mathbb{R}^d : 2^k \epsilon < |x| \leq 2^{k+1}\epsilon\}$, and define

$$g(x) = \sum_{k=0}^{\infty} a_k(x) \quad \text{where} \quad a_k(x) = \frac{1}{(2^k \epsilon)^{d+1}} \chi_{A_k}(x),$$

then we must have $f(x) \leq g(x)$, and hence $\int f \leq \int g$. Since the set A_k is obtained from $\mathcal{A} = \{1 < |x| < 2\}$ by a dilation of factor $2^k \epsilon$, we have by the relative dilation-invariance properties of the Lebesgue measure, that $m(A_k) = (2^k \epsilon)^d m(\mathcal{A})$. Also by Corollary 1.10, we see that

$$\int g = \sum_{k=0}^{\infty} \frac{m(A_k)}{(2^k \epsilon)^{d+1}} = m(\mathcal{A}) \sum_{k=0}^{\infty} \frac{(2^k \epsilon)^d}{(2^k \epsilon)^{d+1}} = \frac{C}{\epsilon},$$

where $C = 2m(\mathcal{A})$. Note that the same dilation-invariance property in fact shows that

$$\int_{|x| \geq \epsilon} \frac{dx}{|x|^{d+1}} = \frac{1}{\epsilon} \int_{|x| \geq 1} \frac{dx}{|x|^{d+1}}.$$

See also the identity (7) below.

Stage four: general case

If f is any real-valued measurable function on $\mathbb{R}^d$, we say that f is **Lebesgue integrable** (or just integrable) if the non-negative measurable function $|f|$ is integrable in the sense of the previous section.

If f is Lebesgue integrable, we give a meaning to its integral as follows. First, we may define

$$f^+(x) = \max(f(x), 0) \quad \text{and} \quad f^-(x) = \max(-f(x), 0),$$

so that both f^+ and f^- are non-negative and $f^+ - f^- = f$. Since $f^{\pm} \leq |f|$, both functions f^+ and f^- are integrable whenever f is, and we then define the **Lebesgue integral** of f by

$$\int f = \int f^+ - \int f^-.$$

In practice one encounters many decompositions $f = f_1 - f_2$, where f_1, f_2 are both non-negative integrable functions, and one would expect that regardless of the decomposition of f, we always have

$$\int f = \int f_1 - \int f_2.$$

In other words, the definition of the integral should be independent of the decomposition $f = f_1 - f_2$. To see why this is so, suppose $f = g_1 - g_2$ is another decomposition where both g_1 and g_2 are non-negative and integrable. Since $f_1 - f_2 = g_1 - g_2$ we have $f_1 + g_2 = g_1 + f_2$; but both sides of this last identity consist of positive measurable functions, so the linearity of the integral in this case yields

$$\int f_1 + \int g_2 = \int g_1 + \int f_2.$$

Since all integrals involved are finite, we find the desired result

$$\int f_1 - \int f_2 = \int g_1 - \int g_2.$$

In considering the above definitions it is useful to keep in mind the following small observations. Both the integrability of f, and the value of its integral are unchanged if we modify f arbitrarily on a set of measure zero. It is therefore useful to adopt the convention that in the context of integration we allow our functions to be undefined on sets of measure zero. Moreover, if f is integrable, then by (v) of Proposition 1.6, it is finite-valued almost everywhere. Thus, availing ourselves of the above convention, we can always add two integrable functions f and g, since the ambiguity of $f + g$, due to the extended values of each, resides in a set of measure zero. Moreover, we note that when speaking of a function f, we are, in effect, also speaking about the collection of all functions that equal f almost everywhere.

Simple applications of the definition and the properties proved previously yield all the elementary properties of the integral:

Proposition 1.11 *The integral of Lebesgue integrable functions is linear, additive, monotonic, and satisfies the triangle inequality.*

We now gather two results which, although instructive in their own right, are also needed in the proof of the next theorem.

Proposition 1.12 *Suppose f is integrable on $\mathbb{R}^d$. Then for every $\epsilon > 0$:*

(i) *There exists a set of finite measure B (a ball, for example) such that*

$$\int_{B^c} |f| < \epsilon.$$

(ii) *There is a $\delta > 0$ such that*

$$\int_E |f| < \epsilon \qquad \text{whenever } m(E) < \delta.$$

The last condition is known as absolute continuity.

Proof. By replacing f with $|f|$ we may assume without loss of generality that $f \geq 0$.

For the first part, let B_N denote the ball of radius N centered at the origin, and note that if $f_N(x) = f(x)\chi_{B_N}(x)$, then $f_N \geq 0$ is measurable, $f_N(x) \leq f_{N+1}(x)$, and $\lim_{N \to \infty} f_N(x) = f(x)$. By the monotone convergence theorem, we must have

$$\lim_{N \to \infty} \int f_N = \int f.$$

In particular, for some large N,

$$0 \leq \int f - \int f\chi_{B_N} < \epsilon,$$

and since $1 - \chi_{B_N} = \chi_{B_N^c}$, this implies $\int_{B_N^c} f < \epsilon$, as we set out to prove.

For the second part, assuming again that $f \geq 0$, we let $f_N(x) = f(x)\chi_{E_N}$ where

$$E_N = \{x : f(x) \leq N\}.$$

Once again, $f_N \geq 0$ is measurable, $f_N(x) \leq f_{N+1}(x)$, and given $\epsilon > 0$ there exists (by the monotone convergence theorem) an integer $N > 0$ such that

$$\int (f - f_N) < \frac{\epsilon}{2}.$$

We now pick $\delta > 0$ so that $N\delta < \epsilon/2$. If $m(E) < \delta$, then

$$\begin{aligned}
\int_E f &= \int_E (f - f_N) + \int_E f_N \\
&\leq \int (f - f_N) + \int_E f_N \\
&\leq \int (f - f_N) + Nm(E) \\
&\leq \frac{\epsilon}{2} + \frac{\epsilon}{2} = \epsilon.
\end{aligned}$$

This concludes the proof of the proposition.

Intuitively, integrable functions should in some sense vanish at infinity since their integrals are finite, and the first part of the proposition attaches a precise meaning to this intuition. One should observe, however, that integrability need not guarantee the more naive pointwise vanishing as $|x|$ becomes large. See Exercise 6.

We are now ready to prove a cornerstone of the theory of Lebesgue integration, the dominated convergence theorem. It can be viewed as a culmination of our efforts, and is a general statement about the interplay between limits and integrals.

Theorem 1.13 *Suppose $\{f_n\}$ is a sequence of measurable functions such that $f_n(x) \to f(x)$ a.e. x, as n tends to infinity. If $|f_n(x)| \le g(x)$, where g is integrable, then*

$$\int |f_n - f| \to 0 \quad \text{as } n \to \infty,$$

and consequently

$$\int f_n \to \int f \quad \text{as } n \to \infty.$$

Proof. For each $N \ge 0$ let $E_N = \{x : |x| \le N, \ g(x) \le N\}$. Given $\epsilon > 0$, we may argue as in the first part of the previous lemma, to see that there exists N so that $\int_{E_N^c} g < \epsilon$. Then the functions $f_n \chi_{E_N}$ are bounded (by N) and supported on a set of finite measure, so that by the bounded convergence theorem, we have

$$\int_{E_N} |f_n - f| < \epsilon, \quad \text{for all large } n.$$

Hence, we obtain the estimate

$$\int |f_n - f| = \int_{E_N} |f_n - f| + \int_{E_N^c} |f_n - f|$$
$$\le \int_{E_N} |f_n - f| + 2\int_{E_N^c} g$$
$$\le \epsilon + 2\epsilon = 3\epsilon$$

for all large n. This proves the theorem.

Complex-valued functions

If f is a complex-valued function on $\mathbb{R}^d$, we may write it as

$$f(x) = u(x) + iv(x),$$

where u and v are real-valued functions called the real and imaginary parts of f, respectively. The function f is measurable if and only if both u and v are measurable. We then say that f is **Lebesgue integrable** if the function $|f(x)| = (u(x)^2 + v(x)^2)^{1/2}$ (which is non-negative) is Lebesgue integrable in the sense defined previously.

It is clear that

$$|u(x)| \leq |f(x)| \quad \text{and} \quad |v(x)| \leq |f(x)|.$$

Also, if $a, b \geq 0$, one has $(a+b)^{1/2} \leq a^{1/2} + b^{1/2}$, so that

$$|f(x)| \leq |u(x)| + |v(x)|.$$

As a result of these simple inequalities, we deduce that a complex-valued function is integrable if and only if both its real and imaginary parts are integrable. Then, the Lebesgue integral of f is defined by

$$\int f(x)\,dx = \int u(x)\,dx + i \int v(x)\,dx.$$

Finally, if E is a measurable subset of $\mathbb{R}^d$, and f is a complex-valued measurable function on E, we say that f is Lebesgue integrable on E if $f\chi_E$ is integrable on $\mathbb{R}^d$, and we define $\int_E f = \int f\chi_E$.

The collection of all complex-valued integrable functions on a measurable subset $E \subset \mathbb{R}^d$ forms a vector space over $\mathbb{C}$. Indeed, if f and g are integrable, then so is $f + g$, since the triangle inequality gives $|(f + g)(x)| \leq |f(x)| + |g(x)|$, and monotonicity of the integral then yields

$$\int_E |f + g| \leq \int_E |f| + \int_E |g| < \infty.$$

Also, it is clear that if $a \in \mathbb{C}$ and if f is integrable, then so is af. Finally, the integral continues to be linear over $\mathbb{C}$.

2 The space L^1 of integrable functions

The fact that the integrable functions form a vector space is an important observation about the algebraic properties of such functions. A fundamental analytic fact is that this vector space is complete in the appropriate norm.

For any integrable function f on $\mathbb{R}^d$ we define the **norm**[2] of f,

$$\|f\| = \|f\|_{L^1} = \|f\|_{L^1(\mathbb{R}^d)} = \int_{\mathbb{R}^d} |f(x)| \, dx.$$

The collection of all integrable functions with the above norm gives a (somewhat imprecise) definition of the space $L^1(\mathbb{R}^d)$. We also note that $\|f\| = 0$ if and only if $f = 0$ almost everywhere (see Proposition 1.6), and this simple property of the norm reflects the practice we have already adopted not to distinguish two functions that agree almost everywhere. With this in mind, we take the precise definition of $L^1(\mathbb{R}^d)$ to be the space of equivalence classes of integrable functions, where we define two functions to be **equivalent** if they agree almost everywhere. Often, however, it is convenient to retain the (imprecise) terminology that an element $f \in L^1(\mathbb{R}^d)$ is an integrable function, even though it is only an equivalence class of such functions. Note that by the above, the norm $\|f\|$ of an element $f \in L^1(\mathbb{R}^d)$ is well-defined by the choice of any integrable function in its equivalence class. Moreover, $L^1(\mathbb{R}^d)$ inherits the property that it is a vector space. This and other straightforward facts are summarized in the following proposition.

Proposition 2.1 *Suppose f and g are two functions in $L^1(\mathbb{R}^d)$.*

(i) $\|af\|_{L^1(\mathbb{R}^d)} = |a| \, \|f\|_{L^1(\mathbb{R}^d)}$ *for all $a \in \mathbb{C}$.*

(ii) $\|f + g\|_{L^1(\mathbb{R}^d)} \leq \|f\|_{L^1(\mathbb{R}^d)} + \|g\|_{L^1(\mathbb{R}^d)}$.

(iii) $\|f\|_{L^1(\mathbb{R}^d)} = 0$ *if and only if $f = 0$ a.e.*

(iv) $d(f, g) = \|f - g\|_{L^1(\mathbb{R}^d)}$ *defines a metric on $L^1(\mathbb{R}^d)$.*

In (iv), we mean that d satisfies the following conditions. First, $d(f, g) \geq 0$ for all integrable functions f and g, and $d(f, g) = 0$ if and only if $f = g$ a.e. Also, $d(f, g) = d(g, f)$, and finally, d satisfies the triangle inequality

$$d(f, g) \leq d(f, h) + d(h, g), \quad \text{for all } f, g, h \in L^1(\mathbb{R}^d).$$

A space V with a metric d is said to be **complete** if for every Cauchy sequence $\{x_k\}$ in V (that is, $d(x_k, x_\ell) \to 0$ as $k, \ell \to \infty$) there exists $x \in V$ such that $\lim_{k \to \infty} x_k = x$ in the sense that

$$d(x_k, x) \to 0, \quad \text{as } k \to \infty.$$

Our main goal of completing the space of Riemann integrable functions will be attained once we have established the next important theorem.

---

[2] In this chapter the only norm we consider is the L^1-norm, so we often write $\|f\|$ for $\|f\|_{L^1}$. Later, we shall have occasion to consider other norms, and then we shall modify our notation accordingly.

Theorem 2.2 (Riesz-Fischer) *The vector space L^1 is complete in its metric.*

Proof. Suppose $\{f_n\}$ is a Cauchy sequence in the norm, so that $\|f_n - f_m\| \to 0$ as $n, m \to \infty$. The plan of the proof is to extract a subsequence of $\{f_n\}$ that converges to f, both pointwise almost everywhere and in the norm.

Under ideal circumstances we would have that the sequence $\{f_n\}$ converges almost everywhere to a limit f, and we would then prove that the sequence converges to f also in the norm. Unfortunately, almost everywhere convergence does not hold for general Cauchy sequences (see Exercise 12). The main point, however, is that if the convergence in the norm is rapid enough, then almost everywhere convergence is a consequence, and this can be achieved by dealing with an appropriate subsequence of the original sequence.

Indeed, consider a subsequence $\{f_{n_k}\}_{k=1}^{\infty}$ of $\{f_n\}$ with the following property:

$$\|f_{n_{k+1}} - f_{n_k}\| \le 2^{-k}, \qquad \text{for all } k \ge 1.$$

The existence of such a subsequence is guaranteed by the fact that $\|f_n - f_m\| \le \epsilon$ whenever $n, m \ge N(\epsilon)$, so that it suffices to take $n_k = N(2^{-k})$.

We now consider the series whose convergence will be seen below,

$$f(x) = f_{n_1}(x) + \sum_{k=1}^{\infty}(f_{n_{k+1}}(x) - f_{n_k}(x))$$

and

$$g(x) = |f_{n_1}(x)| + \sum_{k=1}^{\infty}|f_{n_{k+1}}(x) - f_{n_k}(x)|,$$

and note that

$$\int |f_{n_1}| + \sum_{k=1}^{\infty}\int|f_{n_{k+1}} - f_{n_k}| \le \int |f_{n_1}| + \sum_{k=1}^{\infty}2^{-k} < \infty.$$

So the monotone convergence theorem implies that g is integrable, and since $|f| \le g$, hence so is f. In particular, the series defining f converges almost everywhere, and since the partial sums of this series are precisely the f_{n_k} (by construction of the telescopic series), we find that

$$f_{n_k}(x) \to f(x) \qquad \text{a.e. } x.$$

To prove that $f_{n_k} \to f$ in L^1 as well, we simply observe that $|f - f_{n_k}| \leq g$ for all k, and apply the dominated convergence theorem to get $\|f_{n_k} - f\|_{L^1} \to 0$ as k tends to infinity.

Finally, the last step of the proof consists in recalling that $\{f_n\}$ is Cauchy. Given ϵ, there exists N such that for all $n, m > N$ we have $\|f_n - f_m\| < \epsilon/2$. If n_k is chosen so that $n_k > N$, and $\|f_{n_k} - f\| < \epsilon/2$, then the triangle inequality implies

$$\|f_n - f\| \leq \|f_n - f_{n_k}\| + \|f_{n_k} - f\| < \epsilon$$

whenever $n > N$. Thus $\{f_n\}$ has the limit f in L^1, and the proof of the theorem is complete.

Since every sequence that converges in the norm is a Cauchy sequence in that norm, the argument in the proof of the theorem yields the following.

Corollary 2.3 *If $\{f_n\}_{n=1}^{\infty}$ converges to f in L^1, then there exists a subsequence $\{f_{n_k}\}_{k=1}^{\infty}$ such that*

$$f_{n_k}(x) \to f(x) \quad a.e. \ x.$$

We say that a family $\mathcal{G}$ of integrable functions is **dense** in L^1 if for any $f \in L^1$ and $\epsilon > 0$, there exists $g \in \mathcal{G}$ so that $\|f - g\|_{L^1} < \epsilon$. Fortunately we are familiar with many families that are dense in L^1, and we describe some in the theorem that follows. These are useful when one is faced with the problem of proving some fact or identity involving integrable functions. In this situation a general principle applies: the result is often easier to prove for a more restrictive class of functions (like the ones in the theorem below), and then a density (or limiting) argument yields the result in general.

Theorem 2.4 *The following families of functions are dense in $L^1(\mathbb{R}^d)$:*

(i) *The simple functions.*

(ii) *The step functions.*

(iii) *The continuous functions of compact support.*

Proof. Let f be an integrable function on $\mathbb{R}^d$. First, we may assume that f is real-valued, because we may approximate its real and imaginary parts independently. If this is the case, we may then write $f = f^+ - f^-$, where $f^+, f^- \geq 0$, and it now suffices to prove the theorem when $f \geq 0$.

For (i), Theorem 4.1 in Chapter 1 guarantees the existence of a sequence $\{\varphi_k\}$ of non-negative simple functions that increase to f pointwise. By the dominated convergence theorem (or even simply the monotone convergence theorem) we then have

$$\|f - \varphi_k\|_{L^1} \to 0 \quad \text{as } k \to \infty.$$

Thus there are simple functions that are arbitrarily close to f in the L^1
norm.

For (ii), we first note that by (i) it suffices to approximate simple
functions by step functions. Then, we recall that a simple function is
a finite linear combination of characteristic functions of sets of finite
measure, so it suffices to show that if E is such a set, then there is a
step function ψ so that $\|\chi_E - \psi\|_{L^1}$ is small. However, we now recall
that this argument was already carried out in the proof of Theorem 4.3,
Chapter 1. Indeed, there it is shown that there is an almost disjoint
family of rectangles $\{R_j\}$ with $m(E \triangle \bigcup_{j=1}^{M} R_j) \leq 2\epsilon$. Thus χ_E and $\psi =
\sum_j \chi_{R_j}$ differ at most on a set of measure 2ϵ, and as a result we find
that $\|\chi_E - \psi\|_{L^1} < 2\epsilon$.

By (ii), it suffices to establish (iii) when f is the characteristic function
of a rectangle. In the one-dimensional case, where f is the characteristic
function of an interval $[a, b]$, we may choose a continuous piecewise linear
function g defined by

$$g(x) = \begin{cases} 1 & \text{if } a \leq x \leq b, \\ 0 & \text{if } x \leq a - \epsilon \text{ or } x \geq b + \epsilon, \end{cases}$$

and with g linear on the intervals $[a - \epsilon, a]$ and $[b, b + \epsilon]$. Then $\|f -
g\|_{L^1} < 2\epsilon$. In d dimensions, it suffices to note that the characteristic
function of a rectangle is the product of characteristic functions of intervals. Then, the desired continuous function of compact support is simply
the product of functions like g defined above.

The results above for $L^1(\mathbb{R}^d)$ lead immediately to an extension in which
$\mathbb{R}^d$ can be replaced by any fixed subset E of positive measure. In fact
if E is such a subset, we can define $L^1(E)$ and carry out the arguments
that are analogous to $L^1(\mathbb{R}^d)$. Better yet, we can proceed by extending
any function f on E by setting $\tilde{f} = f$ on E and $\tilde{f} = 0$ on E^c, and defining
$\|f\|_{L^1(E)} = \|\tilde{f}\|_{L^1(\mathbb{R}^d)}$. The analogues of Proposition 2.1 and Theorem 2.2
then hold for the space $L^1(E)$.

Invariance Properties

If f is a function defined on $\mathbb{R}^d$, the **translation** of f by a vector $h \in \mathbb{R}^d$ is the function f_h, defined by $f_h(x) = f(x - h)$. Here we want to examine some basic aspects of translations of integrable functions.

First, there is the translation-invariance of the integral. One way to state this is as follows: if f is an integrable function, then so is f_h and

$$(4) \qquad \int_{\mathbb{R}^d} f(x - h)\, dx = \int_{\mathbb{R}^d} f(x)\, dx.$$

We check this assertion first when $f = \chi_E$, the characteristic function of a measurable set E. Then obviously $f_h = \chi_{E_h}$, where $E_h = \{x + h : x \in E\}$, and thus the assertion follows because $m(E_h) = m(E)$ (see Section 3 in Chapter 1). As a result of linearity, the identity (4) holds for all simple functions. Now if f is non-negative and $\{\varphi_n\}$ is a sequence of simple functions that increase pointwise a.e to f (such a sequence exists by Theorem 4.1 in the previous chapter), then $\{(\varphi_n)_h\}$ is a sequence of simple functions that increase to f_h pointwise a.e, and the monotone convergence theorem implies (4) in this special case. Thus, if f is complex-valued and integrable we see that $\int_{\mathbb{R}^d} |f(x - h)|\, dx = \int_{\mathbb{R}^d} |f(x)|\, dx$, which shows that $f_h \in L^1(\mathbb{R}^d)$ and also $\|f_h\| = \|f\|$. From the definitions, we then conclude that (4) holds whenever $f \in L^1$.

Incidentally, using the relative invariance of Lebesgue measure under dilations and reflections (Section 3, Chapter 1) one can prove in the same way that if $f(x)$ is integrable, so is $f(\delta x)$, $\delta > 0$, and $f(-x)$, and

$$(5)$$
$$\delta^d \int_{\mathbb{R}^d} f(\delta x)\, dx = \int_{\mathbb{R}^d} f(x)\, dx, \qquad \text{while} \qquad \int_{\mathbb{R}^d} f(-x)\, dx = \int_{\mathbb{R}^d} f(x)\, dx.$$

We digress to record for later use two useful consequences of the above invariance properties:

(i) Suppose that f and g are a pair of measurable functions on $\mathbb{R}^d$ so that for some fixed $x \in \mathbb{R}^d$ the function $y \mapsto f(x - y)g(y)$ is integrable. As a consequence, the function $y \mapsto f(y)g(x - y)$ is then also integrable and we have

$$(6) \qquad \int_{\mathbb{R}^d} f(x - y)g(y)\, dy = \int_{\mathbb{R}^d} f(y)g(x - y)\, dy.$$

This follows from (4) and (5) on making the change of variables which replaces y by $x - y$, and noting that this change is a combination of a translation and a reflection.

The integral on the left-hand side is denoted by $(f * g)(x)$ and is defined as the **convolution** of f and g. Thus (6) asserts the commutativity of the convolution product.

(ii) Using (5) one has that for all $\epsilon > 0$

$$(7) \qquad \int_{|x| \geq \epsilon} \frac{dx}{|x|^a} = \epsilon^{-a+d} \int_{|x| \geq 1} \frac{dx}{|x|^a} \qquad \text{whenever } a > d,$$

and

$$(8) \qquad \int_{|x| \leq \epsilon} \frac{dx}{|x|^a} = \epsilon^{-a+d} \int_{|x| \leq 1} \frac{dx}{|x|^a} \qquad \text{whenever } a < d.$$

It can also be seen that the integrals $\int_{|x| \geq 1} \frac{dx}{|x|^a}$ and $\int_{|x| \leq 1} \frac{dx}{|x|^a}$ (respectively, when $a > d$ and $a < d$) are finite by the argument that appears after Corollary 1.10.

Translations and continuity

We shall next examine how continuity properties of f are related to the way the translations f_h vary with h. Note that for any given $x \in \mathbb{R}^d$, the statement that $f_h(x) \to f(x)$ as $h \to 0$ is the same as the continuity of f at the point x.

However, a general f which is integrable may be discontinuous at every x, even when corrected on a set of measure zero; see Exercise 15. Nevertheless, there is an overall continuity that an arbitrary $f \in L^1(\mathbb{R}^d)$ enjoys, one that holds in the norm.

Proposition 2.5 *Suppose $f \in L^1(\mathbb{R}^d)$. Then*

$$\|f_h - f\|_{L^1} \to 0 \qquad \text{as } h \to 0.$$

The proof is a simple consequence of the approximation of integrable functions by continuous functions of compact support as given in Theorem 2.4. In fact for any $\epsilon > 0$, we can find such a function g so that $\|f - g\| < \epsilon$. Now

$$f_h - f = (g_h - g) + (f_h - g_h) - (f - g).$$

However, $\|f_h - g_h\| = \|f - g\| < \epsilon$, while since g is continuous and has compact support we have that clearly

$$\|g_h - g\| = \int_{\mathbb{R}^d} |g(x - h) - g(x)| \, dx \to 0 \qquad \text{as } h \to 0.$$

So if $|h| < \delta$, where δ is sufficiently small, then $\|g_h - g\| < \epsilon$, and as a result $\|f_h - f\| < 3\epsilon$, whenever $|h| < \delta$.

3 Fubini's theorem

In elementary calculus integrals of continuous functions of several variables are often calculated by iterating one-dimensional integrals. We shall now examine this important analytic device from the general point of view of Lebesgue integration in $\mathbb{R}^d$, and we shall see that a number of interesting issues arise.

In general, we may write $\mathbb{R}^d$ as a product

$$\mathbb{R}^d = \mathbb{R}^{d_1} \times \mathbb{R}^{d_2} \quad \text{where } d = d_1 + d_2, \text{ and } d_1, d_2 \geq 1.$$

A point in $\mathbb{R}^d$ then takes the form (x, y), where $x \in \mathbb{R}^{d_1}$ and $y \in \mathbb{R}^{d_2}$. With such a decomposition of $\mathbb{R}^d$ in mind, the general notion of a slice, formed by fixing one variable, becomes natural. If f is a function in $\mathbb{R}^{d_1} \times \mathbb{R}^{d_2}$, the **slice** of f corresponding to $y \in \mathbb{R}^{d_2}$ is the function f^y of the $x \in \mathbb{R}^{d_1}$ variable, given by

$$f^y(x) = f(x, y).$$

Similarly, the slice of f for a fixed $x \in \mathbb{R}^{d_1}$ is $f_x(y) = f(x, y)$.

In the case of a set $E \subset \mathbb{R}^{d_1} \times \mathbb{R}^{d_2}$ we define its **slices** by

$$E^y = \{x \in \mathbb{R}^{d_1} : (x, y) \in E\} \quad \text{and} \quad E_x = \{y \in \mathbb{R}^{d_2} : (x, y) \in E\}.$$

See Figure 1 for an illustration.

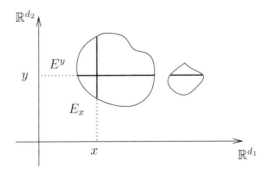

Figure 1. Slices E^y and E_x (for fixed x and y) of a set E

3.1 Statement and proof of the theorem

That the theorem that follows is not entirely straightforward is clear from the first difficulty that arises in its formulation, involving the measurability of the functions and sets in question. In fact, even with the assumption that f is measurable on $\mathbb{R}^d$, it is not necessarily true that the slice f^y is measurable on $\mathbb{R}^{d_1}$ for each y; nor does the corresponding assertion necessarily hold for a measurable set: the slice E^y may not be measurable for each y. An easy example arises in $\mathbb{R}^2$ by placing a one-dimensional non-measurable set on the x-axis; the set E in $\mathbb{R}^2$ has measure zero, but E^y is not measurable for $y = 0$. What saves us is that, nevertheless, measurability holds for almost all slices.

The main theorem is as follows. We recall that by definition all integrable functions are measurable.

Theorem 3.1 *Suppose $f(x, y)$ is integrable on $\mathbb{R}^{d_1} \times \mathbb{R}^{d_2}$. Then for almost every $y \in \mathbb{R}^{d_2}$:*

(i) *The slice f^y is integrable on $\mathbb{R}^{d_1}$.*

(ii) *The function defined by $\int_{\mathbb{R}^{d_1}} f^y(x)\, dx$ is integrable on $\mathbb{R}^{d_2}$.*

Moreover:

(iii) $\displaystyle \int_{\mathbb{R}^{d_2}} \left(\int_{\mathbb{R}^{d_1}} f(x, y)\, dx \right) dy = \int_{\mathbb{R}^d} f.$

Clearly, the theorem is symmetric in x and y so that we also may conclude that the slice f_x is integrable on $\mathbb{R}^{d_2}$ for a.e. x. Moreover, $\int_{\mathbb{R}^{d_2}} f_x(y)\, dy$ is integrable, and

$$\int_{\mathbb{R}^{d_1}} \left(\int_{\mathbb{R}^{d_2}} f(x, y)\, dy \right) dx = \int_{\mathbb{R}^d} f.$$

In particular, Fubini's theorem states that the integral of f on $\mathbb{R}^d$ can be computed by iterating lower-dimensional integrals, and that the iterations can be taken in any order

$$\int_{\mathbb{R}^{d_2}} \left(\int_{\mathbb{R}^{d_1}} f(x, y)\, dx \right) dy = \int_{\mathbb{R}^{d_1}} \left(\int_{\mathbb{R}^{d_2}} f(x, y)\, dy \right) dx = \int_{\mathbb{R}^d} f.$$

We first note that we may assume that f is real-valued, since the theorem then applies to the real and imaginary parts of a complex-valued function. The proof of Fubini's theorem which we give next consists of a sequence of six steps. We begin by letting $\mathcal{F}$ denote the set of integrable functions on $\mathbb{R}^d$ which satisfy all three conclusions in the theorem, and set out to prove that $L^1(\mathbb{R}^d) \subset \mathcal{F}$.

We proceed by first showing that $\mathcal{F}$ is closed under operations such as linear combinations (Step 1) and limits (Step 2). Then we begin to construct families of functions in $\mathcal{F}$. Since any integrable function is the "limit" of simple functions, and simple functions are themselves linear combinations of sets of finite measure, the goal quickly becomes to prove that χ_E belongs to $\mathcal{F}$ whenever E is a measurable subset of $\mathbb{R}^d$ with finite measure. To achieve this goal, we begin with rectangles and work our way up to sets of type G_δ (Step 3), and sets of measure zero (Step 4). Finally, a limiting argument shows that all integrable functions are in $\mathcal{F}$. This will complete the proof of Fubini's theorem.

Step 1. Any finite linear combination of functions in $\mathcal{F}$ also belongs to $\mathcal{F}$.

Indeed, let $\{f_k\}_{k=1}^N \subset \mathcal{F}$. For each k there exists a set $A_k \subset \mathbb{R}^{d_2}$ of measure 0 so that f_k^y is integrable on $\mathbb{R}^{d_1}$ whenever $y \notin A_k$. Then, if $A = \bigcup_{k=1}^N A_k$, the set A has measure 0, and in the complement of A, the y-slice corresponding to any finite linear combination of the f_k is measurable, and also integrable. By linearity of the integral, we then conclude that any linear combination of the f_k's belongs to $\mathcal{F}$.

Step 2. Suppose $\{f_k\}$ is a sequence of measurable functions in $\mathcal{F}$ so that $f_k \nearrow f$ or $f_k \searrow f$, where f is integrable (on $\mathbb{R}^d$). Then $f \in \mathcal{F}$.

By taking $-f_k$ instead of f_k if necessary, we note that it suffices to consider the case of an increasing sequence. Also, we may replace f_k by $f_k - f_1$ and assume that the f_k's are non-negative. Now, we observe that an application of the monotone convergence theorem (Corollary 1.9) yields

$$(9) \qquad \lim_{k \to \infty} \int_{\mathbb{R}^d} f_k(x, y) \, dx \, dy = \int_{\mathbb{R}^d} f(x, y) \, dx \, dy.$$

By assumption, for each k there exists a set $A_k \subset \mathbb{R}^{d_2}$, so that f_k^y is integrable on $\mathbb{R}^{d_1}$ whenever $y \notin A_k$. If $A = \bigcup_{k=1}^\infty A_k$, then $m(A) = 0$ in $\mathbb{R}^{d_2}$, and if $y \notin A$, then f_k^y is integrable on $\mathbb{R}^{d_1}$ for all k, and, by the monotone convergence theorem, we find that

$$g_k(y) = \int_{\mathbb{R}^{d_1}} f_k^y(x) \, dx \quad \text{increases to a limit} \quad g(y) = \int_{\mathbb{R}^{d_1}} f^y(x) \, dx$$

as k tends to infinity. By assumption, each $g_k(y)$ is integrable, so that another application of the monotone convergence theorem yields

$$(10) \qquad \int_{\mathbb{R}^{d_2}} g_k(y) \, dy \to \int_{\mathbb{R}^{d_2}} g(y) \, dy \quad \text{as } k \to \infty.$$

By the assumption that $f_k \in \mathcal{F}$ we have

$$\int_{\mathbb{R}^{d_2}} g_k(y)\, dy = \int_{\mathbb{R}^d} f_k(x, y)\, dx\, dy,$$

and combining this fact with (9) and (10), we conclude that

$$\int_{\mathbb{R}^{d_2}} g(y)\, dy = \int_{\mathbb{R}^d} f(x, y)\, dx\, dy.$$

Since f is integrable, the right-hand integral is finite, and this proves that g is integrable. Consequently $g(y) < \infty$ a.e. y, hence f^y is integrable for a.e. y, and

$$\int_{\mathbb{R}^{d_2}} \left(\int_{\mathbb{R}^{d_1}} f(x, y)\, dx \right) dy = \int_{\mathbb{R}^d} f(x, y)\, dx\, dy.$$

This proves that $f \in \mathcal{F}$ as desired.

$Step$ 3. Any characteristic function of a set E that is a G_δ and of finite measure belongs to $\mathcal{F}$.

We proceed in stages of increasing order of generality.

(a) First suppose E is a bounded open cube in $\mathbb{R}^d$, such that $E = Q_1 \times Q_2$, where Q_1 and Q_2 are open cubes in $\mathbb{R}^{d_1}$ and $\mathbb{R}^{d_2}$, respectively. Then, for each y the function $\chi_E(x, y)$ is measurable in x, and integrable with

$$g(y) = \int_{\mathbb{R}^{d_1}} \chi_E(x, y)\, dx \begin{cases} |Q_1| & \text{if } y \in Q_2, \\ 0 & \text{otherwise.} \end{cases}$$

Consequently, $g = |Q_1| \chi_{Q_2}$ is also measurable and integrable, with

$$\int_{\mathbb{R}^{d_2}} g(y)\, dy = |Q_1|\, |Q_2|.$$

Since we initially have $\int_{\mathbb{R}^d} \chi_E(x, y)\, dx\, dy = |E| = |Q_1|\, |Q_2|$, we deduce that $\chi_E \in \mathcal{F}$.

(b) Now suppose E is a subset of the boundary of some closed cube. Then, since the boundary of a cube has measure 0 in $\mathbb{R}^d$, we have $\int_{\mathbb{R}^d} \chi_E(x, y)\, dx\, dy = 0$.

Next, we note, after an investigation of the various possibilities, that for almost every y, the slice E^y has measure 0 in $\mathbb{R}^{d_1}$, and therefore if $g(y) = \int_{\mathbb{R}^{d_1}} \chi_E(x, y)\, dx$ we have $g(y) = 0$ for a.e. y. As a consequence, $\int_{\mathbb{R}^{d_2}} g(y)\, dy = 0$, and therefore $\chi_E \in \mathcal{F}$.

(c) Suppose now E is a finite union of closed cubes whose interiors are disjoint, $E = \bigcup_{k=1}^{K} Q_k$. Then, if $\tilde{Q}_k$ denotes the interior of Q_k, we may write χ_E as a linear combination of the $\chi_{\tilde{Q}_k}$ and χ_{A_k} where A_k is a subset of the boundary of Q_k for $k = 1, \ldots, K$. By our previous analysis, we know that χ_{Q_k} and χ_{A_k} belong to $\mathcal{F}$ for all k, and since Step 1 guarantees that $\mathcal{F}$ is closed under finite linear combinations, we conclude that $\chi_E \in \mathcal{F}$, as desired.

(d) Next, we prove that if E is open and of finite measure, then $\chi_E \in \mathcal{F}$. This follows from taking a limit in the previous case. Indeed, by Theorem 1.4 in Chapter 1, we may write E as a countable union of almost disjoint closed cubes

$$E = \bigcup_{j=1}^{\infty} Q_j.$$

Consequently, if we let $f_k = \sum_{j=1}^{k} \chi_{Q_j}$, then we note that the functions f_k increase to $f = \chi_E$, which is integrable since $m(E)$ is finite. Therefore, we may conclude by Step 2 that $f \in \mathcal{F}$.

(e) Finally, if E is a G_δ of finite measure, then $\chi_E \in \mathcal{F}$. Indeed, by definition, there exist open sets $\tilde{\mathcal{O}}_1, \tilde{\mathcal{O}}_2, \ldots$, such that

$$E = \bigcap_{k=1}^{\infty} \tilde{\mathcal{O}}_k.$$

Since E has finite measure, there exists an open set $\tilde{\mathcal{O}}_0$ of finite measure with $E \subset \tilde{\mathcal{O}}_0$. If we let

$$\mathcal{O}_k = \mathcal{O}_0 \cap \bigcap_{j=1}^{k} \tilde{\mathcal{O}}_j,$$

then we note that we have a decreasing sequence of open sets of finite measure $\mathcal{O}_1 \supset \mathcal{O}_2 \supset \cdots$ with

$$E = \bigcap_{k=1}^{\infty} \mathcal{O}_k.$$

Therefore, the sequence of functions $f_k = \chi_{\mathcal{O}_k}$ decreases to $f = \chi_E$, and since $\chi_{\mathcal{O}_k} \in \mathcal{F}$ for all k by (d) above, we conclude by Step 2 that χ_E belongs to $\mathcal{F}$.

Step 4. If E has measure 0, then χ_E belongs to $\mathcal{F}$.

Indeed, since E is measurable, we may choose a set G of type G_δ with $E \subset G$ and $m(G) = 0$ (Corollary 3.5, Chapter 1). Since $\chi_G \in \mathcal{F}$ (by the previous step) we find that

$$\int_{\mathbb{R}^{d_2}} \left(\int_{\mathbb{R}^{d_1}} \chi_G(x, y) \, dx \right) dy = \int_{\mathbb{R}^d} \chi_G = 0.$$

Therefore

$$\int_{\mathbb{R}^{d_1}} \chi_G(x, y) \, dx = 0 \quad \text{for a.e. } y.$$

Consequently, the slice G^y has measure 0 for a.e. y. The simple observation that $E^y \subset G^y$ then shows that E^y has measure 0 for a.e. y, and $\int_{\mathbb{R}^{d_1}} \chi_E(x, y) \, dx = 0$ for a.e. y. Therefore,

$$\int_{\mathbb{R}^{d_2}} \left(\int_{\mathbb{R}^{d_1}} \chi_E(x, y) \, dx \right) dy = 0 = \int_{\mathbb{R}^d} \chi_E,$$

and thus $\chi_E \in \mathcal{F}$, as was to be shown.

Step 5. If E is any measurable subset of $\mathbb{R}^d$ with finite measure, then χ_E belongs to $\mathcal{F}$.

To prove this, recall first that there exists a set of finite measure G of type G_δ, with $E \subset G$ and $m(G - E) = 0$. Since

$$\chi_E = \chi_G - \chi_{G-E},$$

and $\mathcal{F}$ is closed under linear combinations, we find that $\chi_E \in \mathcal{F}$, as desired.

Step 6. This is the final step, which consists of proving that if f is integrable, then $f \in \mathcal{F}$.

We note first that f has the decomposition $f = f^+ - f^-$, where both f^+ and f^- are non-negative and integrable, so by Step 1 we may assume that f is itself non-negative. By Theorem 4.1 in the previous chapter, there exists a sequence $\{\varphi_k\}$ of simple functions that increase to f. Since each φ_k is a finite linear combination of characteristic functions of sets with finite measure, we have $\varphi_k \in \mathcal{F}$ by Steps 5 and 1, hence $f \in \mathcal{F}$ by Step 2.

3.2 Applications of Fubini's theorem

Theorem 3.2 *Suppose $f(x, y)$ is a non-negative measurable function on $\mathbb{R}^{d_1} \times \mathbb{R}^{d_2}$. Then for almost every $y \in \mathbb{R}^{d_2}$:*

(i) *The slice f^y is measurable on $\mathbb{R}^{d_1}$.*

(ii) *The function defined by $\int_{\mathbb{R}^{d_1}} f^y(x)\,dx$ is measurable on $\mathbb{R}^{d_2}$.*

Moreover:

(iii) $\displaystyle \int_{\mathbb{R}^{d_2}} \left(\int_{\mathbb{R}^{d_1}} f(x,y)\,dx \right) dy = \int_{\mathbb{R}^d} f(x,y)\,dx\,dy$ *in the extended sense.*

In practice, this theorem is often used in conjunction with Fubini's theorem.[3] Indeed, suppose we are given a measurable function f on $\mathbb{R}^d$ and asked to compute $\int_{\mathbb{R}^d} f$. To justify the use of iterated integration, we first apply the present theorem to $|f|$. Using it, we may freely compute (or estimate) the iterated integrals of the non-negative function $|f|$. If these are finite, Theorem 3.2 guarantees that f is integrable, that is, $\int |f| < \infty$. Then the hypothesis in Fubini's theorem is verified, and we may use that theorem in the calculation of the integral of f.

Proof of Theorem 3.2. Consider the truncations

$$ f_k(x,y) = \begin{cases} f(x,y) & \text{if } |(x,y)| < k \text{ and } f(x,y) < k, \\ 0 & \text{otherwise.} \end{cases} $$

Each f_k is integrable, and by part (i) in Fubini's theorem there exists a set $E_k \subset \mathbb{R}^{d_2}$ of measure 0 such that the slice $f_k^y(x)$ is measurable for all $y \in E_k^c$. Then, if we set $E = \bigcup_k E_k$, we find that $f^y(x)$ is measurable for all $y \in E^c$ and all k. Moreover, $m(E) = 0$. Since $f_k^y \nearrow f^y$, the monotone convergence theorem implies that if $y \notin E$, then

$$ \int_{\mathbb{R}^{d_1}} f_k(x,y)\,dx \ \nearrow \ \int_{\mathbb{R}^{d_1}} f(x,y)\,dx \qquad \text{as } k \to \infty. $$

Again by Fubini's theorem, $\int_{\mathbb{R}^{d_1}} f_k(x,y)\,dx$ is measurable for all $y \in E^c$, hence so is $\int_{\mathbb{R}^{d_1}} f(x,y)\,dx$. Another application of the monotone convergence theorem then gives

(11) $\displaystyle \int_{\mathbb{R}^{d_2}} \left(\int_{\mathbb{R}^{d_1}} f_k(x,y)\,dx \right) dy \to \int_{\mathbb{R}^{d_2}} \left(\int_{\mathbb{R}^{d_1}} f(x,y)\,dx \right) dy.$

By part (iii) in Fubini's theorem we know that

(12) $\displaystyle \int_{\mathbb{R}^{d_2}} \left(\int_{\mathbb{R}^{d_1}} f_k(x,y)\,dx \right) dy = \int_{\mathbb{R}^d} f_k.$

---

[3] Theorem 3.2 was formulated by Tonelli. We will, however, use the short-hand of referring to it, as well as Theorem 3.1 and Corollary 3.3, as Fubini's theorem.

A final application of the monotone convergence theorem directly to f_k also gives

(13)
$$\int_{\mathbb{R}^d} f_k \to \int_{\mathbb{R}^d} f.$$

Combining (11), (12), and (13) completes the proof of Theorem 3.2.

Corollary 3.3 *If E is a measurable set in $\mathbb{R}^{d_1} \times \mathbb{R}^{d_2}$, then for almost every $y \in \mathbb{R}^{d_2}$ the slice*

$$E^y = \{x \in \mathbb{R}^{d_1} : (x, y) \in E\}$$

is a measurable subset of $\mathbb{R}^{d_1}$. Moreover $m(E^y)$ is a measurable function of y and

$$m(E) = \int_{\mathbb{R}^{d_2}} m(E^y) \, dy.$$

This is an immediate consequence of the first part of Theorem 3.2 applied to the function χ_E. Clearly a symmetric result holds for the x-slices in $\mathbb{R}^{d_2}$.

We have thus established the basic fact that if E is measurable on $\mathbb{R}^{d_1} \times \mathbb{R}^{d_2}$, then for almost every $y \in \mathbb{R}^{d_2}$ the slice E^y is measurable in $\mathbb{R}^{d_1}$ (and also the symmetric statement with the roles of x and y interchanged). One might be tempted to think that the converse assertion holds. To see that this is not the case, note that if we let $\mathcal{N}$ denote a non-measurable subset of $\mathbb{R}$, and then define

$$E = [0, 1] \times \mathcal{N} \subset \mathbb{R} \times \mathbb{R},$$

we see that

$$E^y = \begin{cases} [0, 1] & \text{if } y \in \mathcal{N}, \\ \emptyset & \text{if } y \notin \mathcal{N}. \end{cases}$$

Thus E^y is measurable for every y. However, if E were measurable, then the corollary would imply that $E_x = \{y \in \mathbb{R} : (x, y) \in E\}$ is measurable for almost every $x \in \mathbb{R}$, which is not true since E_x is equal to $\mathcal{N}$ for all $x \in [0, 1]$.

A more striking example is that of a set E in the unit square $[0, 1] \times [0, 1]$ that is not measurable, and yet the slices E^y and E_x are measurable with $m(E^y) = 0$ and $m(E_x) = 1$ for each $x, y \in [0, 1]$. The construction of E is based on the existence of a highly paradoxical ordering $\prec$ of the reals, with the property that $\{x : x \prec y\}$ is a countable set for each $y \in \mathbb{R}$. (The construction of this ordering is discussed in Problem 5.) Given this ordering we let

$$E = \{(x, y) \in [0, 1] \times [0, 1], \text{ with } x \prec y\}.$$

Note that for each $y \in [0, 1]$, $E^y = \{x : x \prec y\}$; thus E^y is countable and $m(E^y) = 0$. Similarly $m(E_x) = 1$, because E_x is the complement of a denumerable set in $[0, 1]$. If E were measurable, it would contradict the formula in Corollary 3.3.

In relating a set E to its slices E_x and E^y, matters are straightforward for the basic sets which arise when we consider $\mathbb{R}^d$ as the product $\mathbb{R}^{d_1} \times \mathbb{R}^{d_2}$. These are the **product sets** $E = E_1 \times E_2$, where $E_j \subset \mathbb{R}^{d_j}$.

Proposition 3.4 *If $E = E_1 \times E_2$ is a measurable subset of $\mathbb{R}^d$, and $m_*(E_2) > 0$, then E_1 is measurable.*

Proof. By Corollary 3.3, we know that for a.e. $y \in \mathbb{R}^{d_2}$, the slice function

$$(\chi_{E_1 \times E_2})^y(x) = \chi_{E_1}(x)\chi_{E_2}(y)$$

is measurable as a function of x. In fact, we claim that there is some $y \in E_2$ such that the above slice function is measurable in x; for such a y we would have $\chi_{E_1 \times E_2}(x, y) = \chi_{E_1}(x)$, and this would imply that E_1 is measurable.

To prove the existence of such a y, we use the assumption that $m_*(E_2) > 0$. Indeed, let F denote the set of $y \in \mathbb{R}^{d_2}$ such that the slice E^y is measurable. Then $m(F^c) = 0$ (by the previous corollary). However, $E_2 \cap F$ is not empty because $m_*(E_2 \cap F) > 0$. To see this, note that $E_2 = (E_2 \cap F) \bigcup (E_2 \cap F^c)$, hence

$$0 < m_*(E_2) \le m_*(E_2 \cap F) + m_*(E_2 \cap F^c) = m_*(E_2 \cap F),$$

because $E_2 \cap F^c$ is a subset of a set of measure zero.

To deal with a converse of the above result, we need the following lemma.

Lemma 3.5 *If $E_1 \subset \mathbb{R}^{d_1}$ and $E_2 \subset \mathbb{R}^{d_2}$, then*

$$m_*(E_1 \times E_2) \le m_*(E_1) m_*(E_2),$$

with the understanding that if one of the sets E_j has exterior measure zero, then $m_*(E_1 \times E_2) = 0$.

Proof.   Let $\epsilon > 0$. By definition, we can find cubes $\{Q_k\}_{k=1}^{\infty}$ in $\mathbb{R}^{d_1}$ and $\{Q'_\ell\}_{\ell=1}^{\infty}$ in $\mathbb{R}^{d_2}$ such that

$$E_1 \subset \bigcup_{k=1}^{\infty} Q_k, \quad \text{and} \quad E_2 \subset \bigcup_{\ell=1}^{\infty} Q'_\ell$$

and

$$\sum_{k=1}^{\infty} |Q_k| \leq m_*(E_1) + \epsilon \quad \text{and} \quad \sum_{\ell=1}^{\infty} |Q'_\ell| \leq m_*(E_2) + \epsilon.$$

Since $E_1 \times E_2 \subset \bigcup_{k,\ell=1}^{\infty} Q_k \times Q'_\ell$, the sub-additivity of the exterior measure yields

$$m_*(E_1 \times E_2) \leq \sum_{k,\ell=1}^{\infty} |Q_k \times Q'_\ell|$$

$$= \left(\sum_{k=1}^{\infty} |Q_k| \right) \left(\sum_{\ell=1}^{\infty} |Q'_\ell| \right)$$

$$\leq (m_*(E_1) + \epsilon)(m_*(E_2) + \epsilon).$$

If neither E_1 nor E_2 has exterior measure 0, then from the above we find

$$m_*(E_1 \times E_2) \leq m_*(E_1)\, m_*(E_2) + O(\epsilon),$$

and since ϵ is arbitrary, we must have $m_*(E_1 \times E_2) \leq m_*(E_1)\, m_*(E_2)$.

If for instance $m_*(E_1) = 0$, consider for each positive integer j the set $E_2^j = E_2 \cap \{y \in \mathbb{R}^{d_2} : |y| \leq j\}$. Then, by the above argument, we find that $m_*(E_1 \times E_2^j) = 0$. Since $(E_1 \times E_2^j) \nearrow (E_1 \times E_2)$ as $j \to \infty$, we conclude that $m_*(E_1 \times E_2) = 0$.

Proposition 3.6 *Suppose E_1 and E_2 are measurable subsets of $\mathbb{R}^{d_1}$ and $\mathbb{R}^{d_2}$, respectively. Then $E = E_1 \times E_2$ is a measurable subset of $\mathbb{R}^d$. Moreover,*

$$m(E) = m(E_1)\, m(E_2),$$

with the understanding that if one of the sets E_j has measure zero, then $m(E) = 0$.

Proof.   It suffices to prove that E is measurable, because then the assertion about $m(E)$ follows from Corollary 3.3. Since each set E_j is measurable, there exist sets $G_j \subset \mathbb{R}^{d_j}$ of type G_δ, with $G_j \supset E_j$ and $m_*(G_j - E_j) = 0$ for each $j = 1, 2$. (See Corollary 3.5 in Chapter 1.) Clearly, $G = G_1 \times G_2$ is measurable in $\mathbb{R}^{d_1} \times \mathbb{R}^{d_2}$ and

$$(G_1 \times G_2) - (E_1 \times E_2) \subset ((G_1 - E_1) \times G_2) \cup (G_1 \times (G_2 - E_2)).$$

By the lemma we conclude that $m_*(G - E) = 0$, hence E is measurable.

As a consequence of this proposition we have the following.

Corollary 3.7 *Suppose f is a measurable function on $\mathbb{R}^{d_1}$. Then the function $\tilde{f}$ defined by $\tilde{f}(x, y) = f(x)$ is measurable on $\mathbb{R}^{d_1} \times \mathbb{R}^{d_2}$.*

Proof. To see this, we may assume that f is real-valued, and recall first that if $a \in \mathbb{R}$ and $E_1 = \{x \in \mathbb{R}^{d_1} : f(x) < a\}$, then E_1 is measurable by definition. Since

$$\{(x, y) \in \mathbb{R}^{d_1} \times \mathbb{R}^{d_2} : \tilde{f}(x, y) < a\} = E_1 \times \mathbb{R}^{d_2},$$

the previous proposition shows that $\{\tilde{f}(x, y) < a\}$ is measurable for each $a \in \mathbb{R}$. Thus $\tilde{f}(x, y)$ is a measurable function on $\mathbb{R}^{d_1} \times \mathbb{R}^{d_2}$, as desired.

Finally, we return to an interpretation of the integral that arose first in the calculus. We have in mind the notion that $\int f$ describes the "area" under the graph of f. Here we relate this to the Lebesgue integral and show how it extends to our more general context.

Corollary 3.8 *Suppose $f(x)$ is a non-negative function on $\mathbb{R}^d$, and let*

$$\mathcal{A} = \{(x, y) \in \mathbb{R}^d \times \mathbb{R} : 0 \leq y \leq f(x)\}.$$

Then:

(i) *f is measurable on $\mathbb{R}^d$ if and only if $\mathcal{A}$ is measurable in $\mathbb{R}^{d+1}$.*

(ii) *If the conditions in (i) hold, then*

$$\int_{\mathbb{R}^d} f(x) \, dx = m(\mathcal{A}).$$

Proof. If f is measurable on $\mathbb{R}^d$, then the previous proposition guarantees that the function

$$F(x, y) = y - f(x)$$

is measurable on $\mathbb{R}^{d+1}$, so we immediately see that $\mathcal{A} = \{y \geq 0\} \cap \{F \leq 0\}$ is measurable.

Conversely, suppose that $\mathcal{A}$ is measurable. We note that for each $x \in \mathbb{R}^{d_1}$ the slice $\mathcal{A}_x = \{y \in \mathbb{R} : (x, y) \in \mathcal{A}\}$ is a closed segment, namely $\mathcal{A}_x = [0, f(x)]$. Consequently Corollary 3.3 (with the roles of x and y interchanged) yields the measurability of $m(\mathcal{A}_x) = f(x)$. Moreover

$$m(\mathcal{A}) = \int \chi_{\mathcal{A}}(x, y) \, dx \, dy = \int_{\mathbb{R}^{d_1}} m(\mathcal{A}_x) \, dx = \int_{\mathbb{R}^{d_1}} f(x) \, dx,$$

as was to be shown.

We conclude this section with a useful result.

Proposition 3.9 *If f is a measurable function on $\mathbb{R}^d$, then the function $\tilde{f}(x, y) = f(x - y)$ is measurable on $\mathbb{R}^d \times \mathbb{R}^d$.*

By picking $E = \{z \in \mathbb{R}^d : f(z) < a\}$, we see that it suffices to prove that whenever E is a measurable subset of $\mathbb{R}^d$, then $\tilde{E} = \{(x, y) : x - y \in E\}$ is a measurable subset of $\mathbb{R}^d \times \mathbb{R}^d$.

Note first that if $\mathcal{O}$ is an open set, then $\tilde{\mathcal{O}}$ is also open. Taking countable intersections shows that if E is a G_δ set, then so is $\tilde{E}$. Assume now that $m(\tilde{E}_k) = 0$ for each k, where $\tilde{E}_k = \tilde{E} \cap B_k$ and $B_k = \{|y| < k\}$. Again, take $\mathcal{O}$ to be open in $\mathbb{R}^d$, and let us calculate $m(\tilde{\mathcal{O}} \cap B_k)$. We have that $\chi_{\tilde{\mathcal{O}} \cap B_k} = \chi_{\mathcal{O}}(x - y)\chi_{B_k}(y)$. Hence

$$m(\tilde{\mathcal{O}} \cap B_k) = \int \chi_{\mathcal{O}}(x - y)\chi_{B_k}(y) \, dy \, dx$$

$$= \int \left(\int \chi_{\mathcal{O}}(x - y) \, dx \right) \chi_{B_k}(y) \, dy$$

$$= m(\mathcal{O}) \, m(B_k),$$

by the translation-invariance of the measure. Now if $m(E) = 0$, there is a sequence of open sets $\mathcal{O}_n$ such that $E \subset \mathcal{O}_n$ and $m(\mathcal{O}_n) \to 0$. It follows from the above that $\tilde{E}_k \subset \tilde{\mathcal{O}}_n \cap B_k$ and $m(\tilde{\mathcal{O}}_n \cap B_k) \to 0$ in n for each fixed k. This shows $m(\tilde{E}_k) = 0$, and hence $m(\tilde{E}) = 0$. The proof of the proposition is concluded once we recall that any measurable set E can be written as the difference of a G_δ and a set of measure zero.

4* A Fourier inversion formula

The question of the inversion of the Fourier transform encompasses in effect the problem at the origin of Fourier analysis. This issue involves establishing the validity of the inversion formula for a function f in terms of its Fourier transform $\hat{f}$, that is,

$$(14) \qquad \hat{f}(\xi) = \int_{\mathbb{R}^d} f(x) e^{-2\pi i x \cdot \xi} \, dx,$$

$$(15) \qquad f(x) = \int_{\mathbb{R}^d} \hat{f}(\xi) e^{2\pi i x \cdot \xi} \, d\xi.$$

We have already encountered this problem in Book I in the rudimentary case when in fact both f and $\hat{f}$ were continuous and had rapid (or moderate) decrease at infinity. In Book II we also considered the question in the one-dimensional setting, seen from the viewpoint of complex analysis. The most elegant and useful formulations of Fourier inversion are in terms of the L^2 theory, or in its greatest generality stated in the language of distributions. We shall take up these matters systematically later.[4] It will, nevertheless, be enlightening to digress here to see what our knowledge at this stage teaches us about this problem. We intend to do this by presenting a variant of the inversion formula appropriate for L^1, one that is both simple and adequate in many circumstances.

To begin with, we need to have an idea of what can be said about the Fourier transform of an arbitrary function in $L^1(\mathbb{R}^d)$.

Proposition 4.1 *Suppose $f \in L^1(\mathbb{R}^d)$. Then $\hat{f}$ defined by* (14) *is continuous and bounded on $\mathbb{R}^d$.*

In fact, since $|f(x) e^{-2\pi i x \cdot \xi}| = |f(x)|$, the integral representing $\hat{f}$ converges for each ξ and $\sup_{\xi \in \mathbb{R}^d} |\hat{f}(\xi)| \leq \int_{\mathbb{R}^d} |f(x)| \, dx = \|f\|$. To verify the continuity, note that for every x, $f(x) e^{-2\pi i x \cdot \xi} \to f(x) e^{-2\pi i x \cdot \xi_0}$ as $\xi \to \xi_0$, where ξ_0 is any point in $\mathbb{R}^d$; hence $\hat{f}(\xi) \to \hat{f}(\xi_0)$ by the dominated convergence theorem.

One can assert a little more than the boundedness of $\hat{f}$; namely, one has $\hat{f}(\xi) \to 0$ as $|\xi| \to \infty$, but not much more can be said about the decrease at infinity of $\hat{f}$. (See Exercises 22 and 25.) As a consequence, for general $f \in L^1(\mathbb{R}^d)$ the function $\hat{f}$ is not in $L^1(\mathbb{R}^d)$, and the presumed formula (15) becomes problematical. The following theorem evades this difficulty and yet is useful in a number of situations.

Theorem 4.2 *Suppose $f \in L^1(\mathbb{R}^d)$ and assume also that $\hat{f} \in L^1(\mathbb{R}^d)$. Then the inversion formula* (15) *holds for almost every x.*

An immediate corollary is the uniqueness of the Fourier transform on L^1.

---

[4]The L^2 theory will be dealt with in Chapter 5, and distributions will be studied in Book IV.

Corollary 4.3 *Suppose $\hat{f}(\xi) = 0$ for all ξ. Then $f = 0$ a.e.*

The proof of the theorem requires only that we adapt the earlier arguments carried out for Schwartz functions in Chapter 5 of Book I to the present context. We begin with the "multiplication formula."

Lemma 4.4 *Suppose f and g belong to $L^1(\mathbb{R}^d)$. Then*

$$\int_{\mathbb{R}^d} \hat{f}(\xi) g(\xi)\, d\xi = \int_{\mathbb{R}^d} f(y) \hat{g}(y)\, dy.$$

Note that both integrals converge in view of the proposition above. Consider the function $F(\xi, y) = g(\xi) f(y) e^{-2\pi i \xi \cdot y}$ defined for $(\xi, y) \in \mathbb{R}^d \times \mathbb{R}^d = \mathbb{R}^{2d}$. It is measurable as a function on $\mathbb{R}^{2d}$ in view of Corollary 3.7. We now apply Fubini's theorem to observe first that

$$\int_{\mathbb{R}^d} \int_{\mathbb{R}^d} |F(\xi, y)|\, d\xi\, dy = \int_{\mathbb{R}^d} |g(\xi)|\, d\xi \int_{\mathbb{R}^d} |f(y)|\, dy < \infty.$$

Next, if we evaluate $\int_{\mathbb{R}^d} \int_{\mathbb{R}^d} F(\xi, y)\, d\xi\, dy$ by writing it as $\int_{\mathbb{R}^d} \left(\int_{\mathbb{R}^d} F(\xi, y)\, d\xi \right) dy$ we get the left-hand side of the desired equality. Evaluating the double integral in the reverse order gives as the right-hand side, proving the lemma.

Next we consider the modulated Gaussian, $g(\xi) = e^{-\pi \delta |\xi|^2} e^{2\pi i x \cdot \xi}$, where for the moment δ and x are fixed, with $\delta > 0$ and $x \in \mathbb{R}^d$. An elementary calculation gives[5]

$$\hat{g}(y) = \int_{\mathbb{R}^d} e^{-\pi \delta |\xi|^2} e^{2\pi i (x-y) \cdot \xi}\, d\xi = \delta^{-d/2} e^{-\pi |x-y|^2 / \delta},$$

which we will abbreviate as $K_\delta(x - y)$. We recognize K_δ as a "good kernel" that satisfies:

(i) $\displaystyle\int_{\mathbb{R}^d} K_\delta(y)\, dy = 1$.

(ii) For each $\eta > 0$, $\displaystyle\int_{|y| > \eta} K_\delta(y)\, dy \to 0$ as $\delta \to 0$.

Applying the lemma gives

$$(16) \qquad \int_{\mathbb{R}^d} \hat{f}(\xi) e^{-\pi \delta |\xi|^2} e^{2\pi i x \cdot \xi}\, d\xi = \int_{\mathbb{R}^d} f(y) K_\delta(x - y)\, dy.$$

---

[5]See for example Chapter 6 in Book I.

Note that since $\hat{f} \in L^1(\mathbb{R}^d)$, the dominated convergence theorem shows that the left-hand side of (16) converges to $\int_{\mathbb{R}^d} \hat{f}(\xi)e^{2\pi i x \cdot \xi} \, d\xi$ as $\delta \to 0$, for each x. As for the right-hand side, we make two successive change of variables $y \to y + x$ (a translation), and $y \to -y$ (a reflection), and take into account the corresponding invariance of the integrals (see equations (4) and (5)). Thus the right-hand side becomes $\int_{\mathbb{R}^d} f(x - y)K_\delta(y) \, dy$, and we will prove that this function converges in the L^1-norm to f as $\delta \to 0$. Indeed, we can write the difference as

$$\Delta_\delta(x) = \int_{\mathbb{R}^d} f(x - y)K_\delta(y) \, dy - f(x) = \int_{\mathbb{R}^d} (f(x - y) - f(x))K_\delta(y) \, dy,$$

because of property (i) above. Thus

$$|\Delta_\delta(x)| \leq \int_{\mathbb{R}^d} |f(x - y) - f(x)|K_\delta(y) \, dy.$$

We can now apply Fubini's theorem, recalling that the measurability of $f(x)$ and $f(x - y)$ on $\mathbb{R}^d \times \mathbb{R}^d$ are established in Corollary 3.7 and Proposition 3.9. The result is

$$\|\Delta_\delta\| \leq \int_{\mathbb{R}^d} \|f_y - f\|K_\delta(y) \, dy, \quad \text{where } f_y(x) = f(x - y).$$

Now, for given $\epsilon > 0$ we can find (by Proposition 2.5) $\eta > 0$ so small such that $\|f_y - f\| < \epsilon$ when $|y| < \eta$. Thus

$$\|\Delta_\delta\| \leq \epsilon + \int_{|y|>\eta} \|f_y - f\|K_\delta(y) \, dy \leq \epsilon + 2\|f\| \int_{|y|>\eta} K_\delta(y) \, dy.$$

The first inequality follows by using (i) again; the second holds because $\|f_y - f\| \leq \|f_y\| + \|f\| = 2\|f\|$. Therefore, with the use of (ii), the combination above is $\leq 2\epsilon$ if δ is sufficiently small. To summarize: the right-hand side of (16) converges to f in the L^1-norm as $\delta \to 0$, and thus by Corollary 2.3 there is a subsequence that converges to $f(x)$ almost everywhere, and the theorem is proved.

Note that an immediate consequence of the theorem and the proposition is that if $\hat{f}$ were in L^1, then f could be modified on a set of measure zero to become continuous everywhere. This is of course impossible for the general $f \in L^1(\mathbb{R}^d)$.

5 Exercises

1. Given a collection of sets $F_1, F_2, \ldots, F_n$, construct another collection $F_1^*, F_2^*, \ldots, F_N^*$, with $N = 2^n - 1$, so that $\bigcup_{k=1}^n F_k = \bigcup_{j=1}^N F_j^*$; the collection $\{F_j^*\}$ is disjoint; also

$F_k = \bigcup_{F_j^* \subset F_k} F_j^*$, for every k.

[Hint: Consider the 2^n sets $F_1' \cap F_2' \cap \cdots \cap F_n'$ where each F_k' is either F_k or F_k^c.]

2. In analogy to Proposition 2.5, prove that if f is integrable on $\mathbb{R}^d$ and $\delta > 0$, then $f(\delta x)$ converges to $f(x)$ in the L^1-norm as $\delta \to 1$.

3. Suppose f is integrable on $(-\pi, \pi]$ and extended to $\mathbb{R}$ by making it periodic of period 2π. Show that

$$\int_{-\pi}^{\pi} f(x)\, dx = \int_I f(x)\, dx,$$

where I is any interval in $\mathbb{R}$ of length 2π.

[Hint: I is contained in two consecutive intervals of the form $(k\pi, (k+2)\pi)$.]

4. Suppose f is integrable on $[0, b]$, and

$$g(x) = \int_x^b \frac{f(t)}{t}\, dt \qquad \text{for } 0 < x \le b.$$

Prove that g is integrable on $[0, b]$ and

$$\int_0^b g(x)\, dx = \int_0^b f(t)\, dt.$$

5. Suppose F is a closed set in $\mathbb{R}$, whose complement has finite measure, and let $\delta(x)$ denote the distance from x to F, that is,

$$\delta(x) = d(x, F) = \inf\{|x - y| : y \in F\}.$$

Consider

$$I(x) = \int_{\mathbb{R}} \frac{\delta(y)}{|x - y|^2}\, dy.$$

(a) Prove that δ is continuous, by showing that it satisfies the Lipschitz condition

$$|\delta(x) - \delta(y)| \le |x - y|.$$

(b) Show that $I(x) = \infty$ for each $x \notin F$.

(c) Show that $I(x) < \infty$ for a.e. $x \in F$. This may be surprising in view of the fact that the Lispshitz condition cancels only one power of $|x - y|$ in the integrand of I.

[Hint: For the last part, investigate $\int_F I(x)\,dx$.]

6. Integrability of f on $\mathbb{R}$ does not necessarily imply the convergence of $f(x)$ to 0 as $x \to \infty$.

(a) There exists a positive continuous function f on $\mathbb{R}$ so that f is integrable on $\mathbb{R}$, but yet $\limsup_{x\to\infty} f(x) = \infty$.

(b) However, if we assume that f is uniformly continuous on $\mathbb{R}$ and integrable, then $\lim_{|x|\to\infty} f(x) = 0$.

[Hint: For (a), construct a continuous version of the function equal to n on the segment $[n, n + 1/n^3)$, $n \geq 1$.]

7. Let $\Gamma \subset \mathbb{R}^d \times \mathbb{R}$, $\Gamma = \{(x, y) \in \mathbb{R}^d \times \mathbb{R} : y = f(x)\}$, and assume f is measurable on $\mathbb{R}^d$. Show that Γ is a measurable subset of $\mathbb{R}^{d+1}$, and $m(\Gamma) = 0$.

8. If f is integrable on $\mathbb{R}$, show that $F(x) = \int_{-\infty}^{x} f(t)\,dt$ is uniformly continuous.

9. Tchebychev inequality. Suppose $f \geq 0$, and f is integrable. If $\alpha > 0$ and $E_\alpha = \{x : f(x) > \alpha\}$, prove that

$$m(E_\alpha) \leq \frac{1}{\alpha} \int f.$$

10. Suppose $f \geq 0$, and let $E_{2^k} = \{x : f(x) > 2^k\}$ and $F_k = \{x : 2^k < f(x) \leq 2^{k+1}\}$. If f is finite almost everywhere, then

$$\bigcup_{k=-\infty}^{\infty} F_k = \{f(x) > 0\},$$

and the sets F_k are disjoint.
 Prove that f is integrable if and only if

$$\sum_{k=-\infty}^{\infty} 2^k m(F_k) < \infty, \quad \text{if and only if} \quad \sum_{k=-\infty}^{\infty} 2^k m(E_{2^k}) < \infty.$$

Use this result to verify the following assertions. Let

$$f(x) = \begin{cases} |x|^{-a} & \text{if } |x| \leq 1, \\ 0 & \text{otherwise}, \end{cases} \quad \text{and} \quad g(x) = \begin{cases} |x|^{-b} & \text{if } |x| > 1, \\ 0 & \text{otherwise}. \end{cases}$$

Then f is integrable on $\mathbb{R}^d$ if and only if $a < d$; also g is integrable on $\mathbb{R}^d$ if and only if $b > d$.

11. Prove that if f is integrable on $\mathbb{R}^d$, real-valued, and $\int_E f(x)\,dx \geq 0$ for every measurable E, then $f(x) \geq 0$ a.e. x. As a result, if $\int_E f(x)\,dx = 0$ for every measurable E, then $f(x) = 0$ a.e.

12. Show that there are $f \in L^1(\mathbb{R}^d)$ and a sequence $\{f_n\}$ with $f_n \in L^1(\mathbb{R}^d)$ such that

$$\|f - f_n\|_{L^1} \to 0,$$

but $f_n(x) \to f(x)$ for no x.

[Hint: In $\mathbb{R}$, let $f_n = \chi_{I_n}$, where I_n is an appropriately chosen sequence of intervals with $m(I_n) \to 0$.]

13. Give an example of two measurable sets A and B such that $A + B$ is not measurable.

[Hint: In $\mathbb{R}^2$ take $A = \{0\} \times [0, 1]$ and $B = \mathcal{N} \times \{0\}$.]

14. In Exercise 6 of the previous chapter we saw that $m(B) = v_d r^d$, whenever B is a ball of radius r in $\mathbb{R}^d$ and $v_d = m(B_1)$, with B_1 the unit ball. Here we evaluate the constant v_d.

(a) For $d = 2$, prove using Corollary 3.8 that

$$v_2 = 2 \int_{-1}^{1} (1 - x^2)^{1/2} \, dx,$$

and hence by elementary calculus, that $v_2 = \pi$.

(b) By similar methods, show that

$$v_d = 2v_{d-1} \int_0^1 (1 - x^2)^{(d-1)/2} \, dx.$$

(c) The result is

$$v_d = \frac{\pi^{d/2}}{\Gamma(d/2 + 1)}.$$

Another derivation is in Exercise 5 in Chapter 6 below. Relevant facts about the gamma and beta functions can be found in Chapter 6 of Book II.

15. Consider the function defined over $\mathbb{R}$ by

$$f(x) = \begin{cases} x^{-1/2} & \text{if } 0 < x < 1, \\ 0 & \text{otherwise.} \end{cases}$$

For a fixed enumeration $\{r_n\}_{n=1}^{\infty}$ of the rationals $\mathbb{Q}$, let

$$F(x) = \sum_{n=1}^{\infty} 2^{-n} f(x - r_n).$$

Prove that F is integrable, hence the series defining F converges for almost every $x \in \mathbb{R}$. However, observe that this series is unbounded on every interval, and in fact, any function $\tilde{F}$ that agrees with F a.e is unbounded in any interval.

16. Suppose f is integrable on $\mathbb{R}^d$. If $\delta = (\delta_1, \ldots, \delta_d)$ is a d-tuple of non-zero real numbers, and

$$f^\delta(x) = f(\delta x) = f(\delta_1 x_1, \ldots, \delta_d x_d),$$

show that f^δ is integrable with

$$\int_{\mathbb{R}^d} f^\delta(x)\, dx = |\delta_1|^{-1} \cdots |\delta_d|^{-1} \int_{\mathbb{R}^d} f(x)\, dx.$$

17. Suppose f is defined on $\mathbb{R}^2$ as follows: $f(x, y) = a_n$ if $n \leq x < n+1$ and $n \leq y < n+1$, $(n \geq 0)$; $f(x, y) = -a_n$ if $n \leq x < n+1$ and $n+1 \leq y < n+2$, $(n \geq 0)$; while $f(x, y) = 0$ elsewhere. Here $a_n = \sum_{k \leq n} b_k$, with $\{b_k\}$ a positive sequence such that $\sum_{k=0}^\infty b_k = s < \infty$.

(a) Verify that each slice f^y and f_x is integrable. Also for all x, $\int f_x(y)\, dy = 0$, and hence $\int \left(\int f(x, y)\, dy \right) dx = 0$.

(b) However, $\int f^y(x)\, dx = a_0$ if $0 \leq y < 1$, and $\int f^y(x)\, dx = a_n - a_{n-1}$ if $n \leq y < n+1$ with $n \geq 1$. Hence $y \mapsto \int f^y(x)\, dx$ is integrable on $(0, \infty)$ and

$$\int \left(\int f(x, y)\, dx \right) dy = s.$$

(c) Note that $\int_{\mathbb{R} \times \mathbb{R}} |f(x, y)|\, dx\, dy = \infty$.

18. Let f be a measurable finite-valued function on $[0, 1]$, and suppose that $|f(x) - f(y)|$ is integrable on $[0, 1] \times [0, 1]$. Show that $f(x)$ is integrable on $[0, 1]$.

19. Suppose f is integrable on $\mathbb{R}^d$. For each $\alpha > 0$, let $E_\alpha = \{x : |f(x)| > \alpha\}$. Prove that

$$\int_{\mathbb{R}^d} |f(x)|\, dx = \int_0^\infty m(E_\alpha)\, d\alpha.$$

20. The problem (highlighted in the discussion preceding Fubini's theorem) that certain slices of measurable sets can be non-measurable may be avoided by restricting attention to Borel measurable functions and Borel sets. In fact, prove the following:

Suppose E is a Borel set in $\mathbb{R}^2$. Then for every y, the slice E^y is a Borel set in $\mathbb{R}$.

[Hint: Consider the collection $\mathcal{C}$ of subsets E of $\mathbb{R}^2$ with the property that each slice E^y is a Borel set in $\mathbb{R}$. Verify that $\mathcal{C}$ is a σ-algebra that contains the open sets.]

21. Suppose that f and g are measurable functions on $\mathbb{R}^d$.

(a) Prove that $f(x - y)g(y)$ is measurable on $\mathbb{R}^{2d}$.

(b) Show that if f and g are integrable on $\mathbb{R}^d$, then $f(x - y)g(y)$ is integrable on $\mathbb{R}^{2d}$.

(c) Recall the definition of the convolution of f and g given by

$$(f * g)(x) = \int_{\mathbb{R}^d} f(x - y)g(y)\, dy.$$

Show that $f * g$ is well defined for a.e. x (that is, $f(x - y)g(y)$ is integrable on $\mathbb{R}^d$ for a.e. x).

(d) Show that $f * g$ is integrable whenever f and g are integrable, and that

$$\|f * g\|_{L^1(\mathbb{R}^d)} \le \|f\|_{L^1(\mathbb{R}^d)}\, \|g\|_{L^1(\mathbb{R}^d)},$$

with equality if f and g are non-negative.

(e) The Fourier transform of an integrable function f is defined by

$$\hat{f}(\xi) = \int_{\mathbb{R}^d} f(x)e^{-2\pi i x \cdot \xi}\, dx.$$

Check that $\hat{f}$ is bounded and is a continuous function of ξ. Prove that for each ξ one has

$$\widehat{(f * g)}(\xi) = \hat{f}(\xi)\hat{g}(\xi).$$

22. Prove that if $f \in L^1(\mathbb{R}^d)$ and

$$\hat{f}(\xi) = \int_{\mathbb{R}^d} f(x)e^{-2\pi i x \xi}\, dx,$$

then $\hat{f}(\xi) \to 0$ as $|\xi| \to \infty$. (This is the Riemann-Lebesgue lemma.)
[Hint: Write $\hat{f}(\xi) = \frac{1}{2}\int_{\mathbb{R}^d}[f(x) - f(x - \xi')]e^{-2\pi i x \xi}\, dx$, where $\xi' = \frac{1}{2}\frac{\xi}{|\xi|^2}$, and use Proposition 2.5.]

23. As an application of the Fourier transform, show that there does not exist a function $I \in L^1(\mathbb{R}^d)$ such that

$$f * I = f \quad \text{for all } f \in L^1(\mathbb{R}^d).$$

24. Consider the convolution

$$(f * g)(x) = \int_{\mathbb{R}^d} f(x - y)g(y)\, dy.$$

(a) Show that $f * g$ is uniformly continuous when f is integrable and g bounded.

(b) If in addition g is integrable, prove that $(f * g)(x) \to 0$ as $|x| \to \infty$.

25. Show that for each $\epsilon > 0$ the function $F(\xi) = \frac{1}{(1+|\xi|^2)^\epsilon}$ is the Fourier transform of an L^1 function.

[Hint: With $K_\delta(x) = e^{-\pi|x|^2/\delta}\delta^{-d/2}$ consider $f(x) = \int_0^\infty K_\delta(x)e^{-\pi\delta}\delta^{\epsilon-1}\, d\delta$. Use Fubini's theorem to prove $f \in L^1(\mathbb{R}^d)$, and

$$\hat{f}(\xi) = \int_0^\infty e^{-\pi\delta|\xi|^2} e^{-\pi\delta}\delta^{\epsilon-1}\, d\delta,$$

and evaluate the last integral as $\pi^{-\epsilon}\Gamma(\epsilon)\frac{1}{(1+|\xi|^2)^\epsilon}$. Here $\Gamma(s)$ is the gamma function defined by $\Gamma(s) = \int_0^\infty e^{-t}t^{s-1}\, dt$.]

6 Problems

1. If f is integrable on $[0, 2\pi]$, then $\int_0^{2\pi} f(x)e^{-inx}\, dx \to 0$ as $|n| \to \infty$. Show as a consequence that if E is a measurable subset of $[0, 2\pi]$, then

$$\int_E \cos^2(nx + u_n)\, dx \to \frac{m(E)}{2}, \qquad \text{as } n \to \infty$$

for any sequence $\{u_n\}$.

[Hint: See Exercise 22.]

2. Prove the Cantor-Lebesgue theorem: if

$$\sum_{n=0}^\infty A_n(x) = \sum_{n=0}^\infty (a_n \cos nx + b_n \sin nx)$$

converges for x in a set of positive measure (or in particular for all x), then $a_n \to 0$ and $b_n \to 0$ as $n \to \infty$.

[Hint: Note that $A_n(x) \to 0$ uniformly on a set E of positive measure.]

3. A sequence $\{f_k\}$ of measurable functions on $\mathbb{R}^d$ is **Cauchy in measure** if for every $\epsilon > 0$,

$$m(\{x : |f_k(x) - f_\ell(x)| > \epsilon\}) \to 0 \qquad \text{as } k, \ell \to \infty.$$

We say that $\{f_k\}$ **converges in measure** to a (measurable) function f if for every $\epsilon > 0$

$$m(\{x : |f_k(x) - f(x)| > \epsilon\}) \to 0 \quad \text{as } k \to \infty.$$

This notion coincides with the "convergence in probability" of probability theory.

Prove that if a sequence $\{f_k\}$ of integrable functions converges to f in L^1, then $\{f_k\}$ converges to f in measure. Is the converse true?

We remark that this mode of convergence appears naturally in the proof of Egorov's theorem.

4. We have already seen (in Exercise 8, Chapter 1) that if E is a measurable set in $\mathbb{R}^d$, and L is a linear transformation of $\mathbb{R}^d$ to $\mathbb{R}^d$, then $L(E)$ is also measurable, and if E has measure 0, then so has $L(E)$. The quantitative statement is

$$m(L(E)) = |\det(L)| \, m(E).$$

As a special case, note that the Lebesgue measure is invariant under rotations. (For this special case see also Exercise 26 in the next chapter.)

The above identity can be proved using Fubini's theorem as follows.

(a) Consider first the case $d = 2$, and L a "strictly" upper triangular transformation $x' = x + ay$, $y' = y$. Then

$$\chi_{L(E)}(x, y) = \chi_E(L^{-1}(x, y)) = \chi_E(x - ay, y).$$

Hence

$$m(L(E)) = \int_{\mathbb{R} \times \mathbb{R}} \left(\int \chi_E(x - ay, y) \right) dy$$

$$= \int_{\mathbb{R} \times \mathbb{R}} \left(\int \chi_E(x, y) \, dx \right) dy$$

$$= m(E),$$

by the translation-invariance of the measure.

(b) Similarly $m(L(E)) = m(E)$ if L is strictly lower triangular. In general, one can write $L = L_1 \Delta L_2$, where L_j are strictly (upper and lower) triangular and Δ is diagonal. Thus $m(L(E)) = |\det(L)| m(E)$, if one uses Exercise 7 in Chapter 1.

5. There is an ordering $\prec$ of $\mathbb{R}$ with the property that for each $y \in \mathbb{R}$ the set $\{x \in \mathbb{R} : x \prec y\}$ is at most countable.

The existence of this ordering depends on the **continuum hypothesis**, which asserts: whenever S is an infinite subset of $\mathbb{R}$, then either S is countable, or S has the cardinality of $\mathbb{R}$ (that is, can be mapped bijectively to $\mathbb{R}$).[6]

---

[6]This assertion, formulated by Cantor, is like the well-ordering principle independent of the other axioms of set theory, and so we are also free to accept its validity.

[Hint: Let $\prec$ denote a well-ordering of $\mathbb{R}$, and define the set X by $X = \{y \in \mathbb{R} : \text{the set } \{x : x \prec y\} \text{ is not countable}\}$. If X is empty we are done. Otherwise, consider the smallest element $\bar{y}$ in X, and use the continuum hypothesis.]

3 Differentiation and Integration

The Maximal Problem:
>   The problem is most easily grasped when stated
> in the language of cricket, or any other game in which
> a player compiles a series of scores of which an average
> is recorded.
>
> G. H. Hardy and J. E. Littlewood, 1930

That differentiation and integration are inverse operations was already understood early in the study of the calculus. Here we want to reexamine this basic idea in the framework of the general theory studied in the previous chapters. Our objective is the formulation and proof of the fundamental theorem of the calculus in this setting, and the development of some of the concepts that occur. We shall try to achieve this by answering two questions, each expressing one of the ways of representing the reciprocity between differentiation and integration.

The first problem involved may be stated as follows.

- Suppose f is integrable on $[a, b]$ and F is its indefinite integral $F(x) = \int_a^x f(y)\,dy$. Does this imply that F is differentiable (at least for almost every x), and that $F' = f$?

We shall see that the affirmative answer to this question depends on ideas that have broad application and are not limited to the one-dimensional situation.

For the second question we reverse the order of differentiation and integration.

- What conditions on a function F on $[a, b]$ guarantee that $F'(x)$ exists (for a.e. x), that this function is integrable, and that moreover

$$F(b) - F(a) = \int_a^b F'(x)\,dx \quad ?$$

While this problem will be examined from a narrower perspective than the first, the issues it raises are deep and the consequences entailed are far-reaching. In particular, we shall find that this question is connected to the problem of rectifiability of curves, and as an illustration of this link, we shall establish the general isoperimetric inequality in the plane.

1 Differentiation of the integral

We begin with the first problem, that is, the study of differentiation of the integral. If f is given on $[a, b]$ and integrable on that interval, we let

$$F(x) = \int_a^x f(y)\, dy, \qquad a \le x \le b.$$

To deal with $F'(x)$, we recall the definition of the derivative as the limit of the quotient

$$\frac{F(x+h) - F(x)}{h} \qquad \text{when } h \text{ tends to } 0.$$

We note that this quotient takes the form (say in the case $h > 0$)

$$\frac{1}{h} \int_x^{x+h} f(y)\, dy = \frac{1}{|I|} \int_I f(y)\, dy,$$

where we use the notation $I = (x, x+h)$ and $|I|$ for the length of this interval. At this point, we pause to observe that the above expression is the "average" value of f over I, and that in the limit as $|I| \to 0$, we might expect that these averages tend to $f(x)$. Reformulating the question slightly, we may ask whether

$$\lim_{\substack{|I| \to 0 \\ x \in I}} \frac{1}{|I|} \int_I f(y)\, dy = f(x)$$

holds for suitable points x. In higher dimensions we can pose a similar question, where the averages of f are taken over appropriate sets that generalize the intervals in one dimension. Initially we shall study this problem where the sets involved are the balls B containing x, with their volume $m(B)$ replacing the length $|I|$ of I. Later we shall see that as a consequence of this special case similar results will hold for more general collections of sets, those that have bounded "eccentricity."

With this in mind we restate our first problem in the context of $\mathbb{R}^d$, for all $d \ge 1$.

Suppose f is integrable on $\mathbb{R}^d$. Is it true that

$$\lim_{\substack{m(B) \to 0 \\ x \in B}} \frac{1}{m(B)} \int_B f(y)\, dy = f(x), \qquad \text{for a.e. } x?$$

The limit is taken as the volume of open balls B containing x goes to 0.

We shall refer to this question as the **averaging problem**. We remark that if B is any ball of radius r in $\mathbb{R}^d$, then $m(B) = v_d r^d$, where v_d is the measure of the unit ball. (See Exercise 14 in the previous chapter.)

Note of course that in the special case when f is continuous at x, the limit does converge to $f(x)$. Indeed, given $\epsilon > 0$, there exists $\delta > 0$ such that $|f(x) - f(y)| < \epsilon$ whenever $|x - y| < \delta$. Since

$$f(x) - \frac{1}{m(B)} \int_B f(y)\, dy = \frac{1}{m(B)} \int_B (f(x) - f(y))\, dy,$$

we find that whenever B is a ball of radius $< \delta/2$ that contains x, then

$$\left| f(x) - \frac{1}{m(B)} \int_B f(y)\, dy \right| \leq \frac{1}{m(B)} \int_B |f(x) - f(y)|\, dy < \epsilon,$$

as desired.

The averaging problem has an affirmative answer, but to establish that fact, which is qualitative in nature, we need to make some quantitative estimates bearing on the overall behavior of the averages of f. This will be done in terms of the maximal averages of $|f|$, to which we now turn.

1.1 The Hardy-Littlewood maximal function

The maximal function that we consider below arose first in the one-dimensional situation treated by Hardy and Littlewood. It seems that they were led to the study of this function by toying with the question of how a batsman's score in cricket may best be distributed to maximize his satisfaction. As it turns out, the concepts involved have a universal significance in analysis. The relevant definition is as follows.

If f is integrable on $\mathbb{R}^d$, we define its **maximal function** f^* by

$$f^*(x) = \sup_{x \in B} \frac{1}{m(B)} \int_B |f(y)|\, dy, \qquad x \in \mathbb{R}^d,$$

where the supremum is taken over all balls containing the point x. In other words, we replace the limit in the statement of the averaging problem by a supremum, and f by its absolute value.

The main properties of f^* we shall need are summarized in a theorem.

Theorem 1.1 *Suppose f is integrable on $\mathbb{R}^d$. Then:*

(i) f^* *is measurable.*

(ii) $f^*(x) < \infty$ *for a.e. x.*

(iii) f^* *satisfies*

$$(1) \qquad m(\{x \in \mathbb{R}^d : f^*(x) > \alpha\}) \leq \frac{A}{\alpha} \|f\|_{L^1(\mathbb{R}^d)}$$

for all $\alpha > 0$, where $A = 3^d$, and $\|f\|_{L^1(\mathbb{R}^d)} = \int_{\mathbb{R}^d} |f(x)| \, dx$.

Before we come to the proof we want to clarify the nature of the main conclusion (iii). As we shall observe, one has that $f^*(x) \geq |f(x)|$ for a.e. x; the effect of (iii) is that, broadly speaking, f^* is not much larger than $|f|$. From this point of view, we would have liked to conclude that f^* is integrable, as a result of the assumed integrability of f. However, this is not the case, and (iii) is the best substitute available (see Exercises 4 and 5).

An inequality of the type (1) is called a **weak-type** inequality because it is weaker than the corresponding inequality for the L^1-norms. Indeed, this can be seen from the Tchebychev inequality (Exercise 9 in Chapter 2), which states that for an arbitrary integrable function g,

$$m(\{x : |g(x)| > \alpha\}) \leq \frac{1}{\alpha} \|g\|_{L^1(\mathbb{R}^d)}, \qquad \text{for all } \alpha > 0.$$

We should add that the exact value of A in the inequality (1) is unimportant for us. What matters is that this constant be independent of α and f.

The only simple assertion in the theorem is that f^* is a measurable function. Indeed, the set $E_\alpha = \{x \in \mathbb{R}^d : f^*(x) > \alpha\}$ is open, because if $\overline{x} \in E_\alpha$, there exists a ball B such that $\overline{x} \in B$ and

$$\frac{1}{m(B)} \int_B |f(y)| \, dy > \alpha.$$

Now any point x close enough to $\overline{x}$ will also belong to B; hence $x \in E_\alpha$ as well.

The two other properties of f^* in the theorem are deeper, with (ii) being a consequence of (iii). This follows at once if we observe that

$$\{x : f^*(x) = \infty\} \subset \{x : f^*(x) > \alpha\}$$

for all α. Taking the limit as α tends to infinity, the third property yields $m(\{x : \ f^*(x) = \infty\}) = 0$.

The proof of inequality (1) relies on an elementary version of a Vitali covering argument.[1]

Lemma 1.2 *Suppose $\mathcal{B} = \{B_1, B_2, \ldots, B_N\}$ is a finite collection of open balls in $\mathbb{R}^d$. Then there exists a disjoint sub-collection $B_{i_1}, B_{i_2}, \ldots, B_{i_k}$ of $\mathcal{B}$ that satisfies*

$$m\left(\bigcup_{\ell=1}^{N} B_\ell\right) \leq 3^d \sum_{j=1}^{k} m(B_{i_j}).$$

Loosely speaking, we may always find a disjoint sub-collection of balls that covers a fraction of the region covered by the original collection of balls.

Proof. The argument we give is constructive and relies on the following simple observation: Suppose B and B' are a pair of balls that intersect, with the radius of B' being not greater than that of B. Then B' is contained in the ball $\tilde{B}$ that is concentric with B but with 3 times its radius.

As a first step, we pick a ball B_{i_1} in $\mathcal{B}$ with maximal (that is, largest) radius, and then delete from $\mathcal{B}$ the ball B_{i_1} as well as any balls that intersect B_{i_1}. Thus all the balls that are deleted are contained in the ball $\tilde{B}_{i_1}$ concentric with B_{i_1}, but with 3 times its radius.

The remaining balls yield a new collection $\mathcal{B}'$, for which we repeat the procedure. We pick B_{i_2} with largest radius in $\mathcal{B}'$, and then delete from $\mathcal{B}'$ the ball B_{i_2} and any ball that intersects B_{i_2}. Continuing this way we find, after at most N steps, a collection of disjoint balls $B_{i_1}, B_{i_2}, \ldots, B_{i_k}$.

Finally, to prove that this disjoint collection of balls satisfies the inequality in the lemma, we use the observation made at the beginning of the proof. We let $\tilde{B}_{i_j}$ denote the ball concentric with B_{i_j}, but with 3 times its radius. Since any ball B in $\mathcal{B}$ must intersect a ball B_{i_j} and have equal or smaller radius than B_{i_j}, we must have $B \subset \tilde{B}_{i_j}$, thus

$$m\left(\bigcup_{\ell=1}^{N} B_\ell\right) \leq m\left(\bigcup_{j=1}^{k} \tilde{B}_{i_j}\right) \leq \sum_{j=1}^{k} m(\tilde{B}_{i_j}) = 3^d \sum_{j=1}^{k} m(B_{i_j}).$$

---

[1] We note that the lemma that follows is the first of a series of covering arguments that occur below in the theory of differentiation; see also Lemma 3.9 and its corollary, as well as Lemma 3.5, where the covering assertion is more implicit.

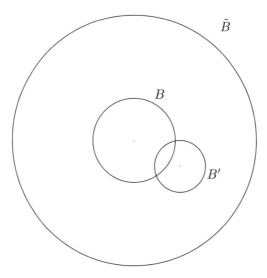

Figure 1. The balls B and $\tilde{B}$

In the last step we have used the fact that in $\mathbb{R}^d$ a dilation of a set by $\delta > 0$ results in the multiplication by δ^d of the Lebesgue measure of this set.

The proof of (iii) in Theorem 1.1 is now in reach. If we let $E_\alpha = \{x : f^*(x) > \alpha\}$, then for each $x \in E_\alpha$ there exists a ball B_x that contains x, and such that

$$\frac{1}{m(B_x)} \int_{B_x} |f(y)|\,dy > \alpha.$$

Therefore, for each ball B_x we have

$$(2) \qquad m(B_x) < \frac{1}{\alpha} \int_{B_x} |f(y)|\,dy.$$

Fix a compact subset K of E_α. Since K is covered by $\bigcup_{x \in E_\alpha} B_x$, we may select a finite subcover of K, say $K \subset \bigcup_{\ell=1}^{N} B_\ell$. The covering lemma guarantees the existence of a sub-collection $B_{i_1}, \ldots, B_{i_k}$ of disjoint balls with

$$(3) \qquad m\left(\bigcup_{\ell=1}^{N} B_\ell\right) \leq 3^d \sum_{j=1}^{k} m(B_{i_j}).$$

Since the balls $B_{i_1}, \ldots, B_{i_k}$ are disjoint and satisfy (2) as well as (3), we find that

$$m(K) \leq m\left(\bigcup_{\ell=1}^N B_\ell\right) \leq 3^d \sum_{j=1}^k m(B_{i_j}) \leq \frac{3^d}{\alpha} \sum_{j=1}^k \int_{B_{i_j}} |f(y)| \, dy$$

$$= \frac{3^d}{\alpha} \int_{\bigcup_{j=1}^k B_{i_j}} |f(y)| \, dy$$

$$\leq \frac{3^d}{\alpha} \int_{\mathbb{R}^d} |f(y)| \, dy.$$

Since this inequality is true for all compact subsets K of E_α, the proof of the weak type inequality for the maximal operator is complete.

1.2 The Lebesgue differentiation theorem

The estimate obtained for the maximal function now leads to a solution of the averaging problem.

Theorem 1.3 *If f is integrable on $\mathbb{R}^d$, then*

$$(4) \qquad \lim_{\substack{m(B) \to 0 \\ x \in B}} \frac{1}{m(B)} \int_B f(y) \, dy = f(x) \qquad \text{for a.e. } x.$$

Proof. It suffices to show that for each $\alpha > 0$ the set

$$E_\alpha = \left\{ x : \limsup_{\substack{m(B) \to 0 \\ x \in B}} \left| \frac{1}{m(B)} \int_B f(y) \, dy - f(x) \right| > 2\alpha \right\}$$

has measure zero, because this assertion then guarantees that the set $E = \bigcup_{n=1}^\infty E_{1/n}$ has measure zero, and the limit in (4) holds at all points of E^c.

We fix α, and recall Theorem 2.4 in Chapter 2, which states that for each $\epsilon > 0$ we may select a continuous function g of compact support with $\|f - g\|_{L^1(\mathbb{R}^d)} < \epsilon$. As we remarked earlier, the continuity of g implies that

$$\lim_{\substack{m(B) \to 0 \\ x \in B}} \frac{1}{m(B)} \int_B g(y) \, dy = g(x), \qquad \text{for all } x.$$

Since we may write the difference $\frac{1}{m(B)} \int_B f(y) \, dy - f(x)$ as

$$\frac{1}{m(B)} \int_B (f(y) - g(y)) \, dy + \frac{1}{m(B)} \int_B g(y) \, dy - g(x) + g(x) - f(x)$$

we find that

$$\limsup_{\substack{m(B)\,\to\,0 \\ x\,\in\,B}} \left| \frac{1}{m(B)} \int_B f(y)\,dy - f(x) \right| \le (f-g)^*(x) + |g(x) - f(x)|,$$

where the symbol $*$ indicates the maximal function. Consequently, if

$$F_\alpha = \{x : (f-g)^*(x) > \alpha\} \quad \text{and} \quad G_\alpha = \{x : |f(x) - g(x)| > \alpha\}$$

then $E_\alpha \subset (F_\alpha \cup G_\alpha)$, because if u_1 and u_2 are positive, then $u_1 + u_2 > 2\alpha$ only if $u_i > \alpha$ for at least one u_i. On the one hand, Tchebychev's inequality yields

$$m(G_\alpha) \le \frac{1}{\alpha} \|f - g\|_{L^1(\mathbb{R}^d)},$$

and on the other hand, the weak type estimate for the maximal function gives

$$m(F_\alpha) \le \frac{A}{\alpha} \|f - g\|_{L^1(\mathbb{R}^d)}.$$

The function g was selected so that $\|f - g\|_{L^1(\mathbb{R}^d)} < \epsilon$. Hence we get

$$m(E_\alpha) \le \frac{A}{\alpha} \epsilon + \frac{1}{\alpha} \epsilon.$$

Since ϵ is arbitrary, we must have $m(E_\alpha) = 0$, and the proof of the theorem is complete.

Note that as an immediate consequence of the theorem applied to $|f|$, we see that $f^*(x) \ge |f(x)|$ for a.e. x, with f^* the maximal function.

We have worked so far under the assumption that f is integrable. This "global" assumption is slightly out of place in the context of a "local" notion like differentiability. Indeed, the limit in Lebesgue's theorem is taken over balls that shrink to the point x, so the behavior of f far from x is irrelevant. Thus, we expect the result to remain valid if we simply assume integrability of f on every ball.

To make this precise, we say that a measurable function f on $\mathbb{R}^d$ is **locally integrable**, if for every ball B the function $f(x)\chi_B(x)$ is integrable. We shall denote by $L^1_{\text{loc}}(\mathbb{R}^d)$ the space of all locally integrable functions. Loosely speaking, the behavior at infinity does not affect the local integrability of a function. For example, the functions $e^{|x|}$ and $|x|^{-1/2}$ are both locally integrable, but not integrable on $\mathbb{R}^d$.

Clearly, the conclusion of the last theorem holds under the weaker assumption that f is locally integrable.

Theorem 1.4 *If $f \in L^1_{\mathrm{loc}}(\mathbb{R}^d)$, then*

$$\lim_{\substack{m(B) \to 0 \\ x \in B}} \frac{1}{m(B)} \int_B f(y)\, dy = f(x), \quad \text{for a.e. } x.$$

Our first application of this theorem yields an interesting insight into the nature of measurable sets. If E is a measurable set and $x \in \mathbb{R}^d$, we say that x is a point of **Lebesgue density** of E if

$$\lim_{\substack{m(B) \to 0 \\ x \in B}} \frac{m(B \cap E)}{m(B)} = 1.$$

Loosely speaking, this condition says that small balls around x are almost entirely covered by E. More precisely, for every $\alpha < 1$ close to 1, and every ball of sufficiently small radius containing x, we have

$$m(B \cap E) \geq \alpha m(B).$$

Thus E covers at least a proportion α of B.

An application of Theorem 1.4 to the characteristic function of E immediately yields the following:

Corollary 1.5 *Suppose E is a measurable subset of $\mathbb{R}^d$. Then:*

(i) *Almost every $x \in E$ is a point of density of E.*

(ii) *Almost every $x \notin E$ is not a point of density of E.*

We next consider a notion that for integrable functions serves as a useful substitute for pointwise continuity.

If f is locally integrable on $\mathbb{R}^d$, the **Lebesgue set** of f consists of all points $\overline{x} \in \mathbb{R}^d$ for which $f(\overline{x})$ is finite and

$$\lim_{\substack{m(B) \to 0 \\ \overline{x} \in B}} \frac{1}{m(B)} \int_B |f(y) - f(\overline{x})|\, dy = 0.$$

At this stage, two simple observations about this definition are in order. First, $\overline{x}$ belongs to the Lebesgue set of f whenever f is continuous at $\overline{x}$. Second, if $\overline{x}$ is in the Lebesgue set of f, then

$$\lim_{\substack{m(B) \to 0 \\ x \in B}} \frac{1}{m(B)} \int_B f(y)\, dy = f(\overline{x}).$$

Corollary 1.6 *If f is locally integrable on $\mathbb{R}^d$, then almost every point belongs to the Lebesgue set of f.*

Proof. An application of Theorem 1.4 to the function $|f(y) - r|$ shows that for each rational r, there exists a set E_r of measure zero, such that

$$\lim_{\substack{m(B) \to 0 \\ x \in B}} \frac{1}{m(B)} \int_B |f(y) - r| \, dy = |f(x) - r| \quad \text{whenever } x \notin E_r.$$

If $E = \bigcup_{r \in \mathbb{Q}} E_r$, then $m(E) = 0$. Now suppose that $\bar{x} \notin E$ and $f(\bar{x})$ is finite. Given $\epsilon > 0$, there exists a rational r such that $|f(\bar{x}) - r| < \epsilon$. Since

$$\frac{1}{m(B)} \int_B |f(y) - f(\bar{x})| \, dy \leq \frac{1}{m(B)} \int_B |f(y) - r| \, dy + |f(\bar{x}) - r|,$$

we must have

$$\limsup_{\substack{m(B) \to 0 \\ \bar{x} \in B}} \frac{1}{m(B)} \int_B |f(y) - f(\bar{x})| \, dy \leq 2\epsilon,$$

and thus $\bar{x}$ is in the Lebesgue set of f. The corollary is therefore proved.

Remark. Recall from the definition in Section 2 of Chapter 2 that elements of $L^1(\mathbb{R}^d)$ are actually equivalence classes, with two functions being equivalent if they differ on a set of measure zero. It is interesting to observe that the set of points where the averages (4) converge to a limit is independent of the representation of f chosen, because

$$\int_B f(y) \, dy = \int_B g(y) \, dy$$

whenever f and g are equivalent. Nevertheless, the Lebesgue set of f depends on the particular representative of f that we consider.

We shall see that the Lebesgue set of a function enjoys a universal property in that at its points the function can be recovered by a wide variety of averages. We will prove this both for averages over sets that generalize balls, and in the setting of approximations to the identity. Note that the theory of differentiation developed so far uses averages over balls, but as we mentioned earlier, one could ask whether similar conclusions hold for other families of sets, such as cubes or rectangles. The answer depends in a fundamental way on the geometric properties of the family in question. For example, we now show that in the case of cubes (and more generally families of sets with bounded "eccentricity") the above results carry over. However, in the case of the family of *all* rectangles the existence of the limit almost everywhere and the weak type inequality fail (see Problem 8).

A collection of sets $\{U_\alpha\}$ is said to **shrink regularly** to $\bar{x}$ (or has **bounded eccentricity** at $\bar{x}$) if there is a constant $c > 0$ such that for each U_α there is a ball B with

$$\bar{x} \in B, \quad U_\alpha \subset B, \quad \text{and} \quad m(U_\alpha) \geq c\,m(B).$$

Thus U_α is contained in B, but its measure is *comparable* to the measure of B. For example, the set of all open cubes containing $\bar{x}$ shrink regularly to $\bar{x}$. However, in $\mathbb{R}^d$ with $d \geq 2$ the collection of all open rectangles containing $\bar{x}$ does not shrink regularly to $\bar{x}$. This can be seen if we consider very thin rectangles.

Corollary 1.7 *Suppose f is locally integrable on $\mathbb{R}^d$. If $\{U_\alpha\}$ shrinks regularly to $\bar{x}$, then*

$$\lim_{\substack{m(U_\alpha) \to 0 \\ x \in U_\alpha}} \frac{1}{m(U_\alpha)} \int_{U_\alpha} f(y)\,dy = f(\bar{x})$$

for every point $\bar{x}$ in the Lebesgue set of f.

The proof is immediate once we observe that if $\bar{x} \in B$ with $U_\alpha \subset B$ and $m(U_\alpha) \geq cm(B)$, then

$$\frac{1}{m(U_\alpha)} \int_{U_\alpha} |f(y) - f(\bar{x})|\,dy \leq \frac{1}{cm(B)} \int_B |f(y) - f(\bar{x})|\,dy.$$

2 Good kernels and approximations to the identity

We shall now turn to averages of functions given as convolutions,[2] which can be written as

$$(f * K_\delta)(x) = \int_{\mathbb{R}^d} f(x - y)K_\delta(y)\,dy.$$

Here f is a general integrable function, which we keep fixed, while the K_δ vary over a specific family of functions, referred to as kernels. Expressions of this kind arise in many questions (for instance, in the Fourier inversion theorem of the previous chapter), and were already discussed in Book I.

In our initial consideration we called these functions "good kernels" if they are integrable and satisfy the following conditions for $\delta > 0$:

---

[2]Some basic properties of convolutions are described in Exercise 21 of the previous chapter.

(i) $\displaystyle\int_{\mathbb{R}^d} K_\delta(x)\,dx = 1.$

(ii) $\displaystyle\int_{\mathbb{R}^d} |K_\delta(x)|\,dx \leq A.$

(iii) For every $\eta > 0$,

$$\int_{|x|\geq\eta} |K_\delta(x)|\,dx \to 0 \quad \text{as } \delta \to 0.$$

Here A is a constant independent of δ.

The main use of these kernels was that whenever f is bounded, then $(f * K_\delta)(x) \to f(x)$ as $\delta \to 0$, at every point of continuity of f. To obtain a similar conclusion, one also valid at all points of the Lebesgue set of f, we need to strengthen somewhat our assumptions on the kernels K_δ. To reflect this situation we adopt a different terminology and refer to the resulting narrower class of kernels as **approximations to the identity**. The assumptions are again that the K_δ are integrable and satisfy conditions (i) but, instead of (ii) and (iii), we assume:

(ii') $|K_\delta(x)| \leq A\delta^{-d}$ for all $\delta > 0$.

(iii') $|K_\delta(x)| \leq A\delta/|x|^{d+1}$ for all $\delta > 0$ and $x \in \mathbb{R}^d$.[3]

We observe that these requirements are stronger and imply the conditions in the definition of good kernels. Indeed, we first prove (ii). For that, we use the second illustration of Corollary 1.10 in Chapter 2, which gives

(5) $$\int_{|x|\geq\epsilon} \frac{dx}{|x|^{d+1}} \leq \frac{C}{\epsilon} \quad \text{for some } C > 0 \text{ and all } \epsilon > 0.$$

Then, using the estimates (ii') and (iii') when $|x| < \delta$ and $|x| \geq \delta$, respectively, yields

$$\int_{\mathbb{R}^d} |K_\delta(x)|\,dx = \int_{|x|<\delta} |K_\delta(x)|\,dx + \int_{|x|\geq\delta} |K_\delta(x)|\,dx$$

$$\leq A \int_{|x|<\delta} \frac{dx}{\delta^d} + A\delta \int_{|x|\geq\delta} \frac{1}{|x|^{d+1}}\,dx$$

$$\leq A' + A'' < \infty.$$

---

[3]Sometimes the condition (iii') is replaced by the requirement $|K_\delta(x)| \leq A\delta^\epsilon/|x|^{d+\epsilon}$ for some fixed $\epsilon > 0$. However, the special case $\epsilon = 1$ suffices in most circumstances.

Finally, the last condition of a good kernel is also verified, since another application of (5) gives

$$\int_{|x|\geq\eta} |K_\delta(x)|\, dx \leq A\delta \int_{|x|\geq\eta} \frac{dx}{|x|^{d+1}}$$
$$\leq \frac{A'\delta}{\eta},$$

and this last expression tends to 0 as $\delta \to 0$.

The term "approximation to the identity" originates in the fact that the mapping $f \mapsto f * K_\delta$ converges to the identity mapping $f \mapsto f$, as $\delta \to 0$, in various senses, as we shall see below. It is also connected with the following heuristics. Figure 2 pictures a typical approximation to the identity: for each $\delta > 0$, the kernel is supported on the set $|x| < \delta$ and has height $1/2\delta$. As δ tends to 0, this family of kernels converges to the

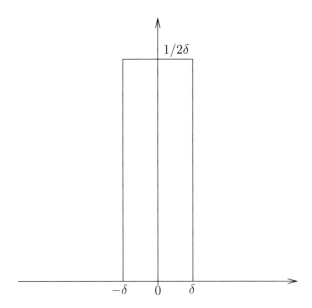

Figure 2. An approximation to the identity

so-called unit mass at the origin or **Dirac delta** "function." The latter is heuristically defined by

$$\mathcal{D}(x) = \begin{cases} \infty & \text{if } x = 0 \\ 0 & \text{if } x \neq 0 \end{cases} \quad \text{and} \quad \int \mathcal{D}(x)\, dx = 1.$$

Since each K_δ integrates to 1, we may say loosely that

$$K_\delta \to \mathcal{D} \quad \text{as } \delta \to 0.$$

If we think of the convolution $f * \mathcal{D}$ as $\int f(x - y)\mathcal{D}(y)\,dy$, the product $f(x - y)\mathcal{D}(y)$ is 0 except when $y = 0$, and the mass of $\mathcal{D}$ is concentrated at $y = 0$, so we may intuitively expect that

$$(f * \mathcal{D})(x) = f(x).$$

Thus $f * \mathcal{D} = f$, and $\mathcal{D}$ plays the role of the identity for convolutions. We should mention that this discussion can be formalized and $\mathcal{D}$ given a precise definition either in terms of Lebesgue-Stieltjes measures, which we take up in Chapter 6, or in terms of "generalized functions" (that is, distributions), which we defer to Book IV.

We now turn to a series of examples of approximations to the identity.

EXAMPLE 1. Suppose φ is a non-negative bounded function in $\mathbb{R}^d$ that is supported on the unit ball $|x| \le 1$, and such that

$$\int_{\mathbb{R}^d} \varphi = 1.$$

Then, if we set $K_\delta(x) = \delta^{-d}\varphi(\delta^{-1}x)$, the family $\{K_\delta\}_{\delta > 0}$ is an approximation to the identity. The simple verification is left to the reader. Important special cases are in the next two examples.

EXAMPLE 2. The Poisson kernel for the upper half-plane is given by

$$\mathcal{P}_y(x) = \frac{1}{\pi}\frac{y}{x^2 + y^2}, \quad x \in \mathbb{R},$$

where the parameter is now $\delta = y > 0$.

EXAMPLE 3. The heat kernel in $\mathbb{R}^d$ is defined by

$$\mathcal{H}_t(x) = \frac{1}{(4\pi t)^{d/2}}e^{-|x|^2/4t}.$$

Here $t > 0$ and we have $\delta = t^{1/2}$. Alternatively, we could set $\delta = 4\pi t$ to make the notation consistent with the specific usage in Chapter 2.

EXAMPLE 4. The Poisson kernel for the disc is

$$\frac{1}{2\pi}P_r(x) = \begin{cases} \dfrac{1}{2\pi}\dfrac{1 - r^2}{1 - 2r\cos x + r^2} & \text{if } |x| \le \pi, \\ 0 & \text{if } |x| > \pi. \end{cases}$$

Here we have $0 < r < 1$ and $\delta = 1 - r$.

EXAMPLE 5. The Fejér kernel is defined by

$$\frac{1}{2\pi} F_N(x) = \begin{cases} \dfrac{1}{2\pi N} \dfrac{\sin^2(Nx/2)}{\sin^2(x/2)} & \text{if } |x| \leq \pi, \\ 0 & \text{if } |x| > \pi, \end{cases}$$

where $\delta = 1/N$.

We note that Examples 2 through 5 have already appeared in Book I.

We now turn to a general result about approximations to the identity that highlights the role of the Lebesgue set.

Theorem 2.1 *If $\{K_\delta\}_{\delta>0}$ is an approximation to the identity and f is integrable on $\mathbb{R}^d$, then*

$$(f * K_\delta)(x) \to f(x) \quad \text{as } \delta \to 0$$

for every x in the Lebesgue set of f. In particular, the limit holds for a.e. x.

Since the integral of each kernel K_δ is equal to 1, we may write

$$(f * K_\delta)(x) - f(x) = \int [f(x - y) - f(x)] \, K_\delta(y) \, dy.$$

Consequently,

$$|(f * K_\delta)(x) - f(x)| \leq \int |f(x - y) - f(x)| \, |K_\delta(y)| \, dy,$$

and it now suffices to prove that the right-hand side tends to 0 as δ goes to 0. The argument we give depends on a simple result that we isolate in the next lemma.

Lemma 2.2 *Suppose that f is integrable on $\mathbb{R}^d$, and that x is a point of the Lebesgue set of f. Let*

$$\mathcal{A}(r) = \frac{1}{r^d} \int_{|y| \leq r} |f(x - y) - f(x)| \, dy, \quad \text{whenever } r > 0.$$

Then $\mathcal{A}(r)$ is a continuous function of $r > 0$, and

$$\mathcal{A}(r) \to 0 \quad \text{as } r \to 0.$$

Moreover, $\mathcal{A}(r)$ is bounded, that is, $\mathcal{A}(r) \leq M$ for some $M > 0$ and all $r > 0$.

Proof. The continuity of $\mathcal{A}(r)$ follows by invoking the absolute continuity in Proposition 1.12 of Chapter 2.

The fact that $\mathcal{A}(r)$ tends to 0 as r tends to 0 follows since x belongs to the Lebesgue set of f, and the measure of a ball of radius r is $v_d r^d$. This and the continuity of $\mathcal{A}(r)$ for $0 < r \leq 1$ show that $\mathcal{A}(r)$ is bounded when $0 < r \leq 1$. To prove that $\mathcal{A}(r)$ is bounded for $r > 1$, note that

$$\mathcal{A}(r) \leq \frac{1}{r^d} \int_{|y| \leq r} |f(x-y)| \, dy + \frac{1}{r^d} \int_{|y| \leq r} |f(x)| \, dy$$
$$\leq r^{-d} \|f\|_{L^1(\mathbb{R}^d)} + v_d |f(x)|,$$

and this concludes the proof of the lemma.

We now return to the proof of the theorem. The key consists in writing the integral over $\mathbb{R}^d$ as a sum of integrals over annuli as follows:

$$\int |f(x-y) - f(x)| \, |K_\delta(y)| \, dy = \int_{|y| \leq \delta} + \sum_{k=0}^{\infty} \int_{2^k \delta < |y| \leq 2^{k+1} \delta}.$$

By using the property (ii') of the approximation to the identity, the first term is estimated by

$$\int_{|y| \leq \delta} |f(x-y) - f(x)| \, |K_\delta(y)| \, dy \leq \frac{c}{\delta^d} \int_{|y| \leq \delta} |f(x-y) - f(x)| \, dy$$
$$\leq c \, \mathcal{A}(\delta).$$

Each term in the sum is estimated similarly, but this time by using property (iii') of approximations to the identity:

$$\int_{2^k \delta < |y| \leq 2^{k+1} \delta} |f(x-y) - f(x)| \, |K_\delta(y)| \, dy$$
$$\leq \frac{c\delta}{(2^k \delta)^{d+1}} \int_{|y| \leq 2^{k+1} \delta} |f(x-y) - f(x)| \, dy$$
$$\leq \frac{c'}{2^k (2^{k+1} \delta)^d} \int_{|y| \leq 2^{k+1} \delta} |f(x-y) - f(x)| \, dy$$
$$\leq c' \, 2^{-k} \mathcal{A}(2^{k+1} \delta).$$

Putting these estimates together, we find that

$$|(f * K_\delta)(x) - f(x)| \leq c \, \mathcal{A}(\delta) + c' \sum_{k=0}^{\infty} 2^{-k} \mathcal{A}(2^{k+1} \delta).$$

Given $\epsilon > 0$, we first choose N so large that $\sum_{k \geq N} 2^{-k} < \epsilon$. Then, by making δ sufficiently small, we have by the lemma

$$A(2^k \delta) < \epsilon/N, \qquad \text{whenever } k = 0, 1, \ldots, N-1.$$

Hence, recalling that $A(r)$ is bounded, we find

$$|(f * K_\delta)(x) - f(x)| \leq C\epsilon$$

for all sufficiently small δ, and the theorem is proved.

In addition to this pointwise result, convolutions with approximations to the identity also provide convergence in the L^1-norm.

Theorem 2.3 *Suppose that f is integrable on $\mathbb{R}^d$ and that $\{K_\delta\}_{\delta > 0}$ is an approximation to the identity. Then, for each $\delta > 0$, the convolution*

$$(f * K_\delta)(x) = \int_{\mathbb{R}^d} f(x - y) K_\delta(y) \, dy$$

is integrable, and

$$\|(f * K_\delta) - f\|_{L^1(\mathbb{R}^d)} \to 0, \qquad \text{as } \delta \to 0.$$

The proof is merely a repetition in a more general context of the argument in the special case where $K_\delta(x) = \delta^{-d/2} e^{-\pi |x|^2/\delta}$ given in Section 4*, Chapter 2, and so will not be repeated.

3 Differentiability of functions

We now take up the second question raised at the beginning of this chapter, that of finding a broad condition on functions F that guarantees the identity

(6) $$F(b) - F(a) = \int_a^b F'(x) \, dx.$$

There are two phenomena that make a general formulation of this identity problematic. First, because of the existence of non-differentiable functions,[4] the right-hand side of (6) might not be meaningful if we merely assumed F was continuous. Second, even if $F'(x)$ existed for every x, the function F' would not necessarily be (Lebesgue) integrable. (See Exercise 12.)

---

[4]In particular, there are continuous nowhere differentiable functions. See Chapter 4 in Book I, or also Chapter 7 below.

How do we deal with these difficulties? One way is by limiting ourselves to those functions F that arise as indefinite integrals (of integrable functions). This raises the issue of how to characterize such functions, and we approach that problem via the study of a wider class, the functions of bounded variation. These functions are closely related to the question of rectifiability of curves, and we start by considering this connection.

3.1 Functions of bounded variation

Let γ be a parametrized curve in the plane given by $z(t) = (x(t), y(t))$, where $a \leq t \leq b$. Here $x(t)$ and $y(t)$ are continuous real-valued functions on $[a, b]$. The curve γ is **rectifiable** if there exists $M < \infty$ such that, for any partition $a = t_0 < t_1 < \cdots < t_N = b$ of $[a, b]$,

$$(7) \qquad \sum_{j=1}^{N} |z(t_j) - z(t_{j-1})| \leq M.$$

By definition, the **length** $L(\gamma)$ of the curve is the supremum over all partitions of the sum on the left-hand side, that is,

$$L(\gamma) = \sup_{a=t_0<t_1<\cdots<t_N=b} \sum_{j=1}^{N} |z(t_j) - z(t_{j-1})|.$$

Alternatively, $L(\gamma)$ is the infimum of all M that satisfy (7). Geometrically, the quantity $L(\gamma)$ is obtained by approximating the curve by polygonal lines and taking the limit of the length of these polygonal lines as the interval $[a, b]$ is partitioned more finely (see the illustration in Figure 3).

Naturally, we may now ask the following questions: What analytic condition on $x(t)$ and $y(t)$ guarantees rectifiability of the curve γ? In particular, must the derivatives of $x(t)$ and $y(t)$ exist? If so, does one have the desired formula

$$L(\gamma) = \int_a^b (x'(t)^2 + y'(t)^2)^{1/2} \, dt?$$

The answer to the first question leads directly to the class of functions of bounded variation, a class that plays a key role in the theory of differentiation.

Suppose $F(t)$ is a complex-valued function defined on $[a, b]$, and $a = t_0 < t_1 < \cdots < t_N = b$ is a partition of this interval. The variation of F

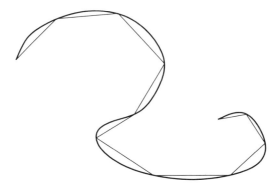

Figure 3. Approximation of a rectifiable curve by polygonal lines

on this partition is defined by

$$\sum_{j=1}^{N} |F(t_j) - F(t_{j-1})|.$$

The function F is said to be of **bounded variation** if the variations of F over all partitions are bounded, that is, there exists $M < \infty$ so that

$$\sum_{j=1}^{N} |F(t_j) - F(t_{j-1})| \leq M$$

for all partitions $a = t_0 < t_1 < \cdots < t_N = b$. In this definition we do not assume that F is continuous; however, when applying it to the case of curves, we will suppose that $F(t) = z(t) = x(t) + iy(t)$ is continuous.

We observe that if a partition $\tilde{\mathcal{P}}$ given by $a = \tilde{t}_0 < \tilde{t}_1 < \cdots < \tilde{t}_M = b$ is a refinement[5] of a partition $\mathcal{P}$ given by $a = t_0 < t_1 < \cdots < t_N = b$, then the variation of F on $\tilde{\mathcal{P}}$ is greater than or equal to the variation of F on $\mathcal{P}$.

Theorem 3.1 *A curve parametrized by $(x(t), y(t))$, $a \leq t \leq b$, is rectifiable if and only if both $x(t)$ and $y(t)$ are of bounded variation.*

The proof is immediate once we observe that if $F(t) = x(t) + iy(t)$, then

$$F(t_j) - F(t_{j-1}) = (x(t_j) - x(t_{j-1})) + i(y(t_j) - y(t_{j-1})),$$

---

[5]We say that a partition $\tilde{\mathcal{P}}$ of $[a, b]$ is a refinement of a partition $\mathcal{P}$ of $[a, b]$ if every point in $\mathcal{P}$ also belongs to $\tilde{\mathcal{P}}$.

and if a and b are real, then $|a + ib| \leq |a| + |b| \leq 2|a + ib|$.

Intuitively, a function of bounded variation cannot oscillate too often with amplitudes that are too large. Some examples should help clarify this assertion.

We first fix some terminology. A real-valued function F defined on $[a, b]$ is **increasing** if $F(t_1) \leq F(t_2)$ whenever $a \leq t_1 \leq t_2 \leq b$. If the inequality is strict, we say that F is **strictly increasing**.

EXAMPLE 1. If F is real-valued, monotonic, and bounded, then F is of bounded variation. Indeed, if for example F is increasing and bounded by M, we see that

$$\sum_{j=1}^{N} |F(t_j) - F(t_{j-1})| = \sum_{j=1}^{N} F(t_j) - F(t_{j-1})$$
$$= F(b) - F(a) \leq 2M.$$

EXAMPLE 2. If F is differentiable at every point, and F' is bounded, then F is of bounded variation. Indeed, if $|F'| \leq M$, the mean value theorem implies

$$|F(x) - F(y)| \leq M|x - y|, \quad \text{for all } x, y \in [a, b],$$

hence $\sum_{j=1}^{N} |F(t_j) - F(t_{j-1})| \leq M(b - a)$. (See also Exercise 23.)

EXAMPLE 3. Let

$$F(x) = \begin{cases} x^a \sin(x^{-b}) & \text{for } 0 < x \leq 1, \\ 0 & \text{if } x = 0. \end{cases}$$

Then F is of bounded variation on $[0, 1]$ if and only if $a > b$ (Exercise 11). Figure 4 illustrates the three cases $a > b$, $a = b$, and $a < b$.

The next result shows that in some sense the first example above exhausts all functions of bounded variation. For its proof, we need the following definitions. The **total variation** of f on $[a, x]$ (where $a \leq x \leq b$) is defined by

$$T_F(a, x) = \sup \sum_{j=1}^{N} |F(t_j) - F(t_{j-1})|,$$

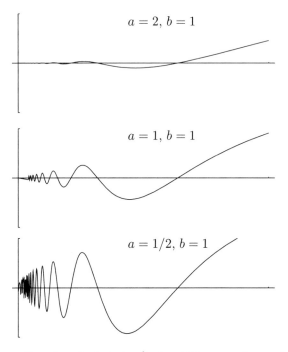

Figure 4. Graphs of $x^a \sin(x^{-b})$ for different values of a and b

where the sup is over all partitions of $[a, x]$. The preceding definition makes sense if F is complex-valued. The succeeding ones require that F is real-valued. In the spirit of the first definition, we say that the **positive variation** of F on $[a, x]$ is

$$P_F(a, x) = \sup \sum_{(+)} F(t_j) - F(t_{j-1}),$$

where the sum is over all j such that $F(t_j) \geq F(t_{j-1})$, and the supremum is over all partitions of $[a, x]$. Finally, the **negative variation** of F on $[a, x]$ is defined by

$$N_F(a, x) = \sup \sum_{(-)} -[F(t_j) - F(t_{j-1})],$$

where the sum is over all j such that $F(t_j) \leq F(t_{j-1})$, and the supremum is over all partitions of $[a, x]$.

Lemma 3.2 *Suppose F is real-valued and of bounded variation on $[a, b]$.*

Then for all $a \leq x \leq b$ one has

$$F(x) - F(a) = P_F(a, x) - N_F(a, x),$$

and

$$T_F(a, x) = P_F(a, x) + N_F(a, x).$$

   Proof.   Given $\epsilon > 0$ there exists a partition $a = t_0 < \cdots < t_N = x$ of $[a, x]$, such that

$$\left| P_F - \sum_{(+)} F(t_j) - F(t_{j-1}) \right| < \epsilon \text{ and } \left| N_F - \sum_{(-)} -[F(t_j) - F(t_{j-1})] \right| < \epsilon.$$

(To see this, it suffices to use the definition to obtain similar estimates for P_F and N_F with possibly different partitions, and then to consider a common refinement of these two partitions.) Since we also note that

$$F(x) - F(a) = \sum_{(+)} F(t_j) - F(t_{j-1}) - \sum_{(-)} -[F(t_j) - F(t_{j-1})],$$

we find that $|F(x) - F(a) - [P_F - N_F]| < 2\epsilon$, which proves the first identity.

   For the second identity, we also note that for any partition of $a = t_0 < \cdots < t_N = x$ of $[a, x]$ we have

$$\sum_{j=1}^{N} |F(t_j) - F(t_{j-1})| = \sum_{(+)} F(t_j) - F(t_{j-1}) + \sum_{(-)} -[F(t_j) - F(t_{j-1})],$$

hence $T_F \leq P_F + N_F$. Also, the above implies

$$\sum_{(+)} F(t_j) - F(t_{j-1}) + \sum_{(-)} -[F(t_j) - F(t_{j-1})] \leq T_F.$$

Once again, one can argue using common refinements of partitions in the definitions of P_F and N_F to deduce the inequality $P_F + N_F \leq T_F$, and the lemma is proved.

Theorem 3.3 *A real-valued function F on $[a, b]$ is of bounded variation if and only if F is the difference of two increasing bounded functions.*

Proof. Clearly, if $F = F_1 - F_2$, where each F_j is bounded and increasing, then F is of bounded variation.

Conversely, suppose F is of bounded variation. Then, we let $F_1(x) = P_F(a, x) + F(a)$ and $F_2(x) = N_F(a, x)$. Clearly, both F_1 and F_2 are increasing, of bounded variation, and by the lemma $F(x) = F_1(x) - F_2(x)$.

Observe that as a consequence, a complex-valued function of bounded variation is a (complex) linear combination of four increasing functions.

Returning to the curve γ parametrized by a continuous function $z(t) = x(t) + iy(t)$, we want to make some comment about its associated length function. Assuming that the curve is rectifiable, we define $L(A, B)$ as the length of the segment of γ that arises as the image of those t for which $A \leq t \leq B$, with $a \leq A \leq B \leq b$. Note that $L(A, B) = T_F(A, B)$, where $F(t) = z(t)$. We see that

$$(8) \qquad L(A, C) + L(C, B) = L(A, B) \quad \text{if } A \leq C \leq B.$$

We also observe that $L(A, B)$ is a continuous function of B (and of A). Since it is an increasing function, to prove its continuity in B from the left, it suffices to see that for each B and $\epsilon > 0$, we can find $B_1 < B$ such that $L(A, B_1) \geq L(A, B) - \epsilon$. We do this by first finding a partition $A = t_0 < t_1 < \cdots < t_N = B$ such that the length of the corresponding polygonal line is $\geq L(A, B) - \epsilon/2$. By continuity of the function $z(t)$, we can find a B_1, with $t_{N-1} < B_1 < B$, such that $|z(B) - z(B_1)| < \epsilon/2$. Now for the refined partition $t_0 < t_1 < \cdots < t_{N-1} < B_1 < B$, the length of the polygonal line is still $\geq L(A, B) - \epsilon/2$. Therefore, the length for the partition $t_0 < t_1 < \cdots < t_{N-1} = B_1$ is $\geq L(A, B) - \epsilon$, and thus $L(A, B_1) \geq L(A, B) - \epsilon$.

To prove continuity from the right at B, let $\epsilon > 0$, pick any $C > B$, and choose a partition $B = t_0 < t_1 < \cdots < t_N = C$ such that $L(B, C) - \epsilon/2 < \sum_{j=0}^{N-1} |z(t_{j+1}) - z(t_j)|$. By considering a refinement of this partition if necessary, we may assume since z is continuous that $|z(t_1) - z(t_0)| < \epsilon/2$. If we denote $B_1 = z(t_1)$, then we get

$$L(B, C) - \epsilon/2 < \epsilon/2 + L(B_1, C).$$

Since $L(B, B_1) + L(B_1, C) = L(B, C)$ we have $L(B, B_1) < \epsilon$, and therefore $L(A, B_1) - L(A, B) < \epsilon$.

Note that what we have observed can be re-stated as follows: if a function of bounded variation is continuous, then so is its total variation.

The next result lies at the heart of the theory of differentiation.

Theorem 3.4 *If F is of bounded variation on $[a, b]$, then F is differentiable almost everywhere.*

In other words, the quotient

$$\lim_{h \to 0} \frac{F(x + h) - F(x)}{h}$$

exists for almost every $x \in [a, b]$. By the previous result, it suffices to consider the case when F is increasing. In fact, we shall first also assume that F is continuous. This makes the argument simpler. As for the general case, we leave that till later. (See Section 3.3.) It will then be instructive to examine the nature of the possible discontinuities of a function of bounded variation, and reduce matters to the case of "jump functions."

We begin with a nice technical lemma of F. Riesz, which has the effect of a covering argument.

Lemma 3.5 *Suppose G is real-valued and continuous on $\mathbb{R}$. Let E be the set of points x such that*

$$G(x + h) > G(x) \quad \text{for some } h = h_x > 0.$$

If E is non-empty, then it must be open, and hence can be written as a countable disjoint union of open intervals $E = \bigcup(a_k, b_k)$. If (a_k, b_k) is a finite interval in this union, then

$$G(b_k) - G(a_k) = 0.$$

Proof. Since G is continuous, it is clear that E is open whenever it is non-empty and can therefore be written as a disjoint union of countably many open intervals (Theorem 1.3 in Chapter 1). If (a_k, b_k) denotes a finite interval in this decomposition, then $a_k \notin E$; therefore we cannot have $G(b_k) > G(a_k)$. We now suppose that $G(b_k) < G(a_k)$. By continuity, there exists $a_k < c < b_k$ so that

$$G(c) = \frac{G(a_k) + G(b_k)}{2},$$

and in fact we may choose c farthest to the right in the interval (a_k, b_k). Since $c \in E$, there exists $d > c$ such that $G(d) > G(c)$. Since $b_k \notin E$, we must have $G(x) \leq G(b_k)$ for all $x \geq b_k$; therefore $d < b_k$. Since $G(d) > G(c)$, there exists (by continuity) $c' > d$ with $c' < b_k$ and $G(c') = G(c)$, which contradicts the fact that c was chosen farthest to the right in (a_k, b_k). This shows that we must have $G(a_k) = G(b_k)$, and the lemma is proved.

Note. This result sometimes carries the name "rising sun lemma" for the following reason. If one thinks of the sun rising from the east (at the right) with the rays of light parallel to the x-axis, then the points $(x, G(x))$ on the graph of G, with $x \in E$, are precisely the points which are in the shade; these points appear in bold in Figure 5.

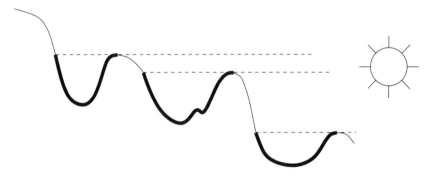

Figure 5. Rising sun lemma

A slight modification of the proof of Lemma 3.5 gives:

Corollary 3.6 *Suppose G is real-valued and continuous on a closed interval $[a, b]$. If E denotes the set of points x in (a, b) so that $G(x + h) > G(x)$ for some $h > 0$, then E is either empty or open. In the latter case, it is a disjoint union of countably many intervals (a_k, b_k), and $G(a_k) = G(b_k)$, except possibly when $a = a_k$, in which case we only have $G(a_k) \leq G(b_k)$.*

For the proof of the theorem, we define the quantity

$$\triangle_h(F)(x) = \frac{F(x + h) - F(x)}{h}.$$

We also consider the four **Dini numbers** at x defined by

$$D^+(F)(x) = \limsup_{\substack{h \to 0 \\ h > 0}} \triangle_h(F)(x)$$

$$D_+(F)(x) = \liminf_{\substack{h \to 0 \\ h > 0}} \triangle_h(F)(x)$$

$$D^-(F)(x) = \limsup_{\substack{h \to 0 \\ h < 0}} \triangle_h(F)(x)$$

$$D_-(F)(x) = \liminf_{\substack{h \to 0 \\ h < 0}} \triangle_h(F)(x).$$

Clearly, one has $D_+ \leq D^+$ and $D_- \leq D^-$. To prove the theorem it suffices to show that

(i) $D^+(F)(x) < \infty$ for a.e. x, and;

(ii) $D^+(F)(x) \leq D_-(F)(x)$ for a.e. x.

Indeed, if these results hold, then by applying (ii) to $-F(-x)$ instead of $F(x)$ we obtain $D^-(F)(x) \leq D_+(F)(x)$ for a.e. x. Therefore

$$D^+ \leq D_- \leq D^- \leq D_+ \leq D^+ < \infty \qquad \text{for a.e. } x.$$

Thus all four Dini numbers are finite and equal almost everywhere, hence $F'(x)$ exists for almost every point x.

We recall that we assume that F is increasing, bounded, and continuous on $[a, b]$. For a fixed $\gamma > 0$, let

$$E_\gamma = \{x : D^+(F)(x) > \gamma\}.$$

First, we assert that E_γ is measurable. (The proof of this simple fact is outlined in Exercise 14.) Next, we apply Corollary 3.6 to the function $G(x) = F(x) - \gamma x$, and note that we then have $E_\gamma \subset \bigcup_k (a_k, b_k)$, where $F(b_k) - F(a_k) \geq \gamma(b_k - a_k)$. Consequently,

$$m(E_\gamma) \leq \sum_k m((a_k, b_k))$$

$$\leq \frac{1}{\gamma} \sum_k F(b_k) - F(a_k)$$

$$\leq \frac{1}{\gamma} (F(b) - F(a)).$$

Therefore $m(E_\gamma) \to 0$ as γ tends to infinity, and since $\{D^+ F(x) < \infty\} \subset E_\gamma$ for all γ, this proves that $D^+ F(x) < \infty$ almost everywhere.

Having fixed real numbers r and R such that $R > r$, we let

$$E = \{x \in [a,b] : D^+(F)(x) > R \quad \text{and} \quad r > D_-(F)(x)\}.$$

We will have shown $D^+(F)(x) \le D_-(F)(x)$ almost everywhere once we prove that $m(E) = 0$, since it then suffices to let R and r vary over the rationals with $R > r$.

To prove that $m(E) = 0$ we may assume that $m(E) > 0$ and arrive at a contradiction. Because $R/r > 1$ we can find an open set $\mathcal{O}$ such that $E \subset \mathcal{O} \subset (a,b)$, yet $m(\mathcal{O}) < m(E) \cdot R/r$.

Now $\mathcal{O}$ can be written as $\bigcup I_n$, with I_n disjoint open intervals. Fix n and apply Corollary 3.6 to the function $G(x) = -F(-x) + rx$ on the interval $-I_n$. Reflecting through the origin again yields an open set $\bigcup_k(a_k, b_k)$ contained in I_n, where the intervals (a_k, b_k) are disjoint, with

$$F(b_k) - F(a_k) \le r(b_k - a_k).$$

However, on each interval (a_k, b_k) we apply Corollary 3.6, this time to $G(x) = F(x) - Rx$. We thus obtain an open set $\mathcal{O}_n = \bigcup_{k,j}(a_{k,j}, b_{k,j})$ of disjoint open intervals $(a_{k,j}, b_{k,j})$ with $(a_{k,j}, b_{k,j}) \subset (a_k, b_k)$ for every j, and

$$F(b_{k,j}) - F(a_{k,j}) \ge R(b_{k,j} - a_{k,j}).$$

Then using the fact that F is increasing we find that

$$m(\mathcal{O}_n) = \sum_{k,j}(b_{k,j} - a_{k,j}) \le \frac{1}{R}\sum_{k,j} F(b_{k,j}) - F(a_{k,j})$$

$$\le \frac{1}{R}\sum_k F(b_k) - F(a_k) \le \frac{r}{R}\sum_k(b_k - a_k)$$

$$\le \frac{r}{R}m(I_n).$$

Note that $\mathcal{O}_n \supset E \cap I_n$, since $D^+F(x) > R$ and $r > D_-F(x)$ for each $x \in E$; of course, $I_n \supset \mathcal{O}_n$. We now sum in n. Therefore

$$m(E) = \sum_n m(E \cap I_n) \le \sum_n m(\mathcal{O}_n) \le \frac{r}{R}\sum m(I_n) = \frac{r}{R}m(\mathcal{O}) < m(E).$$

The strict inequality gives a contradiction and Theorem 3.4 is proved, at least when F is continuous.

Let us see how far we have come regarding (6) if F is a monotonic function.

Corollary 3.7 *If F is increasing and continuous, then F' exists almost everywhere. Moreover F' is measurable, non-negative, and*

$$\int_a^b F'(x)\, dx \leq F(b) - F(a).$$

In particular, if F is bounded on $\mathbb{R}$, then F' is integrable on $\mathbb{R}$.

Proof. For $n \geq 1$, we consider the quotient

$$G_n(x) = \frac{F(x + 1/n) - F(x)}{1/n}.$$

By the previous theorem, we have that $G_n(x) \to F'(x)$ for a.e. x, which shows in particular that F' is measurable and non-negative.

We now extend F as a continuous function on all of $\mathbb{R}$. By Fatou's lemma (Lemma 1.7 in Chapter 2) we know that

$$\int_a^b F'(x)\, dx \leq \liminf_{n \to \infty} \int_a^b G_n(x)\, dx.$$

To complete the proof, it suffices to note that

$$\begin{aligned}
\int_a^b G_n(x)\, dx &= \frac{1}{1/n} \int_a^b F(x + 1/n)\, dx - \frac{1}{1/n} \int_a^b F(x)\, dx \\
&= \frac{1}{1/n} \int_{a+1/n}^{b+1/n} F(y)\, dy - \frac{1}{1/n} \int_a^b F(x)\, dx \\
&= \frac{1}{1/n} \int_b^{b+1/n} F(x)\, dx - \frac{1}{1/n} \int_a^{a+1/n} F(x)\, dx.
\end{aligned}$$

Since F is continuous, the first and second terms converge to $F(b)$ and $F(a)$, respectively, as n goes to infinity, so the proof of the corollary is complete.

We cannot go any farther than the inequality in the corollary if we allow all continuous increasing functions, as is shown by the following important example.

The Cantor-Lebesgue function

The following simple construction yields a continuous function $F : [0, 1] \to [0, 1]$ that is increasing with $F(0) = 0$ and $F(1) = 1$, but $F'(x) = 0$ almost everywhere! Hence F is of bounded variation, but

$$\int_a^b F'(x)\, dx \neq F(b) - F(a).$$

Consider the standard triadic Cantor set $\mathcal{C} \subset [0,1]$ described at the end of Section 1 in Chapter 1, and recall that

$$\mathcal{C} = \bigcap_{k=0}^{\infty} C_k,$$

where each C_k is a disjoint union of 2^k closed intervals. For example, $C_1 = [0, 1/3] \cup [2/3, 1]$. Let $F_1(x)$ be the continuous increasing function on $[0,1]$ that satisfies $F_1(0) = 0$, $F_1(x) = 1/2$ if $1/3 \leq x \leq 2/3$, $F_1(1) = 1$, and F_1 is linear on C_1. Similarly, let $F_2(x)$ be continuous and increasing, and such that

$$F_2(x) = \begin{cases} 0 & \text{if } x = 0, \\ 1/4 & \text{if } 1/9 \leq x \leq 2/9, \\ 1/2 & \text{if } 1/3 \leq x \leq 2/3, \\ 3/4 & \text{if } 7/9 \leq x \leq 8/9, \\ 1 & \text{if } x = 1, \end{cases}$$

and F_2 is linear on C_2. See Figure 6.

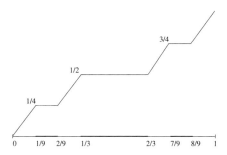

Figure 6. Construction of F_2

This process yields a sequence of continuous increasing functions $\{F_n\}_{n=1}^{\infty}$ such that clearly

$$|F_{n+1}(x) - F_n(x)| \leq 2^{-n-1}.$$

Hence $\{F_n\}_{n=1}^{\infty}$ converges uniformly to a continuous limit F called the **Cantor-Lebesgue function** (Figure 7).[6] By construction, F is increasing, $F(0) = 0$, $F(1) = 1$, and we see that F is constant on each interval of the complement of the Cantor set. Since $m(\mathcal{C}) = 0$, we find that $F'(x) = 0$ almost everywhere, as desired.

---

[6]The reader may check that indeed this function agrees with the one given in Exercise 2 of Chapter 1.

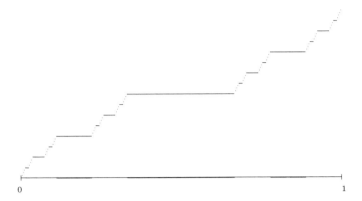

Figure 7. The Cantor-Lebesgue function

The considerations in this section, as well as this last example, show that the assumption of bounded variation guarantees the existence of a derivative almost everywhere, but not the validity of the formula

$$\int_a^b F'(x)\,dx = F(b) - F(a).$$

In the next section, we shall present a condition on a function that will completely settle the problem of establishing the above identity.

3.2 Absolutely continuous functions

A function F defined on $[a, b]$ is **absolutely continuous** if for any $\epsilon > 0$ there exists $\delta > 0$ so that

$$\sum_{k=1}^N |F(b_k) - F(a_k)| < \epsilon \quad \text{whenever} \quad \sum_{k=1}^N (b_k - a_k) < \delta,$$

and the intervals (a_k, b_k), $k = 1, \ldots, N$ are disjoint. Some general remarks are in order.

- From the definition, it is clear that absolutely continuous functions are continuous, and in fact uniformly continuous.

- If F is absolutely continuous on a bounded interval, then it is also of bounded variation on the same interval. Moreover, as is easily seen, its total variation is continuous (in fact absolutely continuous). As a consequence the decomposition of such a function F into two monotonic functions given in Section 3.1 shows that each of these functions is continuous.

- If $F(x) = \int_a^x f(y)\,dy$ where f is integrable, then F is absolutely continuous. This follows at once from (ii) in Proposition 1.12, Chapter 2.

In fact, this last remark shows that absolute continuity is a necessary condition to impose on F if we hope to prove $\int_a^b F'(x)\,dx = F(b) - F(a)$.

Theorem 3.8 *If F is absolutely continuous on $[a,b]$, then $F'(x)$ exists almost everywhere. Moreover, if $F'(x) = 0$ for a.e. x, then F is constant.*

Since an absolutely continuous function is the difference of two continuous monotonic functions, as we have seen above, the existence of $F'(x)$ for a.e. x follows from what we have already proved. To prove that $F'(x) = 0$ a.e. implies F is constant requires a more elaborate version of the covering argument in Lemma 1.2. For the moment we revert to the generality of d dimensions to describe this.

A collection $\mathcal{B}$ of balls $\{B\}$ is said to be a **Vitali covering** of a set E if for every $x \in E$ and any $\eta > 0$ there is a ball $B \in \mathcal{B}$, such that $x \in B$ and $m(B) < \eta$. Thus every point is covered by balls of arbitrarily small measure.

Lemma 3.9 *Suppose E is a set of finite measure and $\mathcal{B}$ is a Vitali covering of E. For any $\delta > 0$ we can find finitely many balls $B_1, \ldots, B_N$ in $\mathcal{B}$ that are disjoint and so that*

$$\sum_{i=1}^{N} m(B_i) \geq m(E) - \delta.$$

Proof. We apply the elementary Lemma 1.2 iteratively, with the aim of exhausting the set E. It suffices to take δ sufficiently small, say $\delta < m(E)$, and using the just cited covering lemma, we can find an initial collection of disjoint balls $B_1, B_2, \ldots, B_{N_1}$ in $\mathcal{B}$ such that $\sum_{i=1}^{N_1} m(B_i) \geq \gamma\delta$. (For simplicity of notation, we have written $\gamma = 3^{-d}$.) Indeed, first we have $m(E') \geq \delta$ for an appropriate compact subset E' of E. Because of the compactness of E', we can cover it by finitely many balls from $\mathcal{B}$, and then the previous lemma allows us to select a disjoint sub-collection of these balls $B_1, B_2, \ldots, B_{N_1}$ such that $\sum_{i=1}^{N_1} m(B_i) \geq \gamma m(E') \geq \gamma\delta$.

With $B_1, \ldots, B_{N_1}$ as our initial sequence of balls, we consider two possibilities: either $\sum_{i=1}^{N_1} m(B_i) \geq m(E) - \delta$ and we are done with $N =$

N_1; or, contrariwise, $\sum_{i=1}^{N_1} m(B_i) < m(E) - \delta$. In the second case, with $E_2 = E - \bigcup_{i=1}^{N_1} \overline{B_i}$, we have $m(E_2) > \delta$ (recall that $m(\overline{B_i}) = m(B_i)$). We then repeat the previous argument, by choosing a compact subset E_2' of E_2 with $m(E_2') \geq \delta$, and by noting that the balls in $\mathcal{B}$ that are *disjoint* from $\bigcup_{i=1}^{N_1} \overline{B_i}$ still cover E_2 and in fact give a Vitali covering for E_2, and hence for E_2'. Thus we can choose a finite disjoint collection of these balls B_i, $N_1 < i \leq N_2$, so that $\sum_{N_1 < i \leq N_2} m(B_i) \geq \gamma\delta$. Therefore, now $\sum_{i=1}^{N_2} m(B_i) \geq 2\gamma\delta$, and the balls B_i, $1 \leq i \leq N_2$, are disjoint.

We again consider two alternatives, whether or not $\sum_{i=1}^{N_2} m(B_i) \geq m(E) - \delta$. In the first case, we are done with $N_2 = N$, and in the second case, we proceed as before. If, continuing this way, we had reached the kth stage and not stopped before then, we would have selected a collection of disjoint balls with the sum of their measures $\geq k\gamma\delta$. In any case, our process achieves the desired goal by the kth stage if $k \geq (m(E) - \delta)/\gamma\delta$, since in this case $\sum_{i=1}^{N_k} m(B_i) \geq m(E) - \delta$.

A simple consequence is the following.

Corollary 3.10 *We can arrange the choice of the balls so that*

$$m(E - \bigcup_{i=1}^{N} B_i) < 2\delta.$$

In fact, let $\mathcal{O}$ be an open set, with $\mathcal{O} \supset E$ and $m(\mathcal{O} - E) < \delta$. Since we are dealing with a Vitali covering of E, we can restrict all of our choices above to balls contained in $\mathcal{O}$. If we do this, then $(E - \bigcup_{i=1}^{N} B_i) \cup \bigcup_{i=1}^{N} B_i \subset \mathcal{O}$, where the union on the left-hand side is a disjoint union. Hence

$$m(E - \bigcup_{i=1}^{N} B_i) \leq m(\mathcal{O}) - m(\bigcup_{i=1}^{N} B_i) \leq m(E) + \delta - (m(E) - \delta) = 2\delta.$$

We now return to the situation on the real line. To complete the proof of the theorem it suffices to show that under its hypotheses we have $F(b) = F(a)$, since if that is proved, we can replace the interval $[a, b]$ by any sub-interval. Now let E be the set of those $x \in (a, b)$ where $F'(x)$ exists and is zero. By our assumption $m(E) = b - a$. Next, momentarily fix $\epsilon > 0$. Since for each $x \in E$ we have

$$\lim_{h \to 0} \left| \frac{F(x+h) - F(x)}{h} \right| = 0,$$

then for *each* $\eta > 0$ we have an open interval $I = (a_x, b_x) \subset [a, b]$ containing x, with

$$|F(b_x) - F(a_x)| \leq \epsilon(b_x - a_x) \quad \text{and} \quad b_x - a_x < \eta.$$

The collection of these intervals forms a Vitali covering of E, and hence by the lemma, for $\delta > 0$, we can select finitely many I_i, $1 \leq i \leq N$, $I_i = (a_i, b_i)$, which are disjoint and such that

$$(9) \qquad \sum_{i=1}^{N} m(I_i) \geq m(E) - \delta = (b - a) - \delta.$$

However, $|F(b_i) - F(a_i)| \leq \epsilon(b_i - a_i)$, and upon adding these inequalities we get

$$\sum_{i=1}^{N} |F(b_i) - F(a_i)| \leq \epsilon(b - a),$$

since the intervals I_i are disjoint and lie in $[a, b]$. Next consider the complement of $\bigcup_{j=1}^{N} I_j$ in $[a, b]$. It consists of finitely many closed intervals $\bigcup_{k=1}^{M} [\alpha_k, \beta_k]$ with total length $\leq \delta$ because of (9). Thus by the absolute continuity of F (if δ is chosen appropriately in terms of ϵ), $\sum_{k=1}^{M} |F(\beta_k) - F(\alpha_k)| \leq \epsilon$. Altogether, then,

$$|F(b) - F(a)| \leq \sum_{i=1}^{N} |F(b_i) - F(a_i)| + \sum_{k=1}^{M} |F(\beta_k) - F(\alpha_k)| \leq \epsilon(b - a) + \epsilon.$$

Since ϵ was positive but otherwise arbitrary, we conclude that $F(b) - F(a) = 0$, which we set out to show.

The culmination of all our efforts is contained in the next theorem. In particular, it resolves our second problem of establishing the reciprocity between differentiation and integration.

Theorem 3.11 *Suppose F is absolutely continuous on $[a, b]$. Then F' exists almost everywhere and is integrable. Moreover,*

$$F(x) - F(a) = \int_a^x F'(y)\, dy, \quad \text{for all } a \leq x \leq b.$$

By selecting $x = b$ we get $F(b) - F(a) = \int_a^b F'(y)\, dy$.

Conversely, if f is integrable on $[a, b]$, then there exists an absolutely continuous function F such that $F'(x) = f(x)$ almost everywhere, and in fact, we may take $F(x) = \int_a^x f(y)\, dy$.

Proof. Since we know that a real-valued absolutely continuous function is the difference of two continuous increasing functions, Corollary 3.7 shows that F' is integrable on $[a, b]$. Now let $G(x) = \int_a^x F'(y)\,dy$. Then G is absolutely continuous; hence so is the difference $G(x) - F(x)$. By the Lebesgue differentiation theorem (Theorem 1.4), we know that $G'(x) = F'(x)$ for a.e. x; hence the difference $F - G$ has derivative 0 almost everywhere. By the previous theorem we conclude that $F - G$ is constant, and evaluating this expression at $x = a$ gives the desired result.

The converse is a consequence of the observation we made earlier, namely that $\int_a^x f(y)\,dy$ is absolutely continuous, and the Lebesgue differentiation theorem, which gives $F'(x) = f(x)$ almost everywhere.

3.3 Differentiability of jump functions

We now examine monotonic functions that are not assumed to be continuous. The resulting analysis will allow us to remove the continuity assumption made earlier in the proof of Theorem 3.4.

As before, we may assume that F is increasing and bounded. In particular, these two conditions guarantee that the limits

$$F(x^-) = \lim_{\substack{y \to x \\ y < x}} F(y) \quad \text{and} \quad F(x^+) = \lim_{\substack{y \to x \\ y > x}} F(y)$$

exist. Then of course $F(x^-) \leq F(x) \leq F(x^+)$, and the function F is continuous at x if $F(x^-) = F(x^+)$; otherwise, we say that it has a jump discontinuity. Fortunately, dealing with these discontinuities is manageable, since there can only be countably many of them.

Lemma 3.12 *A bounded increasing function F on $[a, b]$ has at most countably many discontinuities.*

Proof. If F is discontinuous at x, we may choose a rational number r_x so that $F(x^-) < r_x < F(x^+)$. If f is discontinuous at x and z with $x < z$, we must have $F(x^+) \leq F(z^-)$, hence $r_x < r_z$. Consequently, to each rational number corresponds at most one discontinuity of F, hence F can have at most a countable number of discontinuities.

Now let $\{x_n\}_{n=1}^\infty$ denote the points where F is discontinuous, and let α_n denote the jump of F at x_n, that is, $\alpha_n = F(x_n^+) - F(x_n^-)$. Then

$$F(x_n^+) = F(x_n^-) + \alpha_n$$

and

$$F(x_n) = F(x_n^-) + \theta_n \alpha_n, \quad \text{for some } \theta_n, \text{ with } 0 \leq \theta_n \leq 1.$$

If we let

$$
j_n(x) = \begin{cases} 0 & \text{if } x < x_n, \\ \theta_n & \text{if } x = x_n, \\ 1 & \text{if } x > x_n, \end{cases}
$$

then we define the **jump function** associated to F by

$$
J_F(x) = \sum_{n=1}^{\infty} \alpha_n j_n(x).
$$

For simplicity, and when no confusion is possible, we shall write J instead of J_F.

Our first observation is that if F is bounded, then we must have

$$
\sum_{n=1}^{\infty} \alpha_n \leq F(b) - F(a) < \infty,
$$

and hence the series defining J converges absolutely and uniformly.

Lemma 3.13 *If F is increasing and bounded on $[a, b]$, then:*

(i) $J(x)$ *is discontinuous precisely at the points $\{x_n\}$ and has a jump at x_n equal to that of F.*

(ii) *The difference $F(x) - J(x)$ is increasing and continuous.*

Proof. If $x \neq x_n$ for all n, each j_n is continuous at x, and since the series converges uniformly, J must be continuous at x. If $x = x_N$ for some N, then we write

$$
J(x) = \sum_{n=1}^{N} \alpha_n j_n(x) + \sum_{n=N+1}^{\infty} \alpha_n j_n(x).
$$

By the same argument as above, the series on the right-hand side is continuous at x. Clearly, the finite sum has a jump discontinuity at x_N of size α_N.

For (ii), we note that (i) implies at once that $F - J$ is continuous. Finally, if $y > x$ we have

$$
J(y) - J(x) \leq \sum_{x < x_n \leq y} \alpha_n \leq F(y) - F(x),
$$

where the last inequality follows since F is increasing. Hence

$$F(x) - J(x) \le F(y) - J(y),$$

and the difference $F - J$ is increasing, as desired.

Since we may write $F(x) = [F(x) - J(x)] + J(x)$, our final task is to prove that J is differentiable almost everywhere.

Theorem 3.14 *If J is the jump function considered above, then $J'(x)$ exists and vanishes almost everywhere.*

Proof. Given any $\epsilon > 0$, we note that the set E of those x where

$$(10) \qquad \limsup_{h \to 0} \frac{J(x+h) - J(x)}{h} > \epsilon$$

is a measurable set. (The proof of this little fact is outlined in Exercise 14 below.) Suppose $\delta = m(E)$. We need to show that $\delta = 0$. Now observe that since the series $\sum \alpha_n$ arising in the definition of J converges, then for any η, to be chosen later, we can find an N so large that $\sum_{n > N} \alpha_n < \eta$. We then write

$$J_0(x) = \sum_{n > N} \alpha_n j_n(x),$$

and because of our choice of N we have

$$(11) \qquad\qquad J_0(b) - J_0(a) < \eta.$$

However, $J - J_0$ is a finite sum of terms $\alpha_n j_n(x)$, and therefore the set of points where (10) holds, with J replaced by J_0, differs from E by at most a finite set, the points $\{x_1, x_2, \ldots, x_N\}$. Thus we can find a compact set K, with $m(K) \ge \delta/2$, so that $\limsup_{h \to 0} \frac{J_0(x+h) - J_0(x)}{h} > \epsilon$ for each $x \in K$. Hence there are intervals (a_x, b_x) containing x, $x \in K$, so that $J_0(b_x) - J_0(a_x) > \epsilon(b_x - a_x)$. We can first choose a finite collection of these intervals that covers K, and then apply Lemma 1.2 to select intervals $I_1, I_2, \ldots, I_n$ which are disjoint, and for which $\sum_{j=1}^{n} m(I_j) \ge m(K)/3$. The intervals $I_j = (a_j, b_j)$ of course satisfy

$$J_0(b_j) - J_0(a_j) > \epsilon(b_j - a_j).$$

Now,

$$J_0(b) - J_0(a) \ge \sum_{j=1}^{N} J_0(b_j) - J_0(a_j) > \epsilon \sum (b_j - a_j) \ge \frac{\epsilon}{3} m(K) \ge \frac{\epsilon}{6}\delta.$$

Thus by (11), $\epsilon\delta/6 < \eta$, and since we are free to choose η, it follows that $\delta = 0$ and the theorem is proved.

4 Rectifiable curves and the isoperimetric inequality

We turn to the further study of rectifiable curves and take up first the validity of the formula

$$(12) \qquad L = \int_a^b (x'(t)^2 + y'(t)^2)^{1/2}\, dt,$$

for the length L of the curve parametrized by $(x(t), y(t))$.

We have already seen that rectifiable curves are precisely the curves where, besides the assumed continuity of $x(t)$ and $y(t)$, these functions are of bounded variation. However a simple example shows that formula (12) does not always hold in this context. Indeed, let $x(t) = F(t)$ and $y(t) = F(t)$, where F is the Cantor-Lebesgue function and $0 \leq t \leq 1$. Then this parametrized curve traces out the straight line from $(0,0)$ to $(1,1)$ and has length $\sqrt{2}$, yet $x'(t) = y'(t) = 0$ for a.e. t.

The integral formula expressing the length of L is in fact valid if we assume in addition that the coordinate functions of the parametrization are absolutely continuous.

Theorem 4.1 *Suppose $(x(t), y(t))$ is a curve defined for $a \leq t \leq b$. If both $x(t)$ and $y(t)$ are absolutely continuous, then the curve is rectifiable, and if L denotes its length, we have*

$$L = \int_a^b (x'(t)^2 + y'(t)^2)^{1/2}\, dt.$$

Note that if $F(t) = x(t) + iy(t)$ is absolutely continuous then it is automatically of bounded variation, and hence the curve is rectifiable. The identity (12) is an immediate consequence of the proposition below, which can be viewed as a more precise version of Corollary 3.7 for absolutely continuous functions.

Proposition 4.2 *Suppose F is complex-valued and absolutely continuous on $[a, b]$. Then*

$$T_F(a, b) = \int_a^b |F'(t)|\, dt.$$

In fact, because of Theorem 3.11, for any partition $a = t_0 < t_1 < \cdots < t_N = b$ of $[a, b]$, we have

$$\sum_{j=1}^{N} |F(t_j) - F(t_{j-1})| = \sum_{j=1}^{N} \left| \int_{t_{j-1}}^{t_j} F'(t)\, dt \right|$$

$$\leq \sum_{j=1}^{N} \int_{t_{j-1}}^{t_j} |F'(t)|\, dt$$

$$= \int_{a}^{b} |F'(t)|\, dt.$$

So this proves

$$(13) \qquad T_F(a, b) \leq \int_{a}^{b} |F'(t)|\, dt.$$

To prove the reverse inequality, fix $\epsilon > 0$, and using Theorem 2.4 in Chapter 2 find a step function g on $[a, b]$, such that $F' = g + h$ with $\int_a^b |h(t)|\, dt < \epsilon$. Set $G(x) = \int_a^x g(t)\, dt$, and $H(x) = \int_a^x h(t)\, dt$. Then $F = G + H$, and as is easily seen

$$T_F(a, b) \geq T_G(a, b) - T_H(a, b).$$

However, by (13) $T_H(a, b) < \epsilon$, so that

$$T_F(a, b) \geq T_G(a, b) - \epsilon.$$

Now partition the interval $[a, b]$, as $a = t_0 < \cdots < t_N = b$, so that the step function g is constant on each of the intervals (t_{j-1}, t_j), $j = 1, 2, \ldots, N$. Then

$$T_G(a, b) \geq \sum_{j=1}^{N} |G(t_j) - G(t_{j-1})|$$

$$= \sum_{j=1}^{N} \left| \int_{t_{j-1}}^{t_j} g(t)\, dt \right|$$

$$= \sum \int_{t_{j-1}}^{t_j} |g(t)|\, dt$$

$$= \int_{a}^{b} |g(t)|\, dt.$$

Since $\int_a^b |g(t)|\, dt \geq \int_a^b |F'(t)|\, dt - \epsilon$, we obtain as a consequence that

$$T_F(a, b) \geq \int_a^b |F'(t)|\, dt - 2\epsilon,$$

and letting $\epsilon \to 0$ we establish the assertion and also the theorem.

Now, any curve (viewed as the image of a mapping $t \mapsto z(t)$) can in fact be realized by many different parametrizations. A rectifiable curve, however, has associated to it a unique natural parametrization, the arc-length parametrization. Indeed, let $L(A, B)$ denote the length function (considered in Section 3.1), and for the variable t in $[a, b]$ set $s = s(t) = L(a, t)$. Then $s(t)$, the arc-length, is a continuous increasing function which maps $[a, b]$ to $[0, L]$, where L is the length of the curve. The **arc-length parametrization** of the curve is now given by the pair $\tilde{z}(s) = \tilde{x}(s) + i\tilde{y}(s)$, where $\tilde{z}(s) = z(t)$, for $s = s(t)$. Notice that in this way the function $\tilde{z}(s)$ is well defined on $[0, L]$, since if $s(t_1) = s(t_2)$, $t_1 < t_2$, then in fact $z(t)$ does not vary in the interval $[t_1, t_2]$ and thus $z(t_1) = z(t_2)$. Moreover $|\tilde{z}(s_1) - \tilde{z}(s_2)| \leq |s_1 - s_2|$, for all pairs $s_1, s_2 \in [0, L]$, since the left-hand side of the inequality is the distance between two points on the curve, while the right-hand side is the length of the portion of the curve joining these two points. Also, as s varies from 0 to L, $\tilde{z}(s)$ traces out the same points (in the same order) that $z(t)$ does as t varies from a to b.

Theorem 4.3 *Suppose $(x(t), y(t))$, $a \leq t \leq b$, is a rectifiable curve that has length L. Consider the arc-length parametrization $\tilde{z}(s) = (\tilde{x}(s), \tilde{y}(s))$ described above. Then $\tilde{x}$ and $\tilde{y}$ are absolutely continuous, $|\tilde{z}'(s)| = 1$ for almost every $s \in [0, L]$, and*

$$L = \int_0^L (\tilde{x}'(s)^2 + \tilde{y}'(s)^2)^{1/2}\, ds.$$

Proof. We noted that $|\tilde{z}(s_1) - \tilde{z}(s_2)| \leq |s_1 - s_2|$, so it follows immediately that $\tilde{z}(s)$ is absolutely continuous, hence differentiable almost everywhere. Moreover, this inequality also proves that $|\tilde{z}'(s)| \leq 1$, for almost every s. By definition the total variation of $\tilde{z}$ equals L, and by the previous theorem we must have $L = \int_0^L |\tilde{z}'(s)|\, ds$. Finally, we note that this identity is possible only when $|\tilde{z}'(s)| = 1$ almost everywhere.

4.1* Minkowski content of a curve

The proof we give below of the isoperimetric inequality depends in a key way on the concept of the Minkowski content. While the idea of this content has an interest on its own right, it is particularly relevant for us here. This is because the rectifiability of a curve is tantamount to having (finite) Minkowski content, with that quantity the same as the length of the curve.

We begin our discussion of these matters with several definitions. A curve parametrized by $z(t) = (x(t), y(t))$, $a \leq t \leq b$, is said to be **simple** if the mapping $t \mapsto z(t)$ is injective for $t \in [a, b]$. It is a **closed simple curve** if the mapping $t \mapsto z(t)$ is injective for t in $[a, b)$, and $z(a) = z(b)$. More generally, a curve is **quasi-simple** if the mapping is injective for t in the complement of finitely many points in $[a, b]$.

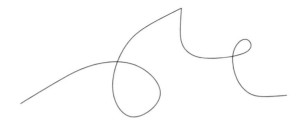

Figure 8. A quasi-simple curve

We shall find it convenient to designate by Γ the pointset traced out by the curve $z(t)$ as t varies in $[a, b]$, that is, $\Gamma = \{z(t) : a \leq t \leq b\}$. For any compact set $K \subset \mathbb{R}^2$ (we take $K = \Gamma$ below), we denote by K^δ the open set that consists of all points at distance (strictly) less than δ from K,

$$K^\delta = \{x \in \mathbb{R}^2 : d(x, K) < \delta\}.$$

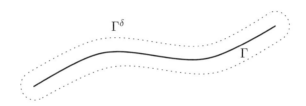

Figure 9. The curve Γ and the set Γ^δ

We then say that the set K has **Minkowski content**[7] if the limit

$$\lim_{\delta \to 0} \frac{m(K^\delta)}{2\delta}$$

exists. When this limit exists, we denote it by $\mathcal{M}(K)$.

Theorem 4.4 *Suppose* $\Gamma = \{z(t), a \leq t \leq b\}$ *is a quasi-simple curve. The Minkowski content of* Γ *exists if and only if* Γ *is rectifiable. When this is the case and* L *is the length of the curve, then* $\mathcal{M}(\Gamma) = L$.

To prove the theorem, we also consider for any compact set K

$$\mathcal{M}^*(K) = \limsup_{\delta \to 0} \frac{m(K^\delta)}{2\delta} \quad \text{and} \quad \mathcal{M}_*(K) = \liminf_{\delta \to 0} \frac{m(K^\delta)}{2\delta}$$

(both taken as extended positive numbers). Of course $\mathcal{M}_*(K) \leq \mathcal{M}^*(K)$. To say that the Minkowski content exists is the same as saying that $\mathcal{M}^*(K) < \infty$ and $\mathcal{M}_*(K) = \mathcal{M}^*(K)$. Their common value is then $\mathcal{M}(K)$.

The theorem just stated is the consequence of two propositions concerning $\mathcal{M}_*(K)$ and $\mathcal{M}^*(K)$. The first is as follows.

Proposition 4.5 *Suppose* $\Gamma = \{z(t), a \leq t \leq b\}$ *is a quasi-simple curve. If* $\mathcal{M}_*(\Gamma) < \infty$, *then the curve is rectifiable, and if* L *denotes its length, then*

$$L \leq \mathcal{M}_*(\Gamma).$$

The proof depends on the following simple observation.

Lemma 4.6 *If* $\Gamma = \{z(t), a \leq t \leq b\}$ *is any curve, and* $\Delta = |z(b) - z(a)|$ *is the distance between its end-points, then* $m(\Gamma^\delta) \geq 2\delta\Delta$.

Proof.   Since the distance function and the Lebesgue measure are invariant under translations and rotations (see Section 3 in Chapter 1 and Problem 4 in Chapter 2) we may transform the situation by an appropriate composition of these motions. Therefore we may assume that the end-points of the curve have been placed on the x-axis, and thus we may suppose that $z(a) = (A, 0)$, $z(b) = (B, 0)$ with $A < B$, and $\Delta = B - A$ (in the case $A = B$ the conclusion is automatically verified).

By the continuity of the function $x(t)$, there is for each x in $[A, B]$ a value $\bar{t}$ in $[a, b]$, such that $x = x(\bar{t})$. Since $\overline{Q} = (x(\bar{t}), y(\bar{t})) \in \Gamma$, the set

---

[7]This is one-dimensional Minkowski content; variants are in Exercise 28 and also in Chapter 7 below.

Γ^δ contains a segment parallel to the y-axis, of length 2δ centered at $\overline{Q}$ lying above x (see Figure 10). In other words the slice $(\Gamma^\delta)_x$ contains the interval $(y(\bar{t}) - \delta, y(\bar{t}) + \delta)$, and hence $m_1((\Gamma^\delta)^x) \geq 2\delta$ (where m_1 is the one-dimensional Lebesgue measure). However by Fubini's theorem

$$m(\Gamma^\delta) = \int_\mathbb{R} m_1((\Gamma^\delta)_x)\, dx \geq \int_A^B m_1((\Gamma^\delta)_x)\, dx \geq 2\delta(B - A) = 2\delta\Delta,$$

and the lemma is proved.

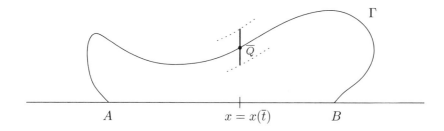

Figure 10. The situation in Lemma 4.6

We now pass to the proof of the proposition. Let us assume first that the curve is simple. Let P be any partition $a = t_0 < t_1 < \cdots < t_N = b$ of the interval $[a, b]$, and let L_P denote the length of the corresponding polygonal line, that is,

$$L_P = \sum_{j=1}^N |z(t_j) - z(t_{j-1})|.$$

For each $\epsilon > 0$, the continuity of $t \mapsto z(t)$ guarantees the existence of N proper closed sub-intervals $I_j = [a_j, b_j]$ of (t_{j-1}, t_j), so that

$$\sum_{j=1}^N |z(b_j) - z(a_j)| \geq L_P - \epsilon.$$

Let Γ_j denote the segment of the curve given by $\Gamma_j = \{z(t); t \in I_j\}$. Since the closed intervals $I_1, \ldots, I_N$ are disjoint, it follows by the simplicity of the curve that the compact sets $\Gamma_1, \Gamma_2, \ldots, \Gamma_N$ are disjoint. However, $\Gamma \supset \bigcup_{j=1}^N \Gamma_j$ and $\Gamma^\delta \supset \bigcup_{j=1}^N (\Gamma_j)^\delta$. Moreover, the disjointness of the Γ_j implies that the sets $(\Gamma_j)^\delta$ are also disjoint for sufficiently small δ. Hence for those δ, the previous lemma applied to each Γ_j gives

$$m(\Gamma^\delta) \geq \sum_{j=1}^{N} m((\Gamma_j)^\delta) \geq 2\delta \sum |z(b_j) - z(a_j)|.$$

As a result, $m(\Gamma^\delta)/(2\delta) \geq L_P - \epsilon$, and a passage to the limit gives $\mathcal{M}_*(\Gamma) \geq L_P - \epsilon$. Since this inequality is true for all partitions P and all $\epsilon > 0$, it implies that the curve is rectifiable and its length does not exceed $\mathcal{M}_*(\Gamma)$.

The proof when the curve is merely quasi-simple is similar, except the partitions P considered must be refined so as to include as partition points those (finitely many) points in whose complement (in $[a, b]$) the mapping $t \mapsto z(t)$ is injective. The details may be left to the reader.

The second proposition is in the reverse direction.

Proposition 4.7 *Suppose $\Gamma = \{z(t), a \leq t \leq b\}$ is a rectifiable curve with length L. Then*

$$\mathcal{M}^*(\Gamma) \leq L.$$

The quantities $\mathcal{M}^*(\Gamma)$ and L are of course independent of the parametrization used; since the curve is rectifiable, it will be convenient to use the arc-length parametrization. Thus we write the curve as $z(s) = (x(s), y(s))$, with $0 \leq s \leq L$, and recall that then $z(s)$ is absolutely continuous and $|z'(s)| = 1$ for a.e. $s \in [0, L]$.

We first fix any $0 < \epsilon < 1$, and find a measurable set $E_\epsilon \subset \mathbb{R}$ and a positive number r_ϵ such that $m(E_\epsilon) < \epsilon$ and

$$(14) \qquad \sup_{0 < |h| < r_\epsilon} \left| \frac{z(s+h) - z(s)}{h} - z'(s) \right| < \epsilon \qquad \text{for all } s \in [0, L] - E_\epsilon.$$

Indeed, for each integer n, let

$$F_n(s) = \sup_{0 < |h| < 1/n} \left| \frac{z(s+h) - z(s)}{h} - z'(s) \right|$$

(where $z(s)$ has been extended outside $[0, L]$, so that $z(s) = z(0)$, when $s < 0$, and $z(s) = z(L)$ when $s > L$). Because $z(s)$ is continuous the supremum of h in the definition of $F_n(s)$ can be replaced by a supremum of countably many measurable functions, and hence each F_n is measurable. However, $F_n(s) \to 0$, as $n \to \infty$ for a.e $s \in [a, b]$. Thus by Egorov's theorem the convergence is uniform outside a set E_ϵ with $m(E_\epsilon) < \epsilon$, and so we merely need to choose $r_\epsilon = 1/n$ for sufficiently large n to establish (14). It will be convenient in what follows to assume, as we may, that $z'(s)$ exists and $|z'(s)| = 1$ for every $s \notin E_\epsilon$.

Now for any $0 < \rho < r_\epsilon$ (with $\rho < 1$), we partition the interval $[0, L]$ into consecutive closed intervals, each of length ρ, (except that the last interval may have length $\leq \rho$). Then there is a total of $N \leq L/\rho + 1$ such intervals that arise. We call these intervals $I_1, I_2, \ldots, I_N$, and divide them into two classes. The first class, those intervals I_j we call "good," are the ones that enjoy the property that $I_j \not\subset E_\epsilon$. The second class, those which are "bad," have the property that $I_j \subset E_\epsilon$. As a result, $\bigcup_{I_j \text{ bad}} I_j \subset E_\epsilon$, hence the union has measure $< \epsilon$.

We have of course that $[0, L] \subset \bigcup_{j=1}^N I_j$, and if we denote by Γ_j the segment of Γ given by $\{z(s) : s \in I_j\}$, then $\Gamma = \bigcup_{j=1}^N \Gamma_j$, and as a result $\Gamma^\delta = \bigcup_{j=1}^N (\Gamma_j)^\delta$ and $m(\Gamma^\delta) \leq \sum_{j=1}^N m((\Gamma_j)^\delta)$.

We consider first the contribution of $m((\Gamma_j)^\delta)$ when I_j is a good interval. Recall that for such $I_j = [a_j, b_j]$ there is an $s_0 \in I_j$ which is not in E_ϵ, and therefore (14) holds for $s = s_0$. Let us now visualize Γ_j by introducing a coordinate system such that $z(s_0) = 0$ and $z'(s_0) = 1$ (which we may assume after a suitable translation and rotation). We maintain the notations $z(s)$ and Γ_j for the so transformed segment of the curve.

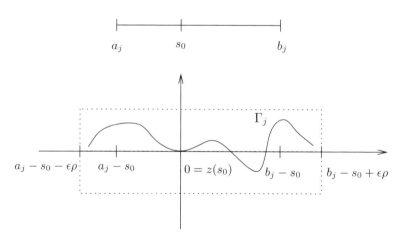

Figure 11. Estimate of $m((\Gamma_j)^\delta)$ for a good interval I_j

Note that as h varies over the interval $[a_j - s_0, b_j - s_0]$, $s_0 + h$ varies over $I_j = [a_j, b_j]$. Therefore Γ_j is contained in the rectangle

$$[a_j - s_0 - \epsilon\rho, b_j - s_0 + \epsilon\rho] \times [-\epsilon\rho, \epsilon\rho],$$

since $|h| \leq \rho < r_\epsilon$ by construction, and $|z(s_0 + h) - h| < \epsilon|h|$ by (14). See Figure 11. Thus $(\Gamma_j)^\delta$ is contained in the rectangle

$$[a_j - s_0 - \epsilon\rho - \delta, b_j - s_0 + \epsilon\rho + \delta] \times [-\epsilon\rho - \delta, \epsilon\rho + \delta],$$

which has measure $\leq (\rho + 2\epsilon\rho + 2\delta)(2\epsilon\rho + 2\delta)$. Therefore, since $\epsilon \leq 1$, we have

$$(15) \qquad m((\Gamma_j)^\delta) \leq 2\delta\rho + O(\epsilon\delta\rho + \delta^2 + \epsilon\rho^2),$$

where the bound arising in O is independent of ϵ, δ, and ρ. This is our desired estimate for the good intervals.

To pass to the remaining intervals we use the fact that $|z(s) - z(s')| \leq |s - s'|$ for all s and s'. Thus in every case Γ_j is contained in a ball (disc) of radius ρ, and hence $(\Gamma_j)^\delta$ is contained in a ball of radius $\rho + \delta$. Therefore we have the crude estimate

$$(16) \qquad m((\Gamma_j)^\delta) = O(\delta^2 + \rho^2).$$

We now sum (15) over the good intervals (of which there are at most $L/\rho + 1$), and (16) over the bad intervals. There are at most $\epsilon/\rho + 1$ of the latter kind, since their union is included in E_ϵ and this set has measure $< \epsilon$. Altogether, then,

$$m(\Gamma^\delta) \leq 2\delta L + 2\delta\rho + O(\epsilon\delta + \delta^2/\rho + \epsilon\rho) + O\left((\epsilon/\rho + 1)(\delta^2 + \rho^2)\right),$$

which simplifies to the inequalities

$$\begin{aligned}
\frac{m(\Gamma^\delta)}{2\delta} &\leq L + O\left(\rho + \epsilon + \frac{\delta}{\rho} + \frac{\epsilon\rho}{\delta} + \frac{\epsilon\delta}{\rho} + \delta + \frac{\rho^2}{\delta}\right) \\
&\leq L + O\left(\rho + \epsilon + \frac{\delta}{\rho} + \frac{\epsilon\rho}{\delta} + \frac{\rho^2}{\delta}\right),
\end{aligned}$$

where in the last line we have used the fact that $\epsilon < 1$ and $\rho < 1$. In order to obtain a favorable estimate from this as $\delta \to 0$, we need to choose ρ (the length of the sub-intervals) very roughly of the same size as δ. An effective choice is $\rho = \delta/\epsilon^{1/2}$. If we fix this choice and restrict our attention to δ for which $0 < \delta < \epsilon^{1/2}r_\epsilon$, then automatically $\rho < r_\epsilon$, as required by (14). Inserting $\rho = \delta/\epsilon^{1/2}$ in the above inequality gives

$$\frac{m(\Gamma^\delta)}{2\delta} \leq L + O\left(\frac{\delta}{\epsilon^{1/2}} + \epsilon + \epsilon^{1/2} + \frac{\delta}{\epsilon}\right),$$

and thus

$$\limsup_{\delta \to 0} \frac{m(\Gamma^\delta)}{2\delta} \le L + O(\epsilon + \epsilon^{1/2}).$$

Now we can let $\epsilon \to 0$ to obtain the desired conclusion $\mathcal{M}^*(\Gamma) \le L$, and the proofs of the proposition and theorem are complete.

4.2* Isoperimetric inequality

The isoperimetric inequality in the plane states, in effect, that among all curves of a given length it is the circle that encloses the maximum area. A simple form of this theorem already appeared in Book I. While the proof given there had the virtue of being brief and elegant, it did suffer several shortcomings. Among them the "area" in the statement was defined indirectly via a technical artifice, and the scope of the conclusion was limited because only relatively smooth curves were considered. Here we want to remedy those defects and deal with a general version of the result.

We suppose that Ω is a bounded open subset of $\mathbb{R}^2$, and that its boundary $\overline{\Omega} - \Omega$, is a rectifiable curve Γ, with length $\ell(\Gamma)$. We do not require that Γ be a simple closed curve. The isoperimetric theorem then asserts the following.

Theorem 4.8 $4\pi\, m(\Omega) \le \ell(\Gamma)^2$.

Proof. For each $\delta > 0$ we consider the outer set

$$\Omega_+(\delta) = \{x \in \mathbb{R}^2 : d(x, \overline{\Omega}) < \delta\},$$

and the inner set

$$\Omega_-(\delta) = \{x \in \mathbb{R}^2 : d(x, \Omega^c) \ge \delta\}.$$

Thus $\Omega_-(\delta) \subset \Omega \subset \Omega_+(\delta)$.
We notice that for $\Gamma^\delta = \{x : d(x, \Gamma) < \delta\}$ we have

$$(17) \qquad \Omega_+(\delta) = \Omega_-(\delta) \cup \Gamma^\delta,$$

and that this union is disjoint. Moreover, if $D(\delta)$ is the open ball (disc) of radius δ centered at the origin, $D(\delta) = \{x \in \mathbb{R}^2,\ |x| < \delta\}$, then clearly

$$(18) \qquad \begin{cases} \Omega_+(\delta) \supset \Omega + D(\delta), \\ \quad\ \Omega \supset \Omega_-(\delta) + D(\delta). \end{cases}$$

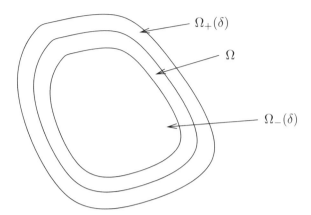

Figure 12. The sets Ω, $\Omega_-(\delta)$ and $\Omega_+(\delta)$

We now apply the Brunn-Minkowski inequality (Theorem 5.1 in Chapter 1) to the first inclusion, and obtain

$$m(\Omega_+(\delta)) \geq (m(\Omega)^{1/2} + m(D(\delta))^{1/2})^2.$$

Since $m(D(\delta)) = \pi\delta^2$ (this standard formula is established in Exercise 14 in the previous chapter), and $(A + B)^2 \geq A^2 + 2AB$ whenever A and B are positive, we find that

$$m(\Omega_+(\delta)) \geq m(\Omega) + 2\pi^{1/2}\delta\, m(\Omega)^{1/2}.$$

Similarly, $m(\Omega) \geq m(\Omega_-(\delta)) + 2\pi^{1/2}\delta\, m(\Omega_-(\delta))^{1/2}$ using the second inclusion in (18), which implies

$$-m(\Omega_-(\delta)) \geq -m(\Omega) + 2\pi^{1/2}\delta\, m(\Omega_-(\delta))^{1/2}.$$

Now by (17)

$$m(\Gamma^\delta) = m(\Omega_+(\delta)) - m(\Omega_-(\delta)),$$

and by the inequalities above, we have

$$m(\Gamma^\delta) \geq 2\pi^{1/2}\delta(m(\Omega)^{1/2} + m(\Omega_-(\delta))^{1/2}).$$

We now divide both sides by 2δ and take the limsup as $\delta \to 0$. This yields

$$\mathcal{M}^*(\Gamma) \geq \pi^{1/2}(2m(\Omega)^{1/2}),$$

since $\Omega_-(\delta) \nearrow \Omega$ as $\delta \to 0$. However, by Proposition 4.7, $\ell(\Gamma) \geq \mathcal{M}^*(\Gamma)$, so

$$\ell(\Gamma) \geq 2\pi^{1/2}m(\Omega)^{1/2},$$

which proves the theorem.

Remark. A similar result holds even without the assumption that the boundary is a (rectifiable) curve. In fact the proof shows that for any bounded open set Ω whose boundary is Γ we have

$$4\pi\,m(\Omega) \leq \mathcal{M}^*(\Gamma)^2.$$

5 Exercises

1. Suppose φ is an integrable function on $\mathbb{R}^d$ with $\int_{\mathbb{R}^d} \varphi(x)\,dx = 1$. Set $K_\delta(x) = \delta^{-d}\varphi(x/\delta)$, $\delta > 0$.

(a) Prove that $\{K_\delta\}_{\delta>0}$ is a family of good kernels.

(b) Assume in addition that φ is bounded and supported in a bounded set. Verify that $\{K_\delta\}_{\delta>0}$ is an approximation to the identity.

(c) Show that Theorem 2.3 (convergence in the L^1-norm) holds for good kernels as well.

2. Suppose $\{K_\delta\}$ is a family of kernels that satisfies:

(i) $|K_\delta(x)| \leq A\delta^{-d}$ for all $\delta > 0$.

(ii) $|K_\delta(x)| \leq A\delta/|x|^{d+1}$ for all $\delta > 0$.

(iii) $\int_{-\infty}^{\infty} K_\delta(x)\,dx = 0$ for all $\delta > 0$.

Thus K_δ satisfies conditions (i) and (ii) of approximations to the identity, but the average value of K_δ is 0 instead of 1. Show that if f is integrable on $\mathbb{R}^d$, then

$$(f * K_\delta)(x) \to 0 \qquad \text{for a.e. } x, \text{ as } \delta \to 0.$$

3. Suppose 0 is a point of (Lebesgue) density of the set $E \subset \mathbb{R}$. Show that for each of the individual conditions below there is an infinite sequence of points $x_n \in E$, with $x_n \neq 0$, and $x_n \to 0$ as $n \to \infty$.

(a) The sequence also satisfies $-x_n \in E$ for all n.

(b) In addition, $2x_n$ belongs to E for all n.

Generalize.

4. Prove that if f is integrable on $\mathbb{R}^d$, and f is not identically zero, then

$$f^*(x) \geq \frac{c}{|x|^d}, \qquad \text{for some } c > 0 \text{ and all } |x| \geq 1.$$

Conclude that f^* is not integrable on $\mathbb{R}^d$. Then, show that the weak type estimate

$$m(\{x : f^*(x) > \alpha\}) \leq c/\alpha$$

for all $\alpha > 0$ whenever $\int |f| = 1$, is best possible in the following sense: if f is supported in the unit ball with $\int |f| = 1$, then

$$m(\{x : f^*(x) > \alpha\}) \geq c'/\alpha$$

for some $c' > 0$ and all sufficiently small α.

[Hint: For the first part, use the fact that $\int_B |f| > 0$ for some ball B.]

5. Consider the function on $\mathbb{R}$ defined by

$$f(x) = \begin{cases} \dfrac{1}{|x|(\log 1/|x|)^2} & \text{if } |x| \leq 1/2, \\ 0 & \text{otherwise.} \end{cases}$$

(a) Verify that f is integrable.

(b) Establish the inequality

$$f^*(x) \geq \frac{c}{|x|(\log 1/|x|)} \qquad \text{for some } c > 0 \text{ and all } |x| \leq 1/2,$$

to conclude that the maximal function f^* is not locally integrable.

6. In one dimension there is a version of the basic inequality (1) for the maximal function in the form of an identity. We define the "one-sided" maximal function

$$f_+^*(x) = \sup_{h>0} \frac{1}{h} \int_x^{x+h} |f(y)| \, dy.$$

If $E_\alpha^+ = \{x \in \mathbb{R} : f_+^*(x) > \alpha\}$, then

$$m(E_\alpha^+) = \frac{1}{\alpha} \int_{E_\alpha^+} |f(y)| \, dy.$$

[Hint: Apply Lemma 3.5 to $F(x) = \int_0^x |f(y)| \, dy - \alpha x$. Then E_α^+ is the union of disjoint intervals (a_k, b_k) with $\int_{a_k}^{b_k} |f(y)| \, dy = \alpha(a_k - b_k)$.]

7. Using Corollary 1.5, prove that if a measurable subset E of $[0,1]$ satisfies $m(E \cap I) \geq \alpha \, m(I)$ for some $\alpha > 0$ and all intervals I in $[0,1]$, then E has measure 1. See also Exercise 28 in Chapter 1.

8. Suppose A is a Lebesgue measurable set in $\mathbb{R}$ with $m(A) > 0$. Does there exist a sequence $\{s_n\}_{n=1}^{\infty}$ such that the complement of $\bigcup_{n=1}^{\infty}(A + s_n)$ in $\mathbb{R}$ has measure zero?

[Hint: For every $\epsilon > 0$, find an interval I_ϵ of length ℓ_ϵ such that $m(A \cap I_\epsilon) \geq (1 - \epsilon)m(I_\epsilon)$. Consider $\bigcup_{k=-\infty}^{\infty}(A + t_k)$, with $t_k = k\ell_\epsilon$. Then vary ϵ.]

9. Let F be a closed subset in $\mathbb{R}$, and $\delta(x)$ the distance from x to F, that is,

$$\delta(x) = d(x, F) = \inf\{|x - y| : y \in F\}.$$

Clearly, $\delta(x + y) \leq |y|$ whenever $x \in F$. Prove the more refined estimate

$$\delta(x + y) = o(|y|) \qquad \text{for a.e. } x \in F,$$

that is, $\delta(x + y)/|y| \to 0$ for a.e. $x \in F$.

[Hint: Assume that x is a point of density of F.]

10. Construct an increasing function on $\mathbb{R}$ whose set of discontinuities is precisely $\mathbb{Q}$.

11. If $a, b > 0$, let

$$f(x) = \begin{cases} x^a \sin(x^{-b}) & \text{for } 0 < x \leq 1, \\ 0 & \text{if } x = 0. \end{cases}$$

Prove that f is of bounded variation in $[0, 1]$ if and only if $a > b$. Then, by taking $a = b$, construct (for each $0 < \alpha < 1$) a function that satisfies the Lipschitz condition of exponent α

$$|f(x) - f(y)| \leq A|x - y|^\alpha$$

but which is not of bounded variation.

[Hint: Note that if $h > 0$, the difference $|f(x + h) - f(x)|$ can be estimated by $C(x + h)^a$, or $C'h/x$ by the mean value theorem. Then, consider two cases, whether $x^{a+1} \geq h$ or $x^{a+1} < h$. What is the relationship between α and a?]

12. Consider the function $F(x) = x^2 \sin(1/x^2)$, $x \neq 0$, with $F(0) = 0$. Show that $F'(x)$ exists for every x, but F' is not integrable on $[-1, 1]$.

13. Show directly from the definition that the Cantor-Lebesgue function is not absolutely continuous.

14. The following measurability issues arose in the discussion of differentiability of functions.

(a) Suppose F is continuous on $[a, b]$. Show that

$$D^+(F)(x) = \limsup_{\substack{h \to 0 \\ h > 0}} \frac{F(x+h) - F(x)}{h}$$

is measurable.

(b) Suppose $J(x) = \sum_{n=1}^{\infty} \alpha_n j_n(x)$ is a jump function as in Section 3.3. Show that

$$\limsup_{h \to 0} \frac{J(x+h) - J(x)}{h}$$

is measurable.

[Hint: For (a), the continuity of F allows one to restrict to countably many h in taking the limsup. For (b), given $k > m$, let $F_{k,m}^N = \sup_{1/k \le |h| \le 1/m} \left| \frac{J_N(x+h) - J_N(x)}{h} \right|$, where $J_N(x) = \sum_{n=1}^{N} \alpha_n j_n(x)$. Note that each $F_{k,m}^N$ is measurable. Then, successively, let $N \to \infty$, $k \to \infty$, and finally $m \to \infty$.]

15. Suppose F is of bounded variation and continuous. Prove that $F = F_1 - F_2$, where both F_1 and F_2 are monotonic and continuous.

16. Show that if F is of bounded variation in $[a, b]$, then:

(a) $\int_a^b |F'(x)|\, dx \le T_F(a, b)$.

(b) $\int_a^b |F'(x)|\, dx = T_F(a, b)$ if and only if F is absolutely continuous.

As a result of (b), the formula $L = \int_a^b |z'(t)|\, dt$ for the length of a rectifiable curve parametrized by z holds if and only if z is absolutely continuous.

17. Prove that if $\{K_\epsilon\}_{\epsilon > 0}$ is a family of approximations to the identity, then

$$\sup_{\epsilon > 0} |(f * K_\epsilon)(x)| \le cf^*(x)$$

for some constant $c > 0$ and all integrable f.

18. Verify the agreement between the two definitions given for the Cantor-Lebesgue function in Exercise 2, Chapter 1 and in Section 3.1 of this chapter.

19. Show that if $f : \mathbb{R} \to \mathbb{R}$ is absolutely continuous, then

(a) f maps sets of measure zero to sets of measure zero.

(b) f maps measurable sets to measurable sets.

20. This exercise deals with functions F that are absolutely continuous on $[a, b]$ and are increasing. Let $A = F(a)$ and $B = F(b)$.

(a) There exists such an F that is in addition strictly increasing, but such that $F'(x) = 0$ on a set of positive measure.

(b) The F in (a) can be chosen so that there is a measurable subset $E \subset [A, B]$, $m(E) = 0$, so that $F^{-1}(E)$ is not measurable.

(c) Prove, however, that for any increasing absolutely continuous F, and E a measurable subset of $[A, B]$, the set $F^{-1}(E) \cap \{F'(x) > 0\}$ is measurable.

[Hint: (a) Let $F(x) = \int_a^x \chi_K(x)\, dx$, where K is the complement of a Cantor-like set C of positive measure. For (b), note that $F(C)$ is a set of measure zero. Finally, for (c) prove first that $m(\mathcal{O}) = \int_{F^{-1}(\mathcal{O})} F'(x)\, dx$ for any open set $\mathcal{O}$.]

21. Let F be absolutely continuous and increasing on $[a, b]$ with $F(a) = A$ and $F(b) = B$. Suppose f is any measurable function on $[A, B]$.

(a) Show that $f(F(x))F'(x)$ is measurable on $[a, b]$. Note: $f(F(x))$ need not be measurable by Exercise 20 (b).

(b) Prove the change of variable formula: If f is integrable on $[A, B]$, then so is $f(F(x))F'(x)$, and

$$\int_A^B f(y)\, dy = \int_a^b f(F(x))F'(x)\, dx.$$

[Hint: Start with the identity $m(\mathcal{O}) = \int_{F^{-1}(\mathcal{O})} F'(x)\, dx$ used in (c) of Exercise 20 above.]

22. Suppose that F and G are absolutely continuous on $[a, b]$. Show that their product FG is also absolutely continuous. This has the following consequences.

(a) Whenever F and G are absolutely continuous in $[a, b]$,

$$\int_a^b F'(x)G(x)\, dx = -\int_a^b F(x)G'(x)\, dx + [F(x)G(x)]_a^b.$$

(b) Let F be absolutely continuous in $[-\pi, \pi]$ with $F(\pi) = F(-\pi)$. Show that if

$$a_n = \frac{1}{2\pi} \int_{-\pi}^{\pi} F(x)e^{-inx}\, dx,$$

such that $F(x) \sim \sum a_n e^{inx}$, then

$$F'(x) \sim \sum in a_n e^{inx}.$$

(c) What happens if $F(-\pi) \neq F(\pi)$? [Hint: Consider $F(x) = x$.]

23. Let F be continuous on $[a, b]$. Show the following.

(a) Suppose $(D^+F)(x) \geq 0$ for every $x \in [a, b]$. Then F is increasing on $[a, b]$.

(b) If $F'(x)$ exists for every $x \in (a, b)$ and $|F'(x)| \leq M$, then $|F(x) - F(y)| \leq M|x - y|$ and F is absolutely continuous.

[Hint: For (a) it suffices to show that $F(b) - F(a) \geq 0$. Assume otherwise. Hence with $G_\epsilon(x) = F(x) - F(a) + \epsilon(x - a)$, for sufficiently small $\epsilon > 0$ we have $G_\epsilon(a) = 0$, but $G_\epsilon(b) < 0$. Now let $x_0 \in [a, b)$ be the greatest value of x_0 such that $G_\epsilon(x_0) \geq 0$. However, $(D^+G_\epsilon)(x_0) > 0$.]

24. Suppose F is an increasing function on $[a, b]$.

(a) Prove that we can write

$$F = F_A + F_C + F_J,$$

where each of the functions F_A, F_C, and F_J is increasing and:

    (i) F_A is absolutely continuous.

    (ii) F_C is continuous, but $F'_C(x) = 0$ for a.e. x.

    (iii) F_J is a jump function.

(b) Moreover, each component F_A, F_C, F_J is uniquely determined up to an additive constant.

The above is the **Lebesgue decomposition** of F. There is a corresponding decomposition for any F of bounded variation.

25. The following shows the necessity of allowing for general exceptional sets of measure zero in the differentiation Theorems 1.4, 3.4, and 3.11. Let E be any set of measure zero in $\mathbb{R}^d$. Show that:

(a) There exists a non-negative integrable f in $\mathbb{R}^d$, so that

$$\liminf_{\substack{m(B) \to 0 \\ x \in B}} \frac{1}{m(B)} \int_B f(y) \, dy = \infty \qquad \text{for each } x \in E.$$

(b) When $d = 1$ this may be restated as follows. There is an increasing absolutely continuous function F so that

$$D_+(F)(x) = D_-(F)(x) = \infty, \qquad \text{for each } x \in E.$$

[Hint: Find open sets $\mathcal{O}_n \supset E$, with $m(\mathcal{O}_n) < 2^{-n}$, and let $f(x) = \sum_{n=1}^{\infty} \chi_{\mathcal{O}_n}(x)$.]

26. An alternative way of defining the exterior measure $m_*(E)$ of an arbitrary set E, as given in Section 2 of Chapter 1, is to replace the coverings of E by cubes with coverings by balls. That is, suppose we define $m_*^{\mathcal{B}}(E)$ as $\inf \sum_{j=1}^{\infty} m(B_j)$, where the infimum is taken over all coverings $E \subset \bigcup_{j=1}^{\infty} B_j$ by open balls. Then $m_*(E) = m_*^{\mathcal{B}}(E)$. (Observe that this result leads to an alternate proof that the Lebesgue measure is invariant under rotations.)

Clearly $m_*(E) \leq m_*^{\mathcal{B}}(E)$. Prove the reverse inequality by showing the following. For any $\epsilon > 0$, there is a collection of balls $\{B_j\}$ such that $E \subset \bigcup_j B_j$ while $\sum_j m(B_j) \leq m_*(E) + \epsilon$. Note also that for any preassigned δ, we can choose the balls to have diameter $< \delta$.

[Hint: Assume first that E is measurable, and pick $\mathcal{O}$ open so that $\mathcal{O} \supset E$ and $m(\mathcal{O} - E) < \epsilon'$. Next, using Corollary 3.10, find balls $B_1, \ldots, B_N$ such that $\sum_{j=1}^{N} m(B_j) \leq m(E) + 2\epsilon'$ and $m(E - \bigcup_{j=1}^{N} B_j) \leq 3\epsilon'$. Finally, cover $E - \bigcup_{j=1}^{N} B_j$ by a union of cubes, the sum of whose measures is $\leq 4\epsilon'$, and replace these cubes by balls that contain them. For the general E, begin by applying the above when E is a cube.]

27. A rectifiable curve has a tangent line at almost all points of the curve. Make this statement precise.

28. A curve in $\mathbb{R}^d$ is a continuous map $t \mapsto z(t)$ of an interval $[a, b]$ into $\mathbb{R}^d$.

(a) State and prove the analogues of the conditions dealing with the rectifiability of curves and their length that are given in Theorems 3.1, 4.1, and 4.3.

(b) Define the (one-dimensional) Minkowski content $\mathcal{M}(K)$ of a compact set in $\mathbb{R}^d$ as the limit (if it exists) of

$$\frac{m(K^\delta)}{m_{d-1}(B(\delta))} \quad \text{as } \delta \to 0,$$

where $m_{d-1}(B(\delta))$ is the measure (in $\mathbb{R}^{d-1}$) of the ball defined by $B(\delta) = \{x \in \mathbb{R}^{d-1}, |x| < \delta\}$. State and prove analogues of Propositions 4.5 and 4.7 for curves in $\mathbb{R}^d$.

29. Let $\Gamma = \{z(t), a \leq t \leq b\}$ be a curve, and suppose it satisfies a Lipschitz condition with exponent α, $1/2 \leq \alpha \leq 1$, that is,

$$|z(t) - z(t')| \leq A|t - t'|^\alpha \quad \text{for all } t, t' \in [a, b].$$

Show that $m(\Gamma^\delta) = O(\delta^{2-1/\alpha})$ for $0 < \delta \leq 1$.

30. A bounded function F is said to be of bounded variation on $\mathbb{R}$ if F is of bounded variation on any finite sub-interval $[a, b]$, and $\sup_{a,b} T_F(a, b) < \infty$.

Prove that such an F enjoys the following two properties:

(a) $\int_{\mathbb{R}} |F(x + h) - F(x)|\, dx \leq A|h|$, for some constant A and all $h \in \mathbb{R}$.

(b) $|\int_{\mathbb{R}} F(x)\varphi'(x)\, dx| \leq A$, where φ ranges over all C^1 functions of bounded support with $\sup_{x \in \mathbb{R}} |\varphi(x)| \leq 1$.

For the converse, and analogues in $\mathbb{R}^d$, see Problem 6^* below.

[Hint: For (a), write $F = F_1 - F_2$, where F_j are monotonic and bounded. For (b), deduce this from (a).]

31. Let F be the Cantor-Lebesgue function described in Section 3.1. Consider the curve that is the graph of F, that is, the curve given by $x(t) = t$ and $y(t) = F(t)$ with $0 \leq t \leq 1$. Prove that the length $L(\overline{x})$ of the segment $0 \leq t \leq \overline{x}$ of the curve is given by $L(\overline{x}) = \overline{x} + F(\overline{x})$. Hence the total length of the curve is 2.

32. Let $f : \mathbb{R} \to \mathbb{R}$. Prove that f satisfies the Lipschitz condition

$$|f(x) - f(y)| \leq M|x - y|$$

for some M and all $x, y \in \mathbb{R}$, if and only if f satisfies the following two properties:

(i) f is absolutely continuous.

(ii) $|f'(x)| \leq M$ for a.e. x.

6 Problems

1. Prove the following variant of the **Vitali covering lemma**: If E is covered in the Vitali sense by a family $\mathcal{B}$ of balls, and $0 < m_*(E) < \infty$, then for every $\eta > 0$ there exists a disjoint collection of balls $\{B_j\}_{j=1}^{\infty}$ in $\mathcal{B}$ such that

$$m_*\left(E / \bigcup_{j=1}^{\infty} B_j \right) = 0 \quad \text{and} \quad \sum_{j=1}^{\infty} |B_j| \leq (1 + \eta)m_*(E).$$

2. The following simple one-dimensional covering lemma can be used in a number of different situations.

Suppose $I_1, I_2, \ldots, I_N$ is a given finite collection of open intervals in $\mathbb{R}$. Then there are two finite sub-collections $I_1', I_2', \ldots, I_K'$, and $I_1'', I_2'', \ldots, I_L''$, so that each sub-collection consists of mutually disjoint intervals and

$$\bigcup_{j=1}^{N} I_j = \bigcup_{k=1}^{K} I_k' \cup \bigcup_{\ell=1}^{L} I_\ell''.$$

Note that, in contrast with Lemma 1.2, the full union is covered and not merely a part.

[Hint: Choose I_1' to be an interval whose left end-point is as far left as possible. Discard all intervals contained in I_1'. If the remaining intervals are disjoint from I_1', select again an interval as far to the left as possible, and call it I_2'. Otherwise choose an interval that intersects I_1', but reaches out to the right as far as possible, and call this interval I_1''. Repeat this procedure.]

3.[*] There is no direct analogue of Problem 2 in higher dimensions. However, a full covering is afforded by the Besicovitch covering lemma. A version of this lemma states that there is an integer N (dependent only on the dimension d) with the following property. Suppose E is any bounded set in $\mathbb{R}^d$ that is covered by a collection $\mathcal{B}$ of balls in the (strong) sense that for each $x \in E$, there is a $B \in \mathcal{B}$ whose center is x. Then, there are N sub-collections $\mathcal{B}_1, \mathcal{B}_2, \ldots, \mathcal{B}_N$ of the original collection $\mathcal{B}$, such that each $\mathcal{B}_j$ is a collection of disjoint balls, and moreover,

$$E \subset \bigcup_{B \in \mathcal{B}'} B, \qquad \text{where } \mathcal{B}' = \mathcal{B}_1 \cup \mathcal{B}_2 \cup \cdots \cup \mathcal{B}_N.$$

4. A real-valued function φ defined on an interval (a, b) is **convex** if the region lying above its graph $\{(x, y) \in \mathbb{R}^2 : y > \varphi(x),\ a \le x \le b\}$ is a convex set, as defined in Section 5*, Chapter 1. Equivalently, φ is convex if

$$\varphi(\theta x_1 + (1 - \theta)x_2) \le \theta \varphi(x_1) + (1 - \theta)\varphi(x_2)$$

for every $x_1, x_2 \in (a, b)$ and $0 \le \theta \le 1$. One can also observe as a consequence that we have the following inequality of the slopes:

$$\frac{\varphi(x + h) - \varphi(x)}{h} \le \frac{\varphi(y) - \varphi(x)}{y - x} \le \frac{\varphi(y) - \varphi(y - h)}{h},$$

whenever $x < y$, $h > 0$, and $x + h < y$.

The following can then be proved.

(a) φ is continuous on (a, b).

(b) φ satisfies a Lipschitz condition of order 1 in any proper closed sub-interval $[a', b']$ of (a, b). Hence φ is absolutely continuous in each sub-interval.

(c) φ' exists at all but an at most denumerable number of points, and $\varphi' = D^+\varphi$ is an increasing function with

$$\varphi(y) - \varphi(x) = \int_x^y \varphi'(t)\, dt.$$

(d) Conversely, if ψ is any increasing function on (a, b), then $\varphi(x) = \int_c^x \psi(t)\, dt$ is a convex function in (a, b) (for $c \in (a, b)$).

5. Suppose that F is continuous on $[a, b]$, $F'(x)$ exists for *every* $x \in (a, b)$, and $F'(x)$ is integrable. Then F is absolutely continuous and

$$F(b) - F(a) = \int_a^b F'(x)\, dx.$$

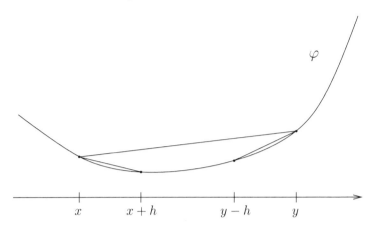

Figure 13. A convex function

[Hint: Assume $F'(x) \geq 0$ for a.e. x. We want to conclude that $F(b) \geq F(a)$. Let E be the set of measure 0 of those x such that $F'(x) < 0$. Then according to Exercise 25, there is a function Φ which is increasing, absolutely continuous, and for which $D^+\Phi(x) = \infty$, $x \in E$. Consider $F + \delta\Phi$, for each δ and apply the result (a) in Exercise 23.]

6.* The following converse to Exercise 30 characterizes functions of bounded variation.

Suppose F is a bounded measurable function on $\mathbb{R}$. If F satisfies either of conditions (a) or (b) in that exercise, then F can be modified on a set of measure zero so as to become a function of bounded variation on $\mathbb{R}$.

Moreover, on $\mathbb{R}^d$ we have the following assertion. Suppose F is a bounded measurable function on $\mathbb{R}^d$. Then the following two conditions on F are equivalent:

(a') $\int_{\mathbb{R}^d} |F(x+h) - F(x)|\, dx \leq A|h|$, for all $h \in \mathbb{R}^d$.

(b') $|\int_{\mathbb{R}^d} F(x)\frac{\partial\varphi}{\partial x_j}\, dx| \leq A$, for all $j = 1, \ldots, d$,

for all $\varphi \in C^1$ that have bounded support, and for which $\sup_{x \in \mathbb{R}^d} |\varphi(x)| \leq 1$.

The class of functions that satisfy either (a') or (b') is the extension to $\mathbb{R}^d$ of the class of functions of bounded variation.

7. Consider the function

$$f_1(x) = \sum_{n=0}^{\infty} 2^{-n} e^{2\pi i 2^n x}.$$

(a) Prove that f_1 satisfies $|f_1(x) - f_1(y)| \leq A_\alpha |x - y|^\alpha$ for each $0 < \alpha < 1$.

(b)* However, f_1 is nowhere differentiable, hence not of bounded variation.

8.* Let $\mathcal{R}$ denote the set of all rectangles in $\mathbb{R}^2$ that contain the origin, and with sides parallel to the coordinate axis. Consider the maximal operator associated to this family, namely

$$f_{\mathcal{R}}^*(x) = \sup_{R \in \mathcal{R}} \frac{1}{m(R)} \int_R |f(x-y)|\, dy.$$

(a) Then, $f \mapsto f_{\mathcal{R}}^*$ does not satisfy the weak type inequality

$$m(\{x : f_{\mathcal{R}}^*(x) > \alpha\}) \leq \frac{A}{\alpha} \|f\|_{L^1}$$

for all $\alpha > 0$, all integrable f, and some $A > 0$.

(b) Using this, one can show that there exists $f \in L^1(\mathbb{R})$ so that for $R \in \mathcal{R}$

$$\limsup_{\mathrm{diam}(R) \to 0} \frac{1}{m(R)} \int_R f(x-y)\, dy = \infty \qquad \text{for almost every } x.$$

Here $\mathrm{diam}(R) = \sup_{x,y \in R} |x-y|$ equals the diameter of the rectangle.

[Hint: For part (a), let B be the unit ball, and consider the function $\varphi(x) = \chi_B(x)/m(B)$. For $\delta > 0$, let $\varphi_\delta(x) = \delta^{-2}\varphi(x/\delta)$. Then

$$(\varphi_\delta)_{\mathcal{R}}^*(x) \to \frac{1}{|x_1||x_2|} \qquad \text{as } \delta \to 0,$$

for every (x_1, x_2), with $x_1 x_2 \neq 0$. If the weak type inequality held, then we would have

$$m(\{|x| \leq 1 : |x_1 x_2|^{-1} > \alpha\}) \leq \frac{A}{\alpha}.$$

This is a contradiction since the left-hand side is of the order of $(\log \alpha)/\alpha$ as α tends to infinity.]

4 Hilbert Spaces: An Introduction

> Born barely 10 years ago, the theory of integral equa-
> tions has attracted wide attention as much as for its
> inherent interest as for the importance of its applica-
> tions. Several of its results are already classic, and no
> one doubts that in a few years every course in analysis
> will devote a chapter to it.
>
> *M. Plancherel, 1912*

There are two reasons that account for the importance of Hilbert spaces. First, they arise as the natural infinite-dimensional generalizations of Euclidean spaces, and as such, they enjoy the familiar properties of orthogonality, complemented by the important feature of completeness. Second, the theory of Hilbert spaces serves both as a conceptual framework and as a language that formulates some basic arguments in analysis in a more abstract setting.

For us the immediate link with integration theory occurs because of the example of the Lebesgue space $L^2(\mathbb{R}^d)$. The related example of $L^2([-\pi, \pi])$ is what connects Hilbert spaces with Fourier series. The latter Hilbert space can also be used in an elegant way to analyze the boundary behavior of bounded holomorphic functions in the unit disc.

A basic aspect of the theory of Hilbert spaces, as in the familiar finite-dimensional case, is the study of their linear transformations. Given the introductory nature of this chapter, we limit ourselves to rather brief discussions of several classes of such operators: unitary mappings, projections, linear functionals, and compact operators.

1 The Hilbert space L^2

A prime example of a Hilbert space is the collection of **square integrable functions** on $\mathbb{R}^d$, which is denoted by $L^2(\mathbb{R}^d)$, and consists of all complex-valued measurable functions f that satisfy

$$\int_{\mathbb{R}^d} |f(x)|^2 \, dx < \infty.$$

The resulting $L^2(\mathbb{R}^d)$-**norm** of f is defined by

$$\|f\|_{L^2(\mathbb{R}^d)} = \left(\int_{\mathbb{R}^d} |f(x)|^2 \, dx \right)^{1/2}.$$

The reader should compare those definitions with these for the space $L^1(\mathbb{R}^d)$ of integrable functions and its norm that were described in Section 2, Chapter 2. A crucial difference is that L^2 has an inner product, which L^1 does not. Some relative inclusion relations between those spaces are taken up in Exercise 5.

The space $L^2(\mathbb{R}^d)$ is naturally equipped with the following inner product:

$$(f, g) = \int_{\mathbb{R}^d} f(x)\overline{g(x)} \, dx, \quad \text{whenever } f, g \in L^2(\mathbb{R}^d),$$

which is intimately related to the L^2-norm since

$$(f, f)^{1/2} = \|f\|_{L^2(\mathbb{R}^d)}.$$

As in the case of integrable functions, the condition $\|f\|_{L^2(\mathbb{R}^d)} = 0$ only implies $f(x) = 0$ almost everywhere. Therefore, we in fact identify functions that are equal almost everywhere, and define $L^2(\mathbb{R}^d)$ as the space of equivalence classes under this identification. However, in practice it is often convenient to think of elements in $L^2(\mathbb{R}^d)$ as functions, and not as equivalence classes of functions.

For the definition of the inner product (f, g) to be meaningful we need to know that $f\overline{g}$ is integrable on $\mathbb{R}^d$ whenever f and g belong to $L^2(\mathbb{R}^d)$. This and other basic properties of the space of square integrable functions are gathered in the next proposition.

In the rest of this chapter we shall denote the L^2-norm by $\| \cdot \|$ (dropping the subscript $L^2(\mathbb{R}^d)$) unless stated otherwise.

Proposition 1.1 *The space $L^2(\mathbb{R}^d)$ has the following properties:*

(i) *$L^2(\mathbb{R}^d)$ is a vector space.*

(ii) *$f(x)\overline{g(x)}$ is integrable whenever $f, g \in L^2(\mathbb{R}^d)$, and the Cauchy-Schwarz inequality holds: $|(f, g)| \le \|f\| \, \|g\|$.*

(iii) *If $g \in L^2(\mathbb{R}^d)$ is fixed, the map $f \mapsto (f, g)$ is linear in f, and also $(f, g) = \overline{(g, f)}$.*

(iv) *The triangle inequality holds: $\|f + g\| \le \|f\| + \|g\|$.*

Proof. If $f, g \in L^2(\mathbb{R}^d)$, then since $|f(x) + g(x)| \leq 2 \max(|f(x)|, |g(x)|)$, we have

$$|f(x) + g(x)|^2 \leq 4(|f(x)|^2 + |g(x)|^2),$$

therefore

$$\int |f + g|^2 \leq 4 \int |f|^2 + 4 \int |g|^2 < \infty,$$

hence $f + g \in L^2(\mathbb{R}^d)$. Also, if $\lambda \in \mathbb{C}$ we clearly have $\lambda f \in L^2(\mathbb{R}^d)$, and part (i) is proved.

To see why $f\overline{g}$ is integrable whenever f and g are in $L^2(\mathbb{R}^d)$, it suffices to recall that for all $A, B \geq 0$, one has $2AB \leq A^2 + B^2$, so that

(1) $$\int |f\overline{g}| \leq \frac{1}{2} \left[\|f\|^2 + \|g\|^2 \right].$$

To prove the Cauchy-Schwarz inequality, we first observe that if either $\|f\| = 0$ or $\|g\| = 0$, then $fg = 0$ is zero almost everywhere, hence $(f, g) = 0$ and the inequality is obvious. Next, if we assume that $\|f\| = \|g\| = 1$, then we get the desired inequality $|(f, g)| \leq 1$. This follows from the fact that $|(f, g)| \leq \int |f\overline{g}|$, and inequality (1). Finally, in the case when both $\|f\|$ and $\|g\|$ are non-zero, we normalize f and g by setting

$$\tilde{f} = f/\|f\| \quad \text{and} \quad \tilde{g} = g/\|g\|,$$

so that $\|\tilde{f}\| = \|\tilde{g}\| = 1$. By our previous observation we then find

$$|(\tilde{f}, \tilde{g})| \leq 1.$$

Multiplying both sides of the above by $\|f\| \|g\|$ yields the Cauchy-Schwarz inequality.

Part (iii) follows from the linearity of the integral.

Finally, to prove the triangle inequality, we use the Cauchy-Schwarz inequality as follows:

$$\begin{aligned}
\|f + g\|^2 &= (f + g, f + g) \\
&= \|f\|^2 + (f, g) + (g, f) + \|g\|^2 \\
&\leq \|f\|^2 + 2 |(f, g)| + \|g\|^2 \\
&\leq \|f\|^2 + 2 \|f\| \|g\| + \|g\|^2 \\
&= (\|f\| + \|g\|)^2,
\end{aligned}$$

and taking square roots completes the argument.

We turn our attention to the notion of a limit in the space $L^2(\mathbb{R}^d)$. The norm on L^2 induces a metric d as follows: if $f, g \in L^2(\mathbb{R}^d)$, then

$$d(f, g) = \|f - g\|_{L^2(\mathbb{R}^d)}.$$

A sequence $\{f_n\} \subset L^2(\mathbb{R}^d)$ is said to be **Cauchy** if $d(f_n, f_m) \to 0$ as $n, m \to \infty$. Moreover, this sequence converges to $f \in L^2(\mathbb{R}^d)$ if $d(f_n, f) \to 0$ as $n \to \infty$.

Theorem 1.2 *The space $L^2(\mathbb{R}^d)$ is complete in its metric.*

In other words, every Cauchy sequence in $L^2(\mathbb{R}^d)$ converges to a function in $L^2(\mathbb{R}^d)$. This theorem, which is in sharp contrast with the situation for Riemann integrable functions, is a graphic illustration of the usefulness of Lebesgue's theory of integration. We elaborate on this point and its relation to Fourier series in Section 3 below.

Proof. The argument given here follows closely the proof in Chapter 2 that L^1 is complete. Let $\{f_n\}_{n=1}^{\infty}$ be a Cauchy sequence in L^2, and consider a subsequence $\{f_{n_k}\}_{k=1}^{\infty}$ of $\{f_n\}$ with the following property:

$$\|f_{n_{k+1}} - f_{n_k}\| \leq 2^{-k}, \qquad \text{for all } k \geq 1.$$

If we now consider the series whose convergence will be seen below,

$$f(x) = f_{n_1}(x) + \sum_{k=1}^{\infty} (f_{n_{k+1}}(x) - f_{n_k}(x))$$

and

$$g(x) = |f_{n_1}(x)| + \sum_{k=1}^{\infty} |(f_{n_{k+1}}(x) - f_{n_k}(x))|,$$

together the partial sums

$$S_K(f)(x) = f_{n_1}(x) + \sum_{k=1}^{K} (f_{n_{k+1}}(x) - f_{n_k}(x))$$

and

$$S_K(g)(x) = |f_{n_1}(x)| + \sum_{k=1}^{K} |f_{n_{k+1}}(x) - f_{n_k}(x)|,$$

then the triangle inequality implies

$$\|S_K(g)\| \leq \|f_{n_1}\| + \sum_{k=1}^{K} \|f_{n_{k+1}} - f_{n_k}\|$$

$$\leq \|f_{n_1}\| + \sum_{k=1}^{K} 2^{-k}.$$

Letting K tend to infinity, and applying the monotone convergence theorem proves that $\int |g|^2 < \infty$, and since $|f| \leq g$, we must have $f \in L^2(\mathbb{R}^d)$.

In particular, the series defining f converges almost everywhere, and since (by construction of the telescopic series) the $(K-1)^{\text{th}}$ partial sum of this series is precisely f_{n_K}, we find that

$$f_{n_k}(x) \to f(x) \qquad \text{a.e. } x.$$

To prove that $f_{n_k} \to f$ in $L^2(\mathbb{R}^d)$ as well, we simply observe that $|f - S_K(f)|^2 \leq (2g)^2$ for all K, and apply the dominated convergence theorem to get $\|f_{n_k} - f\| \to 0$ as k tends to infinity.

Finally, the last step of the proof consists of recalling that $\{f_n\}$ is Cauchy. Given ϵ, there exists N such that for all $n, m > N$ we have $\|f_n - f_m\| < \epsilon/2$. If n_k is chosen so that $n_k > N$, and $\|f_{n_k} - f\| < \epsilon/2$, then the triangle inequality implies

$$\|f_n - f\| \leq \|f_n - f_{n_k}\| + \|f_{n_k} - f\| < \epsilon$$

whenever $n > N$. This concludes the proof of the theorem.

An additional useful property of $L^2(\mathbb{R}^d)$ is contained in the following theorem.

Theorem 1.3 *The space $L^2(\mathbb{R}^d)$ is **separable**, in the sense that there exists a countable collection $\{f_k\}$ of elements in $L^2(\mathbb{R}^d)$ such that their linear combinations are dense in $L^2(\mathbb{R}^d)$.*

Proof. Consider the family of functions of the form $r\chi_R(x)$, where r is a complex number with rational real and imaginary parts, and R is a rectangle in $\mathbb{R}^d$ with rational coordinates. We claim that finite linear combinations of these type of functions are dense in $L^2(\mathbb{R}^d)$.

Suppose $f \in L^2(\mathbb{R}^d)$ and let $\epsilon > 0$. Consider for each $n \geq 1$ the function g_n defined by

$$g_n(x) = \begin{cases} f(x) & \text{if } |x| \leq n \text{ and } |f(x)| \leq n, \\ 0 & \text{otherwise.} \end{cases}$$

Then $|f - g_n|^2 \leq 4|f|^2$ and $g_n(x) \to f(x)$ almost everywhere.[1] The dominated convergence theorem implies that $\|f - g_n\|^2_{L^2(\mathbb{R}^d)} \to 0$ as n tends to infinity; therefore we have

$$\|f - g_N\|_{L^2(\mathbb{R}^d)} < \epsilon/2 \quad \text{for some } N.$$

Let $g = g_N$, and note that g is a bounded function supported on a bounded set; thus $g \in L^1(\mathbb{R}^d)$. We may now find a step function φ so that $|\varphi| \leq N$ and $\int |g - \varphi| < \epsilon^2/16N$ (Theorem 2.4, Chapter 2). By replacing the coefficients and rectangles that appear in the canonical form of φ by complex numbers with rational real and imaginary parts, and rectangles with rational coordinates, we may find a ψ with $|\psi| \leq N$ and $\int |g - \psi| < \epsilon^2/8N$. Finally, we note that

$$\int |g - \psi|^2 \leq 2N \int |g - \psi| < \epsilon^2/4.$$

Consequently $\|g - \psi\| < \epsilon/2$, therefore $\|f - \psi\| < \epsilon$, and the proof is complete.

The example $L^2(\mathbb{R}^d)$ possesses all the characteristic properties of a Hilbert space, and motivates the definition of the abstract version of this concept.

2 Hilbert spaces

A set $\mathcal{H}$ is a **Hilbert space** if it satisfies the following:

(i) $\mathcal{H}$ is a vector space over $\mathbb{C}$ (or $\mathbb{R}$).[2]

(ii) $\mathcal{H}$ is equipped with an inner product $(\cdot, \cdot)$, so that

- $f \mapsto (f, g)$ is linear on $\mathcal{H}$ for every fixed $g \in \mathcal{H}$,
- $(f, g) = \overline{(g, f)}$,
- $(f, f) \geq 0$ for all $f \in \mathcal{H}$.

We let $\|f\| = (f, f)^{1/2}$.

(iii) $\|f\| = 0$ if and only if $f = 0$.

---

[1] By definition $f \in L^2(\mathbb{R}^d)$ implies that $|f|^2$ is integrable, hence $f(x)$ is finite for a.e x.
[2] At this stage we consider both cases, where the scalar field can be either $\mathbb{C}$ or $\mathbb{R}$. However, in many applications, such as in the context of Fourier analysis, one deals primarily with Hilbert spaces over $\mathbb{C}$.

(iv) The Cauchy-Schwarz and triangle inequalities hold

$$|(f,g)| \leq \|f\| \, \|g\| \quad \text{and} \quad \|f+g\| \leq \|f\| + \|g\|$$

for all $f, g \in \mathcal{H}$.

(v) $\mathcal{H}$ is complete in the metric $d(f,g) = \|f - g\|$.

(vi) $\mathcal{H}$ is separable.

We make two comments about the definition of a Hilbert space. First, the Cauchy-Schwarz and triangle inequalities in (iv) are in fact easy consequences of assumptions (i) and (ii). (See Exercise 1.) Second, we make the requirement that $\mathcal{H}$ be separable because that is the case in most applications encountered. That is not to say that there are no interesting non-separable examples; one such example is described in Problem 2.

Also, we remark that in the context of a Hilbert space we shall often write $\lim_{n\to\infty} f_n = f$ or $f_n \to f$ to mean that $\lim_{n\to\infty} \|f_n - f\| = 0$, which is the same as $d(f_n, f) \to 0$.

We give some examples of Hilbert spaces.

EXAMPLE 1. If E is a measurable subset of $\mathbb{R}^d$ with $m(E) > 0$, we let $L^2(E)$ denote the space of square integrable functions that are supported on E,

$$L^2(E) = \left\{ f \text{ supported on } E, \text{ so that } \int_E |f(x)|^2 \, dx < \infty \right\}.$$

The inner product and norm on $L^2(E)$ are then

$$(f,g) = \int_E f(x)\overline{g(x)} \, dx \quad \text{and} \quad \|f\| = \left(\int_E |f(x)|^2 \, dx \right)^{1/2}.$$

Once again, we consider two elements of $L^2(E)$ to be equivalent if they differ only on a set of measure zero; this guarantees that $\|f\| = 0$ implies $f = 0$. The properties (i) through (vi) follow from these of $L^2(\mathbb{R}^d)$ proved above.

EXAMPLE 2. A simple example is the finite-dimensional complex Euclidean space. Indeed,

$$\mathbb{C}^N = \{(a_1, \ldots, a_N) : a_k \in \mathbb{C}\}$$

becomes a Hilbert space when equipped with the inner product

$$\sum_{k=1}^{N} a_k \overline{b_k},$$

where $a = (a_1, \ldots, a_N)$ and $b = (b_1, \ldots, b_N)$ are in $\mathbb{C}^N$. The norm is then

$$\|a\| = \left(\sum_{k=1}^{N} |a_k|^2 \right)^{1/2}.$$

One can formulate in the same way the real Hilbert space $\mathbb{R}^N$.

EXAMPLE 3. An infinite-dimensional analogue of the above example is the space $\ell^2(\mathbb{Z})$. By definition

$$\ell^2(\mathbb{Z}) = \left\{ (\ldots, a_{-2}, a_{-1}, a_0, a_1, \ldots) : a_i \in \mathbb{C}, \sum_{n=-\infty}^{\infty} |a_n|^2 < \infty \right\}.$$

If we denote infinite sequences by a and b, the inner product and norm on $\ell^2(\mathbb{Z})$ are

$$(a, b) = \sum_{k=-\infty}^{\infty} a_k \overline{b_k} \quad \text{and} \quad \|a\| = \left(\sum_{k=-\infty}^{\infty} |a_k|^2 \right)^{1/2}.$$

We leave the proof that $\ell^2(\mathbb{Z})$ is a Hilbert space as Exercise 4.

While this example is very simple, it will turn out that all infinite-dimensional (separable) Hilbert spaces are $\ell^2(\mathbb{Z})$ in disguise.

Also, a slight variant of this space is $\ell^2(\mathbb{N})$, where we take only one-sided sequences, that is,

$$\ell^2(\mathbb{N}) = \left\{ (a_1, a_2, \ldots) : a_i \in \mathbb{C}, \sum_{n=1}^{\infty} |a_n|^2 < \infty \right\}.$$

The inner product and norm are then defined in the same way with the sums extending from $n = 1$ to ∞.

A characteristic feature of a Hilbert space is the notion of orthogonality. This aspect, with its rich geometric and analytic consequences, distinguishes Hilbert spaces from other normed vector spaces. We now describe some of these properties.

2.1 Orthogonality

Two elements f and g in a Hilbert space $\mathcal{H}$ with inner product $(\cdot, \cdot)$ are **orthogonal** or **perpendicular** if

$$(f, g) = 0, \quad \text{and we then write } f \perp g.$$

The first simple observation is that the usual theorem of Pythagoras holds in the setting of abstract Hilbert spaces:

Proposition 2.1 *If $f \perp g$, then $\|f + g\|^2 = \|f\|^2 + \|g\|^2$.*

Proof. It suffices to note that $(f, g) = 0$ implies $(g, f) = 0$, and therefore

$$\|f + g\|^2 = (f + g, f + g) = \|f\|^2 + (f, g) + (g, f) + \|g\|^2$$
$$= \|f\|^2 + \|g\|^2.$$

A finite or countably infinite subset $\{e_1, e_2, \ldots\}$ of a Hilbert space $\mathcal{H}$ is **orthonormal** if

$$(e_k, e_\ell) = \begin{cases} 1 & \text{when } k = \ell, \\ 0 & \text{when } k \neq \ell. \end{cases}$$

In other words, each e_k has unit norm and is orthogonal to e_ℓ whenever $\ell \neq k$.

Proposition 2.2 *If $\{e_k\}_{k=1}^{\infty}$ is orthonormal, and $f = \sum a_k e_k \in \mathcal{H}$ where the sum is finite, then*

$$\|f\|^2 = \sum |a_k|^2.$$

The proof is a simple application of the Pythagorean theorem.

Given an orthonormal subset $\{e_1, e_2, \ldots\} = \{e_k\}_{k=1}^{\infty}$ of $\mathcal{H}$, a natural problem is to determine whether this subset spans all of $\mathcal{H}$, that is, whether finite linear combinations of elements in $\{e_1, e_2, \ldots\}$ are dense in $\mathcal{H}$. If this is the case, we say that $\{e_k\}_{k=1}^{\infty}$ is an **orthonormal basis** for $\mathcal{H}$. If we are in the presence of an orthonormal basis, we might expect that any $f \in \mathcal{H}$ takes the form

$$f = \sum_{k=1}^{\infty} a_k e_k,$$

for some constants $a_k \in \mathbb{C}$. In fact, taking the inner product of both sides with e_j, and recalling that $\{e_k\}$ is orthonormal yields (formally)

$$(f, e_j) = a_j.$$

This question is motivated by Fourier series. In fact, a good insight into the theorem below is afforded by considering the case where $\mathcal{H}$ is $L^2([-\pi, \pi])$ with inner product $(f, g) = \frac{1}{2\pi} \int_{-\pi}^{\pi} f(x)\overline{g(x)} \, dx$, and the orthonormal set $\{e_k\}_{k=1}^{\infty}$ is merely a relabeling of the exponentials $\{e^{inx}\}_{n=-\infty}^{\infty}$.

Adapting the notation used in Fourier series, we write $f \sim \sum_{k=1}^{\infty} a_k e_k$, where $a_j = (f, e_j)$ for all j.

In the next theorem, we provide four equivalent characterizations that $\{e_k\}$ is an orthonormal basis for $\mathcal{H}$.

Theorem 2.3 *The following properties of an orthonormal set $\{e_k\}_{k=1}^{\infty}$ are equivalent.*

(i) *Finite linear combinations of elements in $\{e_k\}$ are dense in $\mathcal{H}$.*

(ii) *If $f \in \mathcal{H}$ and $(f, e_j) = 0$ for all j, then $f = 0$.*

(iii) *If $f \in \mathcal{H}$, and $S_N(f) = \sum_{k=1}^{N} a_k e_k$, where $a_k = (f, e_k)$, then $S_N(f) \to f$ as $N \to \infty$ in the norm.*

(iv) *If $a_k = (f, e_k)$, then $\|f\|^2 = \sum_{k=1}^{\infty} |a_k|^2$.*

Proof. We prove that each property implies the next, with the last one implying the first.

We begin by assuming (i). Given $f \in \mathcal{H}$ with $(f, e_j) = 0$ for all j, we wish to prove that $f = 0$. By assumption, there exists a sequence $\{g_n\}$ of elements in $\mathcal{H}$ that are finite linear combinations of elements in $\{e_k\}$, and such that $\|f - g_n\|$ tends to 0 as n goes to infinity. Since $(f, e_j) = 0$ for all j, we must have $(f, g_n) = 0$ for all n; therefore an application of the Cauchy-Schwarz inequality gives

$$\|f\|^2 = (f, f) = (f, f - g_n) \leq \|f\| \, \|f - g_n\| \qquad \text{for all } n.$$

Letting $n \to \infty$ proves that $\|f\|^2 = 0$; hence $f = 0$, and (i) implies (ii).

Now suppose that (ii) is verified. For $f \in \mathcal{H}$ we define

$$S_N(f) = \sum_{k=1}^{N} a_k e_k, \qquad \text{where } a_k = (f, e_k),$$

and prove first that $S_N(f)$ converges to some element $g \in \mathcal{H}$. Indeed, one notices that the definition of a_k implies $(f - S_N(f)) \perp S_N(f)$, so the Pythagorean theorem and Proposition 2.2 give

$$(2) \quad \|f\|^2 = \|f - S_N(f)\|^2 + \|S_N(f)\|^2 = \|f - S_N(f)\|^2 + \sum_{k=1}^{N} |a_k|^2.$$

Hence $\|f\|^2 \geq \sum_{k=1}^{N} |a_k|^2$, and letting N tend to infinity we obtain **Bessel's inequality**

$$\sum_{k=1}^{\infty} |a_k|^2 \leq \|f\|^2,$$

which implies that the series $\sum_{k=1}^{\infty} |a_k|^2$ converges. Therefore, $\{S_N(f)\}_{N=1}^{\infty}$ forms a Cauchy sequence in $\mathcal{H}$ since

$$\|S_N(f) - S_M(f)\|^2 = \sum_{k=M+1}^{N} |a_k|^2 \quad \text{whenever } N > M.$$

Since $\mathcal{H}$ is complete, there exists $g \in \mathcal{H}$ such that $S_N(f) \to g$ as N tends to infinity.

Fix j, and note that for all sufficiently large N, $(f - S_N(f), e_j) = a_j - a_j = 0$. Since $S_N(f)$ tends to g, we conclude that

$$(f - g, e_j) = 0 \quad \text{for all } j.$$

Hence $f = g$ by assumption (ii), and we have proved that $f = \sum_{k=1}^{\infty} a_k e_k$.

Now assume that (iii) holds. Observe from (2) that we immediately get in the limit as N goes to infinity

$$\|f\|^2 = \sum_{k=1}^{\infty} |a_k|^2.$$

Finally, if (iv) holds, then again from (2) we see that $\|f - S_N(f)\|$ converges to 0. Since each $S_N(f)$ is a finite linear combination of elements in $\{e_k\}$, we have completed the circle of implications, and the theorem is proved.

In particular, a closer look at the proof shows that Bessel's inequality holds for any orthonormal family $\{e_k\}$. In contrast, the identity

$$\|f\|^2 = \sum_{k=1}^{\infty} |a_k|^2, \quad \text{where } a_k = (f, e_k),$$

which is called **Parseval's identity**, holds if and only if $\{e_k\}_{k=1}^{\infty}$ is also an orthonormal basis.

Now we turn our attention to the existence of a basis.

Theorem 2.4 *Any Hilbert space has an orthonormal basis.*

The first step in the proof of this fact is to recall that (by definition) a Hilbert space $\mathcal{H}$ is separable. Hence, we may choose a countable collection of elements $\mathcal{F} = \{h_k\}$ in $\mathcal{H}$ so that finite linear combinations of elements in $\mathcal{F}$ are dense in $\mathcal{H}$.

We start by recalling a definition already used in the case of finite-dimensional vector spaces. Finitely many elements $g_1, \ldots, g_N$ are said to be **linearly independent** if whenever

$$a_1 g_1 + \cdots + a_N g_N = 0 \quad \text{for some complex numbers } a_i,$$

then $a_1 = a_2 = \cdots = a_N = 0$. In other words, no element g_j is a linear combination of the others. In particular, we note that none of the g_j can be 0. We say that a countable family of elements is **linearly independent** if all finite subsets of this family are linearly independent.

If we next successively disregard the elements h_k that are linearly dependent on the previous elements $h_1, h_2, \ldots, h_{k-1}$, then the resulting collection $h_1 = f_1, f_2, \ldots, f_k, \ldots$ consists of linearly independent elements, whose finite linear combinations are the same as those given by $h_1, h_2, \ldots, h_k, \ldots$, and hence these linear combinations are also dense in $\mathcal{H}$.

The proof of the theorem now follows from an application of a familiar construction called the **Gram-Schmidt process**. Given a finite family of elements $\{f_1, \ldots, f_k\}$ we call the **span** of this family the set of all elements which are finite linear combinations of the elements $\{f_1, \ldots, f_k\}$. We denote the span of $\{f_1, \ldots, f_k\}$ by $\mathrm{Span}(\{f_1, \ldots, f_k\})$.

We now construct a sequence of orthonormal vectors $e_1, e_2, \ldots$ such that $\mathrm{Span}(\{e_1, \ldots, e_n\}) = \mathrm{Span}(\{f_1, \ldots, f_n\})$ for all $n \geq 1$. We do this by induction.

By the linear independence hypothesis, $f_1 \neq 0$, so we may take $e_1 = f_1 / \|f_1\|$. Next, assume that orthonormal vectors $e_1, \ldots, e_k$ have been found such that $\mathrm{Span}(\{e_1, \ldots, e_k\}) = \mathrm{Span}(\{f_1, \ldots, f_k\})$ for a given k. We then try e'_{k+1} as $f_{k+1} + \sum_{j=1}^{k} a_j e_j$. To have $(e'_{k+1}, e_j) = 0$ requires that $a_j = -(f_{k+1}, e_j)$, and this choice of a_j for $1 \leq j \leq k$ assures that e'_{k+1} is orthogonal to $e_1, \ldots, e_k$. Moreover our linear independence hypothesis assures that $e'_{k+1} \neq 0$; hence we need only "renormalize" and take $e_{k+1} = e'_{k+1}/\|e'_{k+1}\|$ to complete the inductive step. With this we have found an orthonormal basis for $\mathcal{H}$

Note that we have implicitly assumed that the number of linearly independent elements $f_1, f_2, \ldots$ is infinite. In the case where there are only N linearly independent vectors $f_1, \ldots, f_N$, then $e_1, \ldots, e_N$ constructed in the same way also provide an orthonormal basis for $\mathcal{H}$. These two cases are differentiated in the following definition. If $\mathcal{H}$ is a Hilbert space with an orthonormal basis consisting of finitely many elements, then we say that $\mathcal{H}$ is **finite-dimensional**. Otherwise $\mathcal{H}$ is said to be **infinite-dimensional**.

2.2 Unitary mappings

A correspondence between two Hilbert spaces that preserves their structure is a unitary transformation. More precisely, suppose we are given two Hilbert spaces $\mathcal{H}$ and $\mathcal{H}'$ with respective inner products $(\cdot, \cdot)_{\mathcal{H}}$ and $(\cdot, \cdot)_{\mathcal{H}'}$, and the corresponding norms $\| \cdot \|_{\mathcal{H}}$ and $\| \cdot \|_{\mathcal{H}'}$. A mapping $U : \mathcal{H} \to \mathcal{H}'$ between these space is called **unitary** if:

(i) U is linear, that is, $U(\alpha f + \beta g) = \alpha U(f) + \beta U(g)$.

(ii) U is a bijection.

(iii) $\|Uf\|_{\mathcal{H}'} = \|f\|_{\mathcal{H}}$ for all $f \in \mathcal{H}$.

Some observations are in order. First, since U is bijective it must have an inverse $U^{-1} : \mathcal{H}' \to \mathcal{H}$ that is also unitary. Part (iii) above also implies that if U is unitary, then

$$(Uf, Ug)_{\mathcal{H}'} = (f, g)_{\mathcal{H}} \quad \text{for all } f, g \in \mathcal{H}.$$

To see this, it suffices to "polarize," that is, to note that for any vector space (say over $\mathbb{C}$) with inner product $(\cdot, \cdot)$ and norm $\| \cdot \|$, we have

$$(F, G) = \frac{1}{4} \left[\|F + G\|^2 - \|F - G\|^2 + i \left(\|\frac{F}{i} + G\|^2 - \|\frac{F}{i} - G\|^2 \right) \right]$$

whenever F and G are elements of the space.

The above leads us to say that the two Hilbert spaces $\mathcal{H}$ and $\mathcal{H}'$ are **unitarily equivalent** or **unitarily isomorphic** if there exists a unitary mapping $U : \mathcal{H} \to \mathcal{H}'$. Clearly, unitary isomorphism of Hilbert spaces is an equivalence relation.

With this definition we are now in a position to give precise meaning to the statement we made earlier that all infinite-dimensional Hilbert spaces are the same and in that sense $\ell^2(\mathbb{Z})$ in disguise.

Corollary 2.5 *Any two infinite-dimensional Hilbert spaces are unitarily equivalent.*

Proof. If $\mathcal{H}$ and $\mathcal{H}'$ are two infinite-dimensional Hilbert spaces, we may select for each an orthonormal basis, say

$$\{e_1, e_2, \ldots\} \subset \mathcal{H} \quad \text{and} \quad \{e_1', e_2', \ldots\} \subset \mathcal{H}'.$$

Then, consider the mapping defined as follows: if $f = \sum_{k=1}^{\infty} a_k e_k$, then

$$U(f) = g, \quad \text{where} \quad g = \sum_{k=1}^{\infty} a_k e_k'.$$

Clearly, the mapping U is both linear and invertible. Moreover, by Parseval's identity, we must have

$$\|Uf\|_{\mathcal{H}'}^2 = \|g\|_{\mathcal{H}'}^2 = \sum_{k=1}^{\infty} |a_k|^2 = \|f\|_{\mathcal{H}}^2,$$

and the corollary is proved.

Consequently, all infinite-dimensional Hilbert spaces are unitarily equivalent to $\ell^2(\mathbb{N})$, and thus, by relabeling, to $\ell^2(\mathbb{Z})$. By similar reasoning we also have the following:

Corollary 2.6 *Any two finite-dimensional Hilbert spaces are unitarily equivalent if and only if they have the same dimension.*

Thus every finite-dimensional Hilbert space over $\mathbb{C}$ (or over $\mathbb{R}$) is equivalent with $\mathbb{C}^d$ (or $\mathbb{R}^d$), for some d.

2.3 Pre-Hilbert spaces

Although Hilbert spaces arise naturally, one often starts with a **pre-Hilbert space** instead, that is, a space $\mathcal{H}_0$ that satisfies all the defining properties of a Hilbert space except (v); in other words $\mathcal{H}_0$ is not assumed to be complete. A prime example arose implicitly early in the study of Fourier series with the space $\mathcal{H}_0 = \mathcal{R}$ of Riemann integrable functions on $[-\pi, \pi]$ with the usual inner product; we return to this below. Other examples appear in the next chapter in the study of the solutions of partial differential equations.

Fortunately, every pre-Hilbert space $\mathcal{H}_0$ can be completed.

Proposition 2.7 *Suppose we are given a pre-Hilbert space $\mathcal{H}_0$ with inner product $(\cdot,\cdot)_0$. Then we can find a Hilbert space $\mathcal{H}$ with inner product $(\cdot,\cdot)$ such that*

(i) $\mathcal{H}_0 \subset \mathcal{H}$.

(ii) $(f,g)_0 = (f,g)$ whenever $f, g \in \mathcal{H}_0$.

(iii) $\mathcal{H}_0$ is dense in $\mathcal{H}$.

A Hilbert space satisfying properties like $\mathcal{H}$ in the above proposition is called a **completion** of $\mathcal{H}_0$. We shall only sketch the construction of $\mathcal{H}$, since it follows closely Cantor's familiar method of obtaining the real numbers as the completion of the rationals in terms of Cauchy sequences of rationals.

Indeed, consider the collection of all Cauchy sequences $\{f_n\}$ with $f_n \in \mathcal{H}_0$, $1 \leq n < \infty$. One defines an equivalence relation in this collection by saying that $\{f_n\}$ is equivalent to $\{f'_n\}$ if $f_n - f'_n$ converges to 0 as $n \to \infty$. The collection of equivalence classes is then taken to be $\mathcal{H}$. One then easily verifies that $\mathcal{H}$ inherits the structure of a vector space, with an inner product (f,g) defined as $\lim_{n\to\infty}(f_n, g_n)$, where $\{f_n\}$ and $\{g_n\}$ are Cauchy sequences in $\mathcal{H}_0$, representing, respectively, the elements f and g in $\mathcal{H}$. Next, if $f \in \mathcal{H}_0$ we take the sequence $\{f_n\}$, with $f_n = f$ for all n, to represent f as an element of $\mathcal{H}$, giving $\mathcal{H}_0 \subset \mathcal{H}$. To see that $\mathcal{H}$ is complete, let $\{F^k\}_{k=1}^{\infty}$ be a Cauchy sequence in $\mathcal{H}$, with each F^k represented by $\{f_n^k\}_{n=1}^{\infty}$, $f_n^k \in \mathcal{H}_0$. If we define $F \in \mathcal{H}$ as represented by the sequence $\{f_n\}$ with $f_n = f_{N(n)}^n$, where $N(n)$ is so that $|f_{N(n)}^n - f_j^n| \leq 1/n$ for $j \geq N(n)$, then we note that $F^k \to F$ in $\mathcal{H}$.

One can also observe that the completion $\mathcal{H}$ of $\mathcal{H}_0$ is unique up to isomorphism. (See Exercise 14.)

3 Fourier series and Fatou's theorem

We have already seen an interesting relation between Hilbert spaces and some elementary facts about Fourier series. Here we want to pursue this idea and also connect it with complex analysis.

When considering Fourier series, it is natural to begin by turning to the broader class of all integrable functions on $[-\pi, \pi]$. Indeed, note that $L^2([-\pi,\pi]) \subset L^1([-\pi,\pi])$, by the Cauchy-Schwarz inequality, since the interval $[-\pi, \pi]$ has finite measure. Thus, if $f \in L^1([-\pi,\pi])$ and $n \in \mathbb{Z}$, we define the nth **Fourier coefficient** of f by

$$a_n = \frac{1}{2\pi} \int_{-\pi}^{\pi} f(x)e^{-inx}\, dx.$$

The **Fourier series** of f is then formally $\sum_{n=-\infty}^{\infty} a_n e^{inx}$, and we write

$$f(x) \sim \sum_{n=-\infty}^{\infty} a_n e^{inx}$$

to indicate that the sum on the right is the Fourier series of the function on the left. The theory developed thus far provides the natural generalization of some earlier results obtained in Book I.

Theorem 3.1 *Suppose f is integrable on $[-\pi, \pi]$.*

(i) *If $a_n = 0$ for all n, then $f(x) = 0$ for a.e. x.*

(ii) *$\sum_{n=-\infty}^{\infty} a_n r^{|n|} e^{inx}$ tends to $f(x)$ for a.e. x, as $r \to 1$, $r < 1$.*

The second conclusion is the almost everywhere "Abel summability" to f of its Fourier series. Note that since $|a_n| \leq \frac{1}{2\pi} \int_{-\pi}^{\pi} |f(x)| \, dx$, the series $\sum a_n r^{|n|} e^{inx}$ converges absolutely and uniformly for each r, $0 \leq r < 1$.

Proof. The first conclusion is an immediate consequence of the second. To prove the latter we recall the identity

$$\sum_{n=-\infty}^{\infty} r^{|n|} e^{iny} = P_r(y) = \frac{1 - r^2}{1 - 2r \cos y + r^2}$$

for the Poisson kernel; see Book I, Chapter 2. Starting with our given $f \in L^1([-\pi, \pi])$ we extend it as a function on $\mathbb{R}$ by making it periodic of period 2π.[3] We then claim that for every x

$$(3) \qquad \sum_{n=-\infty}^{\infty} a_n r^{|n|} e^{inx} = \frac{1}{2\pi} \int_{-\pi}^{\pi} f(x - y) P_r(y) \, dy.$$

Indeed, by the dominated convergence theorem the right-hand side equals

$$\sum r^{|n|} \frac{1}{2\pi} \int_{-\pi}^{\pi} f(x - y) e^{iny} \, dy.$$

Moreover, for each x and n

$$\int_{-\pi}^{\pi} f(x - y) e^{iny} \, dy = \int_{-\pi+x}^{\pi+x} f(y) e^{in(x-y)} \, dy$$

$$= e^{inx} \int_{-\pi}^{\pi} f(y) e^{-iny} \, dy = e^{inx} 2\pi a_n.$$

---

[3] Note that we may without loss of generality assume that $f(\pi) = f(-\pi)$ so as to make the periodic extension unambiguous.

The first equality follows by translation invariance (see Section 3, Chapter 2), and the second since $\int_{-\pi}^{\pi} F(y)\,dy = \int_I F(y)\,dy$ whenever F is periodic of period 2π and I is an interval of length 2π (Exercise 3, Chapter 2). With these observations, the identity (3) is established. We can now invoke the facts about approximations to the identity (Theorem 2.1 and Example 4, Chapter 3) to conclude that the left-hand side of (3) tends to $f(x)$ at every point of the Lebesgue set of f, hence almost everywhere. (To be correct, the hypotheses of the theorem require that f be integrable on all of $\mathbb{R}$. We can achieve this for our periodic function by setting f equal to zero outside $[-2\pi, 2\pi]$, and then (3) still holds for this modified f, whenever $x \in [-\pi, \pi]$.)

We return to the more restrictive setting of L^2. We express the essential conclusions of Theorem 2.3 in the context of Fourier series. With $f \in L^2([-\pi, \pi])$, we write as before $a_n = \frac{1}{2\pi} \int_{-\pi}^{\pi} f(x)e^{-inx}\,dx$.

Theorem 3.2 *Suppose $f \in L^2([-\pi, \pi])$. Then:*

(i) *We have Parseval's relation*

$$\sum_{n=-\infty}^{\infty} |a_n|^2 = \frac{1}{2\pi} \int_{-\pi}^{\pi} |f(x)|^2\,dx.$$

(ii) *The mapping $f \mapsto \{a_n\}$ is a unitary correspondence between $L^2([-\pi, \pi])$ and $\ell^2(\mathbb{Z})$.*

(iii) *The Fourier series of f converges to f in the L^2-norm, that is,*

$$\frac{1}{2\pi} \int_{-\pi}^{\pi} |f(x) - S_N(f)(x)|^2\,dx \to 0 \qquad as\ N \to \infty,$$

where $S_N(f) = \sum_{|n| \leq N} a_n e^{inx}$.

To apply the previous results, we let $\mathcal{H} = L^2([-\pi, \pi])$ with inner product $(f, g) = \frac{1}{2\pi} \int_{-\pi}^{\pi} f(x)\overline{g(x)}\,dx$, and take the orthonormal set $\{e_k\}_{k=1}^{\infty}$ to be the exponentials $\{e^{inx}\}_{n=-\infty}^{\infty}$, with $k = 1$ when $n = 0$, $k = 2n$ for $n > 0$, and $k = 2|n| - 1$ for $n < 0$.

By the previous result, assertion (ii) of Theorem 2.3 holds and thus all the other conclusions hold. We therefore have Parseval's relation, and from (iv) we conclude that $\|f - S_N(f)\|^2 = \sum_{|n|>N} |a_n|^2 \to 0$ as $N \to \infty$. Similarly, if $\{a_n\} \in \ell^2(\mathbb{Z})$ is given, then $\|S_N(f) - S_M(f)\|^2 \to 0$, as $N, M \to \infty$. Hence the completeness of L^2 guarantees that there is an $f \in L^2$ such that $\|f - S_N(f)\| \to 0$, and one verifies directly that f

has $\{a_n\}$ as its Fourier coefficients. Thus we deduce that the mapping $f \mapsto \{a_n\}$ is onto and hence unitary. This is a key conclusion that holds in the setting on L^2 and was not valid in an earlier context of Riemann integrable functions. In fact the space $\mathcal{R}$ of such functions on $[-\pi, \pi]$ is not complete in the norm, containing as it does the continuous functions, but $\mathcal{R}$ is itself restricted to bounded functions.

3.1 Fatou's theorem

Fatou's theorem is a remarkable result in complex analysis. Its proof combines elements of Hilbert spaces, Fourier series, and deeper ideas of differentiation theory, and yet none of these notions appear in its statement. The question that Fatou's theorem answers may be put simply as follows.

> Suppose $F(z)$ is holomorphic in the unit disc $\mathbb{D} = \{z \in \mathbb{C} : |z| < 1\}$. What are conditions on F that guarantee that $F(z)$ will converge, in an appropriate sense, to boundary values $F(e^{i\theta})$ on the unit circle?

In general a holomorphic function in the unit disc can behave quite erratically near the boundary. It turns out, however, that imposing a simple boundedness condition is enough to obtain a strong conclusion.

If F is a function defined in the unit disc $\mathbb{D}$, we say that F has a **radial limit** at the point $-\pi \le \theta \le \pi$ on the circle, if the limit

$$\lim_{\substack{r \to 1 \\ r < 1}} F(re^{i\theta})$$

exists.

Theorem 3.3 *A bounded holomorphic function $F(re^{i\theta})$ on the unit disc has radial limits at almost every θ.*

Proof. We know that $F(z)$ has a power series expansion $\sum_{n=0}^{\infty} a_n z^n$ in $\mathbb{D}$ that converges absolutely and uniformly whenever $z = re^{i\theta}$ and $r < 1$. In fact, for $r < 1$ the series $\sum_{n=0}^{\infty} a_n r^n e^{in\theta}$ is the Fourier series of the function $F(re^{i\theta})$, that is,

$$a_n r^n = \frac{1}{2\pi} \int_{-\pi}^{\pi} F(re^{i\theta}) e^{-in\theta} \, d\theta \quad \text{when } n \ge 0,$$

and the integral vanishes when $n < 0$. (See also Chapter 3, Section 7 in Book II).

We pick M so that $|F(z)| \leq M$, for all $z \in \mathbb{D}$. By Parseval's identity

$$\sum_{n=0}^{\infty} |a_n|^2 r^{2n} = \frac{1}{2\pi} \int_{-\pi}^{\pi} |F(re^{i\theta})|^2 d\theta \quad \text{for each } 0 \leq r < 1.$$

Letting $r \to 1$ one sees that $\sum |a_n|^2$ converges (and is $\leq M^2$). We now let $F(e^{i\theta})$ be the L^2-function whose Fourier coefficients are a_n when $n \geq 0$, and 0 when $n < 0$. Hence by conclusion (ii) in Theorem 3.1

$$\sum_{n=0}^{\infty} a_n r^n e^{in\theta} \to F(e^{i\theta}), \quad \text{for a.e } \theta,$$

concluding the proof of the theorem.

If we examine the argument given above we see that the same conclusion holds for a larger class of functions. In this connection, we define the **Hardy space** $H^2(\mathbb{D})$ to consist of all holomorphic functions F on the unit disc $\mathbb{D}$ that satisfy

$$\sup_{0 \leq r < 1} \frac{1}{2\pi} \int_{-\pi}^{\pi} |F(re^{i\theta})|^2 \, d\theta < \infty.$$

We also define the "norm" for functions F in this class, $\|F\|_{H^2(\mathbb{D})}$, to be the square root of the above quantity.

One notes that if F is bounded, then $F \in H^2(\mathbb{D})$, and moreover the conclusion of the existence of radial limits almost everywhere holds for any $F \in H^2(\mathbb{D})$, by the same argument given for the bounded case.[4] Finally, one notes that $F \in H^2(\mathbb{D})$ if and only if $F(z) = \sum_{n=0}^{\infty} a_n z^n$ with $\sum_{n=0}^{\infty} |a_n|^2 < \infty$; moreover, $\sum_{n=0}^{\infty} |a_n|^2 = \|F\|_{H^2(\mathbb{D})}^2$. This states in particular that $H^2(\mathbb{D})$ is in fact a Hilbert space that can be viewed as the "subspace" $\ell^2(\mathbb{Z}^+)$ of $\ell^2(\mathbb{Z})$, consisting of all $\{a_n\} \in \ell^2(\mathbb{Z})$, with $a_n = 0$ when $n < 0$.

Some general considerations of subspaces and their concomitant orthogonal projections will be taken up next.

4 Closed subspaces and orthogonal projections

A **linear subspace** S (or simply subspace) of $\mathcal{H}$ is a subset of $\mathcal{H}$ that satisfies $\alpha f + \beta g \in S$ whenever $f, g \in S$ and α, β are scalars. In other words, S is also a vector space. For example in $\mathbb{R}^3$, lines passing through

---

[4] An even more general statement is given in Problem 5*.

the origin and planes passing through the origin are the one-dimensional and two-dimensional subspaces, respectively.

The subspace S is **closed** if whenever $\{f_n\} \subset S$ converges to some $f \in \mathcal{H}$, then f also belongs to S. In the case of finite-dimensional Hilbert spaces, every subspace is closed. This is, however, not true in the general case of infinite-dimensional Hilbert spaces. For instance, as we have already indicated, the subspace of Riemann integrable functions in $L^2([-\pi, \pi])$ is not closed, nor is the subspace obtained by fixing a basis and taking all vectors that are finite linear combinations of these basis elements. It is useful to note that every closed subspace S of $\mathcal{H}$ is itself a Hilbert space, with the inner product on S that which is inherited from $\mathcal{H}$. (For the separability of S, see Exercise 11.)

Next, we show that a closed subspace enjoys an important characteristic property of Euclidean geometry.

Lemma 4.1 *Suppose S is a closed subspace of $\mathcal{H}$ and $f \in \mathcal{H}$. Then:*

(i) *There exists a (unique) element $g_0 \in S$ which is closest to f, in the sense that*

$$\|f - g_0\| = \inf_{g \in S} \|f - g\|.$$

(ii) *The element $f - g_0$ is perpendicular to S, that is,*

$$(f - g_0, g) = 0 \quad \text{for all } g \in S.$$

The situation in the lemma can be visualized as in Figure 1.

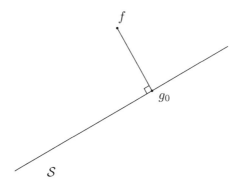

Figure 1. Nearest element to f in S

Proof. If $f \in \mathcal{S}$, then we choose $f = g_0$, and there is nothing left to prove. Otherwise, we let $d = \inf_{g \in \mathcal{S}} \|f - g\|$, and note that we must have $d > 0$ since $f \notin \mathcal{S}$ and $\mathcal{S}$ is closed. Consider a sequence $\{g_n\}_{n=1}^{\infty}$ in $\mathcal{S}$ such that

$$\|f - g_n\| \to d \quad \text{as } n \to \infty.$$

We claim that $\{g_n\}$ is a Cauchy sequence whose limit will be the desired element g_0. In fact, it would suffice to show that a subsequence of $\{g_n\}$ converges, and this is immediate in the finite-dimensional case because a closed ball is compact. However, in general this compactness fails, as we shall see in Section 6, and so a more intricate argument is needed at this point.

To prove our claim, we use the **parallelogram law**, which states that in a Hilbert space $\mathcal{H}$

$$(4) \quad \|A + B\|^2 + \|A - B\|^2 = 2 \left[\|A\|^2 + \|B\|^2\right] \quad \text{for all } A, B \in \mathcal{H}.$$

The simple verification of this equality, which consists of writing each norm in terms of the inner product, is left to the reader. Putting $A = f - g_n$ and $B = f - g_m$ in the parallelogram law, we find

$$\|2f - (g_n + g_m)\|^2 + \|g_m - g_n\|^2 = 2 \left[\|f - g_n\|^2 + \|f - g_m\|^2\right].$$

However $\mathcal{S}$ is a subspace, so the quantity $\frac{1}{2}(g_n + g_m)$ belongs to $\mathcal{S}$, hence

$$\|2f - (g_n + g_m)\| = 2\|f - \frac{1}{2}(g_n + g_m)\| \geq 2d.$$

Therefore

$$\|g_m - g_n\|^2 = 2 \left[\|f - g_n\|^2 + \|f - g_m\|^2\right] - \|2f - (g_n + g_m)\|^2$$
$$\leq 2 \left[\|f - g_n\|^2 + \|f - g_m\|^2\right] - 4d^2.$$

By construction, we know that $\|f - g_n\| \to d$ and $\|f - g_m\| \to d$ as $n, m \to \infty$, so the above inequality implies that $\{g_n\}$ is a Cauchy sequence. Since $\mathcal{H}$ is complete and $\mathcal{S}$ closed, the sequence $\{g_n\}$ must have a limit g_0 in $\mathcal{S}$, and then it satisfies $d = \|f - g_0\|$.

We prove that if $g \in \mathcal{S}$, then $g \perp (f - g_0)$. For each ϵ (positive or negative), consider the perturbation of g_0 defined by $g_0 - \epsilon g$. This element belongs to $\mathcal{S}$, hence

$$\|f - (g_0 - \epsilon g)\|^2 \geq \|f - g_0\|^2.$$

Since $\|f - (g_0 - \epsilon g)\|^2 = \|f - g_0\|^2 + \epsilon^2\|g\|^2 + 2\epsilon\,\mathrm{Re}(f - g_0, g)$, we find that

$$(5) \qquad\qquad 2\epsilon\,\mathrm{Re}(f - g_0, g) + \epsilon^2\|g\|^2 \geq 0.$$

If $\mathrm{Re}(f - g_0, g) < 0$, then taking ϵ small and positive contradicts (5). If $\mathrm{Re}(f - g_0, g) > 0$, a contradiction also follows by taking ϵ small and negative. Thus $\mathrm{Re}(f - g_0, g) = 0$. By considering the perturbation $g_0 - i\epsilon g$, a similar argument gives $\mathrm{Im}(f - g_0, g) = 0$, and hence $(f - g_0, g) = 0$.

Finally, the uniqueness of g_0 follows from the above observation about orthogonality. Suppose $\tilde{g}_0$ is another point in $\mathcal{S}$ that minimizes the distance to f. By taking $g = g_0 - \tilde{g}_0$ in our last argument we find $(f - g_0) \perp (g_0 - \tilde{g}_0)$, and the Pythagorean theorem gives

$$\|f - \tilde{g}_0\|^2 = \|f - g_0\|^2 + \|g_0 - \tilde{g}_0\|^2.$$

Since by assumption $\|f - \tilde{g}_0\|^2 = \|f - g_0\|^2$, we conclude that $\|g_0 - \tilde{g}_0\| = 0$, as desired.

Using the lemma, we may now introduce a useful concept that is another expression of the notion of orthogonality. If $\mathcal{S}$ is a subspace of a Hilbert space $\mathcal{H}$, we define the **orthogonal complement** of $\mathcal{S}$ by

$$\mathcal{S}^\perp = \{f \in \mathcal{H} : (f, g) = 0 \quad \text{for all } g \in \mathcal{S}\}.$$

Clearly, $\mathcal{S}^\perp$ is also a subspace of $\mathcal{H}$, and moreover $\mathcal{S} \cap \mathcal{S}^\perp = \{0\}$. To see this, note that if $f \in \mathcal{S} \cap \mathcal{S}^\perp$, then f must be orthogonal to itself; thus $0 = (f, f) = \|f\|$, and therefore $f = 0$. Moreover, $\mathcal{S}^\perp$ is itself a closed subspace. Indeed, if $f_n \to f$, then $(f_n, g) \to (f, g)$ for every g, by the Cauchy-Schwarz inequality. Hence if $(f_n, g) = 0$ for all $g \in \mathcal{S}$ and all n, then $(f, g) = 0$ for all those g.

Proposition 4.2 *If $\mathcal{S}$ is a closed subspace of a Hilbert space $\mathcal{H}$, then*

$$\mathcal{H} = \mathcal{S} \oplus \mathcal{S}^\perp.$$

The notation in the proposition means that every $f \in \mathcal{H}$ can be written uniquely as $f = g + h$, where $g \in \mathcal{S}$ and $h \in \mathcal{S}^\perp$; we say that $\mathcal{H}$ is the **direct sum** of $\mathcal{S}$ and $\mathcal{S}^\perp$. This is equivalent to saying that any f in $\mathcal{H}$ is the sum of two elements, one in $\mathcal{S}$, the other in $\mathcal{S}^\perp$, and that $\mathcal{S} \cap \mathcal{S}^\perp$ contains only 0.

The proof of the proposition relies on the previous lemma giving the closest element of f in $\mathcal{S}$. In fact, for any $f \in \mathcal{H}$, we choose g_0 as in the lemma and write

$$f = g_0 + (f - g_0).$$

By construction $g_0 \in \mathcal{S}$, and the lemma implies $f - g_0 \in \mathcal{S}^{\perp}$, and this shows that f is the sum of an element in $\mathcal{S}$ and one in $\mathcal{S}^{\perp}$. To prove that this decomposition is unique, suppose that

$$f = g + h = \tilde{g} + \tilde{h} \quad \text{where } g, \tilde{g} \in \mathcal{S} \text{ and } h, \tilde{h} \in \mathcal{S}^{\perp}.$$

Then, we must have $g - \tilde{g} = \tilde{h} - h$. Since the left-hand side belongs to $\mathcal{S}$ while the right-hand side belongs to $\mathcal{S}^{\perp}$ the fact that $\mathcal{S} \cap \mathcal{S}^{\perp} = \{0\}$ implies $g - \tilde{g} = 0$ and $\tilde{h} - h = 0$. Therefore $g = \tilde{g}$ and $h = \tilde{h}$ and the uniqueness is established.

With the decomposition $\mathcal{H} = \mathcal{S} \oplus \mathcal{S}^{\perp}$ one has the natural projection onto $\mathcal{S}$ defined by

$$P_{\mathcal{S}}(f) = g, \quad \text{where } f = g + h \text{ and } g \in \mathcal{S}, h \in \mathcal{S}^{\perp}.$$

The mapping $P_{\mathcal{S}}$ is called the **orthogonal projection** onto $\mathcal{S}$ and satisfies the following simple properties:

(i) $f \mapsto P_{\mathcal{S}}(f)$ is linear,

(ii) $P_{\mathcal{S}}(f) = f$ whenever $f \in \mathcal{S}$,

(iii) $P_{\mathcal{S}}(f) = 0$ whenever $f \in \mathcal{S}^{\perp}$,

(iv) $\|P_{\mathcal{S}}(f)\| \leq \|f\|$ for all $f \in \mathcal{H}$.

Property (i) means that $P_{\mathcal{S}}(\alpha f_1 + \beta f_2) = \alpha P_{\mathcal{S}}(f_1) + \beta P_{\mathcal{S}}(f_2)$, whenever $f_1, f_2 \in \mathcal{H}$ and α and β are scalars.

It will be useful to observe the following. Suppose $\{e_k\}$ is a (finite or infinite) collection of orthonormal vectors in $\mathcal{H}$. Then the orthogonal projection P in the closure of the subspace spanned by $\{e_k\}$ is given by $P(f) = \sum_k (f, e_k) e_k$. In case the collection is infinite, the sum converges in the norm of $\mathcal{H}$.

We illustrate this with two examples that arise in Fourier analysis.

EXAMPLE 1. On $L^2([-\pi, \pi])$, recall that if $f(\theta) \sim \sum_{n=-\infty}^{\infty} a_n e^{in\theta}$ then the partial sums of the Fourier series are

$$S_N(f)(\theta) = \sum_{n=-N}^{N} a_n e^{in\theta}.$$

Therefore, the partial sum operator S_N consists of the projection onto the closed subspace spanned by $\{e_{-N}, \ldots, e_N\}$.

The sum S_N can be realized as a convolution

$$S_N(f)(\theta) = \frac{1}{2\pi} \int_{-\pi}^{\pi} D_N(\theta - \varphi) f(\varphi) \, d\varphi,$$

where $D_N(\theta) = \sin((N + 1/2)\theta)/\sin(\theta/2)$ is the **Dirichlet kernel**.

EXAMPLE 2. Once again, consider $L^2([-\pi, \pi])$ and let $\mathcal{S}$ denote the subspace that consists of all $F \in L^2([-\pi, \pi])$ with

$$F(\theta) \sim \sum_{n=0}^{\infty} a_n e^{in\theta}.$$

In other words, $\mathcal{S}$ is the space of square integrable functions whose Fourier coefficients a_n vanish for $n < 0$. From the proof of Fatou's theorem, this implies that $\mathcal{S}$ can be identified with the Hardy space $H^2(\mathbb{D})$, where $\mathbb{D}$ is the unit disc, and so is a closed subspace unitarily isomorphic to $\ell^2(\mathbb{Z}^+)$. Therefore, using this identification, if P denotes the orthogonal projection from $L^2([-\pi, \pi])$ to $\mathcal{S}$, we may also write $P(f)(z)$ for the element corresponding to $H^2(\mathbb{D})$, that is,

$$P(f)(z) = \sum_{n=0}^{\infty} a_n z^n.$$

Given $f \in L^2([-\pi, \pi])$, we define the **Cauchy integral** of f by

$$C(f)(z) = \frac{1}{2\pi i} \int_{\gamma} \frac{f(\zeta)}{\zeta - z} \, d\zeta,$$

where γ denotes the unit circle and z belongs to the unit disc. Then we have the identity

$$P(f)(z) = C(f)(z), \quad \text{for all } z \in \mathbb{D}.$$

Indeed, since $f \in L^2$ it follows by the Cauchy-Schwarz inequality that $f \in L^1([-\pi, \pi])$, and therefore we may interchange the sum and integral in the following calculation (recall $|z| < 1$):

$$P(f)(z) = \sum_{n=0}^{\infty} a_n z^n = \sum_{n=0}^{\infty} \left(\frac{1}{2\pi} \int_{-\pi}^{\pi} f(e^{i\theta}) e^{-in\theta} d\theta \right) z^n$$

$$= \frac{1}{2\pi} \int_{-\pi}^{\pi} f(e^{i\theta}) \sum_{n=0}^{\infty} (e^{-i\theta} z)^n d\theta$$

$$= \frac{1}{2\pi} \int_{-\pi}^{\pi} \frac{f(e^{i\theta})}{1 - e^{-i\theta} z} d\theta$$

$$= \frac{1}{2\pi i} \int_{-\pi}^{\pi} \frac{f(e^{i\theta})}{e^{i\theta} - z} i e^{i\theta} d\theta$$

$$= C(f)(z).$$

5 Linear transformations

The focus of analysis in Hilbert spaces is largely the study of their linear transformations. We have already encountered two classes of such transformations, the unitary mappings and the orthogonal projections. There are two other important classes we shall deal with in this chapter in some detail: the "linear functionals" and the "compact operators," and in particular those that are symmetric.

Suppose $\mathcal{H}_1$ and $\mathcal{H}_2$ are two Hilbert spaces. A mapping $T : \mathcal{H}_1 \to \mathcal{H}_2$ is a **linear transformation** (also called **linear operator** or **operator**) if

$$T(af + bg) = aT(f) + bT(g) \quad \text{for all scalars } a, b \text{ and } f, g \in \mathcal{H}_1.$$

Clearly, linear operators satisfy $T(0) = 0$.

We shall say that a linear operator $T : \mathcal{H}_1 \to \mathcal{H}_2$ is **bounded** if there exists $M > 0$ so that

$$(6) \qquad \qquad \|T(f)\|_{\mathcal{H}_2} \leq M \|f\|_{\mathcal{H}_1}.$$

The **norm** of T is denoted by $\|T\|_{\mathcal{H}_1 \to \mathcal{H}_2}$ or simply $\|T\|$ and defined by

$$\|T\| = \inf M,$$

where the infimum is taken over all M so that (6) holds. A trivial example is given by the **identity operator** I, with $I(f) = f$. It is of course a unitary operator and a projection, with $\|I\| = 1$.

In what follows we shall generally drop the subscripts attached to the norms of elements of a Hilbert space, when this causes no confusion.

Lemma 5.1 $\|T\| = \sup\{|(Tf, g)| : \|f\| \leq 1, \|g\| \leq 1\}$, *where of course* $f \in \mathcal{H}_1$ *and* $g \in \mathcal{H}_2$.

Proof. If $\|T\| \leq M$, the Cauchy-Schwarz inequality gives

$$|(Tf, g)| \leq M \qquad \text{whenever } \|f\| \leq 1 \text{ and } \|g\| \leq 1;$$

thus $\sup\{|(Tf, g)| : \|f\| \leq 1, \|g\| \leq 1\} \leq \|T\|$.

Conversely, if $\sup\{|(Tf, g)| : \|f\| \leq 1, \|g\| \leq 1\} \leq M$, we claim that $\|Tf\| \leq M\|f\|$ for all f. If f or Tf is zero, there is nothing to prove. Otherwise, $f' = f/\|f\|$ and $g' = Tf/\|Tf\|$ have norm 1, so by assumption

$$|(Tf', g')| \leq M.$$

But since $|(Tf', g')| = \|Tf\|/\|f\|$ this gives $\|Tf\| \leq M\|f\|$, and the lemma is proved.

A linear transformation T is **continuous** if $T(f_n) \to T(f)$ whenever $f_n \to f$. Clearly, linearity implies that T is continuous on all of $\mathcal{H}_1$ if and only if it is continuous at the origin. In fact, the conditions of being bounded or continuous are equivalent.

Proposition 5.2 *A linear operator* $T : \mathcal{H}_1 \to \mathcal{H}_2$ *is bounded if and only if it is continuous.*

Proof. If T is bounded, then $\|T(f) - T(f_n)\|_{\mathcal{H}_2} \leq M\|f - f_n\|_{\mathcal{H}_1}$, hence T is continuous. Conversely, suppose that T is continuous but not bounded. Then for each n there exists $f_n \neq 0$ such that $\|T(f_n)\| \geq n\|f_n\|$. The element $g_n = f_n/(n\|f_n\|)$ has norm $1/n$, hence $g_n \to 0$. Since T is continuous at 0, we must have $T(g_n) \to 0$, which contradicts the fact that $\|T(g_n)\| \geq 1$. This proves the proposition.

In the rest of this chapter we shall assume that all linear operators are bounded, hence continuous. It is noteworthy to recall that any linear operator between finite-dimensional Hilbert spaces is necessarily continuous.

5.1 Linear functionals and the Riesz representation theorem

A **linear functional** ℓ is a linear transformation from a Hilbert space $\mathcal{H}$ to the underlying field of scalars, which we may assume to be the complex numbers,

$$\ell : \mathcal{H} \to \mathbb{C}.$$

Of course, we view $\mathbb{C}$ as a Hilbert space equipped with its standard norm, the absolute value.

A natural example of a linear functional is provided by the inner product on $\mathcal{H}$. Indeed, for fixed $g \in \mathcal{H}$, the map

$$\ell(f) = (f, g)$$

is linear, and also bounded by the Cauchy-Schwarz inequality. Indeed,

$$|(f, g)| \leq M\|f\|, \quad \text{where } M = \|g\|.$$

Moreover, $\ell(g) = M\|g\|$ so we have $\|\ell\| = \|g\|$. The remarkable fact is that this example is exhaustive, in the sense that every *continuous* linear functional on a Hilbert space arises as an inner product. This is the so-called Riesz representation theorem.

Theorem 5.3 *Let ℓ be a continuous linear functional on a Hilbert space $\mathcal{H}$. Then, there exists a unique $g \in \mathcal{H}$ such that*

$$\ell(f) = (f, g) \quad \text{for all } f \in \mathcal{H}.$$

Moreover, $\|\ell\| = \|g\|$.

Proof. Consider the subspace of $\mathcal{H}$ defined by

$$\mathcal{S} = \{f \in \mathcal{H} : \ell(f) = 0\}.$$

Since ℓ is continuous the subspace $\mathcal{S}$, which is called the **null-space** of ℓ, is closed. If $\mathcal{S} = \mathcal{H}$, then $\ell = 0$ and we take $g = 0$. Otherwise $\mathcal{S}^\perp$ is non-trivial and we may pick any $h \in \mathcal{S}^\perp$ with $\|h\| = 1$. With this choice of h we determine g by setting $g = \overline{\ell(h)}h$. Thus if we let $u = \ell(f)h - \ell(h)f$, then $u \in \mathcal{S}$, and therefore $(u, h) = 0$. Hence

$$0 = (\ell(f)h - \ell(h)f, h) = \ell(f)(h, h) - (f, \overline{\ell(h)}h).$$

Since $(h, h) = 1$, we find that $\ell(f) = (f, g)$ as desired.

At this stage we record the following remark for later use. Let $\mathcal{H}_0$ be a pre-Hilbert space whose completion is $\mathcal{H}$. Suppose ℓ_0 is a linear functional on $\mathcal{H}_0$ which is bounded, that is, $|\ell_0(f)| \leq M\|f\|$ for all $f \in$

$\mathcal{H}_0$. Then ℓ_0 has an extension ℓ to a bounded linear functional on $\mathcal{H}$, with $|\ell(f)| \leq M\|f\|$ for all $f \in \mathcal{H}$. This extension is also unique. To see this, one merely notes that $\{\ell_0(f_n)\}$ is a Cauchy sequence whenever the vectors $\{f_n\}$ belong to $\mathcal{H}_0$, and $f_n \to f$ in $\mathcal{H}$, as $n \to \infty$. Thus we may define $\ell(f)$ as $\lim_{n\to\infty} \ell_0(f_n)$. The verification of the asserted properties of ℓ is then immediate. (This result is a special case of the extension Lemma 1.3 in the next chapter.)

5.2 Adjoints

The first application of the Riesz representation theorem is to determine the existence of the "adjoint" of a linear transformation.

Proposition 5.4 *Let $T : \mathcal{H} \to \mathcal{H}$ be a bounded linear transformation. There exists a unique bounded linear transformation T^* on $\mathcal{H}$ so that:*

(i) $(Tf, g) = (f, T^*g)$,

(ii) $\|T\| = \|T^*\|$,

(iii) $(T^*)^* = T$.

The linear operator $T^* : \mathcal{H} \to \mathcal{H}$ satisfying the above conditions is called the **adjoint** of T.

To prove the existence of an operator satisfying (i) above, we observe that for each fixed $g \in \mathcal{H}$, the linear functional $\ell = \ell_g$, defined by

$$\ell(f) = (Tf, g),$$

is bounded. Indeed, since T is bounded one has $\|Tf\| \leq M\|f\|$; hence the Cauchy-Schwarz inequality implies that

$$|\ell(f)| \leq \|Tf\| \, \|g\| \leq B\|f\|,$$

where $B = M\|g\|$. Consequently, the Riesz representation theorem guarantees the existence of a unique $h \in \mathcal{H}$, $h = h_g$, such that

$$\ell(f) = (f, h).$$

Then we define $T^*g = h$, and note that the association $T^* : g \mapsto h$ is linear and satisfies (i).

The fact that $\|T\| = \|T^*\|$ follows at once from (i) and Lemma 5.1:

$$\begin{aligned} \|T\| &= \sup\{|(Tf, g)| : \|f\| \leq 1, \|g\| \leq 1\} \\ &= \sup\{|(f, T^*g)| : \|f\| \leq 1, \|g\| \leq 1\} \\ &= \|T^*\|. \end{aligned}$$

To prove (iii), note that $(Tf, g) = (f, T^*g)$ for all f and g if and only if $(T^*f, g) = (f, Tg)$ for all f and g, as one can see by taking complex conjugates and reversing the roles of f and g.

We record here a few additional remarks.

(a) In the special case when $T = T^*$ (we say that T is **symmetric**), then

(7) $$\|T\| = \sup\{|(Tf, f)| : \|f\| = 1\}.$$

This should be compared to Lemma 5.1, which holds for any linear operator. To establish (7), let $M = \sup\{|(Tf, f)| : \|f\| = 1\}$. By Lemma 5.1 it is clear that $M \leq \|T\|$. Conversely, if f and g belong on $\mathcal{H}$, then one has the following "polarization" identity which is easy to verify

$$(Tf, g) = \frac{1}{4}[(T(f + g), f + g) - (T(f - g), f - g) \\ + i(T(f + ig), f + ig) - i(T(f - ig), f - ig)].$$

For any $h \in \mathcal{H}$, the quantity (Th, h) is real, because $T = T^*$, hence $(Th, h) = (h, T^*h) = (h, Th) = \overline{(Th, h)}$. Consequently

$$\mathrm{Re}(Tf, g) = \frac{1}{4}[(T(f + g), f + g) - (T(f - g), f - g)].$$

Now $|(Th, h)| \leq M\|h\|^2$, so $|\mathrm{Re}(Tf, g)| \leq \frac{M}{4}[\|f + g\|^2 + \|f - g\|^2]$, and an application of the parallelogram law (4) then implies

$$|\mathrm{Re}(Tf, g)| \leq \frac{M}{2}[\|f\|^2 + \|g\|^2].$$

So if $\|f\| \leq 1$ and $\|g\| \leq 1$, then $|\mathrm{Re}(Tf, g)| \leq M$. In general, we may replace g by $e^{i\theta}g$ in the last inequality to find that whenever $\|f\| \leq 1$ and $\|g\| \leq 1$, then $|(Tf, g)| \leq M$, and invoking Lemma 5.1 once again gives the result, $\|T\| \leq M$.

(b) Let us note that if T and S are bounded linear transformations of $\mathcal{H}$ to itself, then so is their product TS, defined by $(TS)(f) = T(S(f))$. Moreover we have automatically $(TS)^* = S^*T^*$; in fact, $(TSf, g) = (Sf, T^*g) = (f, S^*T^*g)$.

(c) One can also exhibit a natural connection between linear transformations on a Hilbert space and their associated bilinear forms. Suppose first that T is a bounded operator in $\mathcal{H}$. Define the corresponding bilinear form B by

(8) $$B(f, g) = (Tf, g).$$

Note that B is linear in f and conjugate linear in g. Also by the Cauchy-Schwarz inequality $|B(f,g)| \leq M\|f\|\,\|g\|$, where $M = \|T\|$. Conversely if B is linear in f, conjugate linear in g and satisfies $|B(f,g)| \leq M\|f\|\,\|g\|$, there is a unique linear transformation so that (8) holds with $M = \|T\|$. This can be proved by the argument of Proposition 5.4; the details are left to the reader.

5.3 Examples

Having presented the elementary facts about Hilbert spaces, we now digress to describe briefly the background of some of the early developments of the theory. A motivating problem of considerable interest was that of the study of the "eigenfunction expansion" of a differential operator L. A particular case, that of a Sturm-Liouville operator, arises on an interval $[a, b]$ of $\mathbb{R}$ with L defined by

$$L = \frac{d^2}{dx^2} - q(x),$$

where q is a given real-valued function. The question is then that of expanding an "arbitrary" function in terms of the eigenfunctions φ, that is those functions that satisfy $L(\varphi) = \mu\varphi$ for some $\mu \in \mathbb{R}$. The classical example of this is that of Fourier series, where $L = d^2/dx^2$ on the interval $[-\pi, \pi]$ with each exponential e^{inx} an eigenfunction of L with eigenvalue $\mu = -n^2$.

When made precise in the "regular" case, the problem for L can be resolved by considering an associated "integral operator" T defined on $L^2([a, b])$ by

$$T(f)(x) = \int_a^b K(x, y) f(y)\, dy,$$

with the property that for suitable f,

$$LT(f) = f.$$

It turns out that a key feature that makes the study of T tractable is a certain compactness it enjoys. We now pass to the definitions and elaboration of some of these ideas, and begin by giving two relevant illustrations of classes of operators on Hilbert spaces.

Infinite diagonal matrix

Suppose $\{\varphi_k\}_{k=1}^\infty$ is an orthonormal basis of $\mathcal{H}$. Then, a linear transformation $T : \mathcal{H} \to \mathcal{H}$ is said to be **diagonalized** with respect to the basis

$\{\varphi_k\}$ if

$$T(\varphi_k) = \lambda_k \varphi_k, \qquad \text{where } \lambda_k \in \mathbb{C} \text{ for all } k.$$

In general, a *non-zero* element φ is called an **eigenvector** of T with **eigenvalue** λ if $T\varphi = \lambda\varphi$. So the φ_k above are eigenvectors of T, and the numbers λ_k are the corresponding eigenvalues.

So if

$$f \sim \sum_{k=1}^{\infty} a_k \varphi_k \qquad \text{then} \qquad Tf \sim \sum_{k=1}^{\infty} a_k \lambda_k \varphi_k.$$

The sequence $\{\lambda_k\}$ is called the **multiplier sequence** corresponding to T.

In this case, one can easily verify the following facts:

- $\|T\| = \sup_k |\lambda_k|$.

- T^* corresponds to the sequence $\{\overline{\lambda}_k\}$; hence $T = T^*$ if and only if the λ_k are real.

- T is unitary if and only if $|\lambda_k| = 1$ for all k.

- T is an orthogonal projection if and only if $\lambda_k = 0$ or 1 for all k.

As a particular example, consider $\mathcal{H} = L^2([-\pi, \pi])$, and assume that every $f \in L^2([-\pi, \pi])$ is extended to $\mathbb{R}$ by periodicity, so that $f(x + 2\pi) = f(x)$ for all $x \in \mathbb{R}$. Let $\varphi_k(x) = e^{ikx}$ for $k \in \mathbb{Z}$. For a fixed $h \in \mathbb{R}$ the operator U_h defined by

$$U_h(f)(x) = f(x + h)$$

is unitary with $\lambda_k = e^{ikh}$. Hence

$$U_h(f) \sim \sum_{k=-\infty}^{\infty} a_k \lambda_k e^{ikx} \qquad \text{if} \qquad f \sim \sum_{k=-\infty}^{\infty} a_k e^{ikx}.$$

Integral operators, and in particular, Hilbert-Schmidt operators

Let $\mathcal{H} = L^2(\mathbb{R}^d)$. If we can define an operator $T : \mathcal{H} \to \mathcal{H}$ by the formula

$$T(f)(x) = \int_{\mathbb{R}^d} K(x, y) f(y) \, dy \qquad \text{whenever } f \in L^2(\mathbb{R}^d),$$

we say that the operator T is an **integral operator** and K is its associated **kernel**.

In fact, it was the problem of invertibility related to such operators, and more precisely the question of solvability of the equation $f - Tf = g$ for given g, that initiated the study of Hilbert spaces. These equations were then called "integral equations."

In general a bounded linear transformation cannot be expressed as an (absolutely convergent) integral operator. However, there is an interesting class for which this is possible and which has a number of other worthwhile properties: **Hilbert-Schmidt operators**, those with a kernel K that belongs to $L^2(\mathbb{R}^d \times \mathbb{R}^d)$.

Proposition 5.5 *Let T be a Hilbert-Schmidt operator on $L^2(\mathbb{R}^d)$ with kernel K.*

(i) *If $f \in L^2(\mathbb{R}^d)$, then for almost every x the function $y \mapsto K(x,y)f(y)$ is integrable.*

(ii) *The operator T is bounded from $L^2(\mathbb{R}^d)$ to itself, and*

$$\|T\| \leq \|K\|_{L^2(\mathbb{R}^d \times \mathbb{R}^d)},$$

where $\|K\|_{L^2(\mathbb{R}^d \times \mathbb{R}^d)}$ is the L^2-norm of K on $\mathbb{R}^d \times \mathbb{R}^d = \mathbb{R}^{2d}$.

(iii) *The adjoint T^* has kernel $\overline{K(y,x)}$.*

Proof. By Fubini's theorem we know that for almost every x, the function $y \mapsto |K(x,y)|^2$ is integrable. Then, part (i) follows directly from an application of the Cauchy-Schwarz inequality.

For (ii), we make use again of the Cauchy-Schwarz inequality as follows

$$\left| \int K(x,y)f(y)\,dy \right| \leq \int |K(x,y)||f(y)|\,dy$$

$$\leq \left(\int |K(x,y)|^2\,dy \right)^{1/2} \left(\int |f(y)|^2\,dy \right)^{1/2}.$$

Therefore, squaring this and integrating in x yields

$$\|Tf\|_{L^2(\mathbb{R}^d)}^2 \leq \int \left(\int |K(x,y)|^2 dy \int |f(y)|^2 dy \right) dx$$

$$= \|K\|_{L^2(\mathbb{R}^d \times \mathbb{R}^d)}^2 \|f\|_{L^2(\mathbb{R}^d)}^2.$$

Finally, part (iii) follows by writing out (Tf, g) in terms of a double integral, and then interchanging the order of integration, as is permissible by Fubini's theorem.

Hilbert-Schmidt operators can be defined analogously for the Hilbert space $L^2(E)$, where E is a measurable subset of $\mathbb{R}^d$. We leave it to the reader to formulate an prove the analogue of Proposition 5.5 that holds in this case.

Hilbert-Schmidt operators enjoy another important property: they are compact. We will now discuss this feature in more detail.

6 Compact operators

We shall use the notion of sequential compactness in a Hilbert space $\mathcal{H}$: a set $X \subset \mathcal{H}$ is **compact** if for every sequence $\{f_n\}$ in X, there exists a subsequence $\{f_{n_k}\}$ that converges in the norm to an element *in* X.

Let $\mathcal{H}$ denote a Hilbert space, and B the closed unit ball in $\mathcal{H}$,

$$B = \{f \in \mathcal{H} : \|f\| \leq 1\}.$$

A well-known result in elementary real analysis says that in a finite-dimensional Euclidean space, a closed and bounded set is compact. However, this does not carry over to the infinite-dimensional case. The fact is that in this case the unit ball, while closed and bounded, is not compact. To see this, consider the sequence $\{f_n\} = \{e_n\}$, where the e_n are orthonormal. By the Pythagorean theorem, $\|e_n - e_m\|^2 = 2$ if $n \neq m$, so no subsequence of the $\{e_n\}$ can converge.

In the infinite-dimensional case we say that a linear operator $T : \mathcal{H} \to \mathcal{H}$ is **compact** if the closure of

$$T(B) = \{g \in \mathcal{H} : g = T(f) \text{ for some } f \in B\}$$

is a compact set. Equivalently, an operator T is compact if, whenever $\{f_k\}$ is a bounded sequence in $\mathcal{H}$, there exists a subsequence $\{f_{n_k}\}$ so that Tf_{n_k} converges. Note that a compact operator is automatically bounded.

Note that by what has been said, a linear transformation is in general not compact (take for instance the identity operator!). However, if T is of **finite rank**, which means that its range is finite-dimensional, then it is automatically compact. It turns out that dealing with compact operators provides us with the closest analogy to the usual theorems of (finite-dimensional) linear algebra. Some relevant analytic properties of compact operators are given by the proposition below.

Proposition 6.1 *Suppose T is a bounded linear operator on $\mathcal{H}$.*

(i) *If S is compact on $\mathcal{H}$, then ST and TS are also compact.*

(ii) *If $\{T_n\}$ is a family of compact linear operators with $\|T_n - T\| \to 0$ as n tends to infinity, then T is compact.*

(iii) *Conversely, if T is compact, there is a sequence $\{T_n\}$ of operators of finite rank such that $\|T_n - T\| \to 0$.*

(iv) *T is compact if and only if T^* is compact.*

Proof. Part (i) is immediate. For part (ii) we use a diagonalization argument. Suppose $\{f_k\}$ is a bounded sequence in $\mathcal{H}$. Since T_1 is compact, we may extract a subsequence $\{f_{1,k}\}_{k=1}^{\infty}$ of $\{f_k\}$ such that $T_1(f_{1,k})$ converges. From $\{f_{1,k}\}$ we may find a subsequence $\{f_{2,k}\}_{k=1}^{\infty}$ such that $T_2(f_{2,k})$ converges, and so on. If we let $g_k = f_{k,k}$, then we claim $\{T(g_k)\}$ is a Cauchy sequence. We have

$$\|T(g_k) - T(g_\ell)\| \leq \|T(g_k) - T_m(g_k)\| + \|T_m(g_k) - T_m(g_\ell)\| + $$
$$+ \|T_m(g_\ell) - T(g_\ell)\|.$$

Since $\|T - T_m\| \to 0$ and $\{g_k\}$ is bounded, we can make the first and last term each $< \epsilon/3$ for some large m independent of k and ℓ. With this fixed m, we note that by construction $\|T_m(g_k) - T_m(g_\ell)\| < \epsilon/3$ for all large k and ℓ. This proves our claim; hence $\{T(g_k)\}$ converges in $\mathcal{H}$.

To prove (iii) let $\{e_k\}_{k=1}^{\infty}$ be a basis of $\mathcal{H}$ and let Q_n be the orthogonal projection on the subspace spanned by the e_k with $k > n$. Then clearly $Q_n(g) \sim \sum_{k>n} a_k e_k$ whenever $g \sim \sum_{k=1}^{\infty} a_k e_k$, and $\|Q_n g\|^2$ is a decreasing sequence that tends to 0 as $n \to \infty$ for any $g \in \mathcal{H}$. We claim that $\|Q_n T\| \to 0$ as $n \to \infty$. If not, there is a $c > 0$ so that $\|Q_n T\| \geq c$, and hence for each n we can find f_n, with $\|f_n\| = 1$ so that $\|Q_n T f_n\| \geq c$. Now by compactness of T, choosing an appropriate subsequence $\{f_{n_k}\}$, we have $T f_{n_k} \to g$ for some g. But $Q_{n_k}(g) = Q_{n_k} T f_{n_k} - Q_{n_k}(T f_{n_k} - g)$, and hence we conclude that $\|Q_{n_k}(g)\| \geq c/2$, for large k. This contradiction shows that $\|Q_n T\| \to 0$. So if P_n is the complementary projection on the finite-dimensional space spanned by $e_1, \ldots, e_n$, $I = P_n + Q_n$, then $\|Q_n T\| \to 0$ means that $\|P_n T - T\| \to 0$. Since each $P_n T$ is of finite rank, assertion (iii) is established.

Finally, if T is compact the fact that $\|P_n T - T\| \to 0$ implies $\|T^* P_n - T^*\| \to 0$, and clearly $T^* P_n$ is again of finite rank. Thus we need only appeal to the second conclusion to prove the last.

We now state two further observations about compact operators.

- If T can be diagonalized with respect to some basis $\{\varphi_k\}$ of eigenvectors and corresponding eigenvalues $\{\lambda_k\}$, then T is compact if and only if $|\lambda_k| \to 0$. See Exercise 25.

- Every Hilbert-Schmidt operator is compact.

To prove the second point, recall that a Hilbert-Schmidt operator is given on $L^2(\mathbb{R}^d)$ by

$$T(f)(x) = \int_{\mathbb{R}^d} K(x,y) f(y)\, dy, \quad \text{where } K \in L^2(\mathbb{R}^d \times \mathbb{R}^d).$$

If $\{\varphi_k\}_{k=1}^\infty$ denotes an orthonormal basis for $L^2(\mathbb{R}^d)$, then the collection $\{\varphi_k(x)\varphi_\ell(y)\}_{k,\ell \geq 1}$ is an orthonormal basis for $L^2(\mathbb{R}^d \times \mathbb{R}^d)$; the proof of this simple fact is outlined in Exercise 7. As a result

$$K(x,y) \sim \sum_{k,\ell=1}^\infty a_{k\ell} \varphi_k(x)\varphi_\ell(y), \quad \text{with } \sum_{k,\ell} |a_{k\ell}|^2 < \infty.$$

We define an operator

$$T_n f(x) = \int_{\mathbb{R}^d} K_n(x,y) f(y) dy, \quad \text{where } K_n(x,y) = \sum_{k,\ell=1}^n a_{k\ell}\varphi_k(x)\varphi_\ell(y).$$

Then, each T_n has finite-dimensional range, hence is compact. Moreover,

$$\|K - K_n\|_{L^2(\mathbb{R}^d \times \mathbb{R}^d)}^2 = \sum_{k \geq n \text{ or } \ell \geq n} |a_{k\ell}|^2 \to 0 \quad \text{as } n \to \infty.$$

By Proposition 5.5, $\|T - T_n\| \leq \|K - K_n\|_{L^2(\mathbb{R}^d \times \mathbb{R}^d)}$, so we can conclude the proof that T is compact by appealing to Proposition 6.1.

The climax of our efforts regarding compact operators is the infinite-dimensional version of the familiar diagonalization theorem in linear algebra for symmetric matrices. Using a similar terminology, we say that a bounded linear operator T is **symmetric** if $T^* = T$. (These operators are also called "self-adjoint" or "Hermitian.")

Theorem 6.2 (Spectral theorem) *Suppose T is a compact symmetric operator on a Hilbert space $\mathcal{H}$. Then there exists an (orthonormal) basis $\{\varphi_k\}_{k=1}^\infty$ of $\mathcal{H}$ that consists of eigenvectors of T. Moreover, if*

$$T\varphi_k = \lambda_k \varphi_k,$$

then $\lambda_k \in \mathbb{R}$ and $\lambda_k \to 0$ as $k \to \infty$.

Conversely, every operator of the above form is compact and symmetric.
The collection $\{\lambda_k\}$ is called the **spectrum** of T.

Lemma 6.3 *Suppose T is a bounded symmetric linear operator on a Hilbert space $\mathcal{H}$.*

(i) *If λ is an eigenvalue of T, then λ is real.*

(ii) *If f_1 and f_2 are eigenvectors corresponding to two distinct eigenvalues, then f_1 and f_2 are orthogonal.*

Proof. To prove (i), we first choose a non-zero eigenvector f such that $T(f) = \lambda f$. Since T is symmetric (that is, $T = T^*$), we find that

$$\lambda(f, f) = (Tf, f) = (f, Tf) = (f, \lambda f) = \overline{\lambda}(f, f),$$

where we have used in the last equality the fact that the inner product is conjugate linear in the second variable. Since $f \neq 0$, we must have $\lambda = \overline{\lambda}$ and hence $\lambda \in \mathbb{R}$.

For (ii), suppose f_1 and f_2 have eigenvalues λ_1 and λ_2, respectively. By the previous argument both λ_1 and λ_2 are real, and we note that

$$\begin{aligned}
\lambda_1(f_1, f_2) &= (\lambda_1 f_1, f_2) \\
&= (Tf_1, f_2) \\
&= (f_1, Tf_2) \\
&= (f_1, \lambda_2 f_2) \\
&= \lambda_2(f_1, f_2).
\end{aligned}$$

Since by assumption $\lambda_1 \neq \lambda_2$ we must have $(f_1, f_2) = 0$ as desired.

For the next lemma note that every non-zero element of the null-space of $T - \lambda I$ is an eigenvector with eigenvalue λ.

Lemma 6.4 *Suppose T is compact, and $\lambda \neq 0$. Then the dimension of the null space of $T - \lambda I$ is finite. Moreover, the eigenvalues of T form at most a denumerable set $\lambda_1, \ldots, \lambda_k, \ldots$, with $\lambda_k \to 0$ as $k \to \infty$. More specifically, for each $\mu > 0$, the linear space spanned by the eigenvectors corresponding to the eigenvalues λ_k with $|\lambda_k| > \mu$ is finite-dimensional.*

Proof. Let V_λ denote the null-space of $T - \lambda I$, that is, the eigenspace of T corresponding to λ. If V_λ is not finite-dimensional, there exists a countable sequence of orthonormal vectors $\{\varphi_k\}$ in V_λ. Since T is compact, there exists a subsequence $\{\varphi_{n_k}\}$ such that $T(\varphi_{n_k})$ converges.

But since $T(\varphi_{n_k}) = \lambda \varphi_{n_k}$ and $\lambda \neq 0$, we conclude that φ_{n_k} converges, which is a contradiction since $\|\varphi_{n_k} - \varphi_{n_{k'}}\|^2 = 2$ if $k \neq k'$.

The rest of the lemma follows if we can show that for each $\mu > 0$, there are only finitely many eigenvalues whose absolute values are greater than μ. We argue again by contradiction. Suppose there are infinitely many distinct eigenvalues whose absolute values are greater than μ, and let $\{\varphi_k\}$ be a corresponding sequence of eigenvectors. Since the eigenvalues are distinct, we know from the previous lemma that $\{\varphi_k\}$ is orthogonal, and after normalization, we may assume that this set of eigenvectors is orthonormal. One again, since T is compact, we may find a subsequence so that $T(\varphi_{n_k})$ converges, and since

$$T(\varphi_{n_k}) = \lambda_{n_k} \varphi_{n_k}$$

the fact that $|\lambda_{n_k}| > \mu$ leads to a contradiction, since $\{\varphi_k\}$ is an orthonormal set and thus $\|\lambda_{n_k} \varphi_{n_k} - \lambda_{n_j} \varphi_{n_j}\|^2 = \lambda_{n_k}^2 + \lambda_{n_j}^2 \geq 2\mu^2$.

Lemma 6.5 *Suppose $T \neq 0$ is compact and symmetric. Then either $\|T\|$ or $-\|T\|$ is an eigenvalue of T.*

Proof. By the observation (7) made earlier, either

$$\|T\| = \sup\{(Tf, f) : \|f\| = 1\} \quad \text{or} \quad -\|T\| = \inf\{(Tf, f) : \|f\| = 1\}.$$

We assume the first case, that is,

$$\lambda = \|T\| = \sup\{(Tf, f) : \|f\| = 1\},$$

and prove that λ is an eigenvalue of T. (The proof of the other case is similar.)

We pick a sequence $\{f_n\} \subset \mathcal{H}$ such that $\|f_n\| = 1$ and $(Tf_n, f_n) \to \lambda$. Since T is compact, we may assume also (by passing to a subsequence of $\{f_n\}$ if necessary) that $\{Tf_n\}$ converges to a limit $g \in \mathcal{H}$. We claim that g is an eigenvector of T with eigenvalue λ. To see this, we first observe that $Tf_n - \lambda f_n \to 0$ because

$$\begin{aligned}
\|Tf_n - \lambda f_n\|^2 &= \|Tf_n\|^2 - 2\lambda(Tf_n, f_n) + \lambda^2 \|f_n\|^2 \\
&\leq \|T\|^2 \|f_n\|^2 - 2\lambda(Tf_n, f_n) + \lambda^2 \|f_n\|^2 \\
&\leq 2\lambda^2 - 2\lambda(Tf_n, f_n) \to 0.
\end{aligned}$$

Since $Tf_n \to g$, we must have $\lambda f_n \to g$, and since T is continuous, this implies that $\lambda Tf_n \to Tg$. This proves that $\lambda g = Tg$. Finally, we must have $g \neq 0$, for otherwise $\|T_n f_n\| \to 0$, hence $(Tf_n, f_n) \to 0$, and $\lambda = \|T\| = 0$, which is a contradiction.

We are now equipped with the necessary tools to prove the spectral theorem. Let $\mathcal{S}$ denote the closure of the linear space spanned by all eigenvectors of T. By Lemma 6.5, the space $\mathcal{S}$ is non-empty. The goal is to prove that $\mathcal{S} = \mathcal{H}$. If not, then since

$$(9) \qquad\qquad\qquad \mathcal{S} \oplus \mathcal{S}^\perp = \mathcal{H},$$

$\mathcal{S}^\perp$ would be non-empty. We will have reached a contradiction once we show that $\mathcal{S}^\perp$ contains an eigenvector of T. First, we note that T respects the decomposition (9). In other words, if $f \in \mathcal{S}$ then $Tf \in \mathcal{S}$, which follows from the definitions. Also, if $g \in \mathcal{S}^\perp$ then $Tg \in \mathcal{S}^\perp$. This is because T is symmetric and maps $\mathcal{S}$ to itself, and hence

$$(Tg, f) = (g, Tf) = 0 \quad \text{whenever } g \in \mathcal{S}^\perp \text{ and } f \in \mathcal{S}.$$

Now consider the operator T_1, which by definition is the restriction of T to the subspace $\mathcal{S}^\perp$. The closed subspace $\mathcal{S}^\perp$ inherits its Hilbert space structure from $\mathcal{H}$. We see immediately that T_1 is also a compact and symmetric operator on this Hilbert space. Moreover, if $\mathcal{S}^\perp$ is non-empty, the lemma implies that T_1 has a non-zero eigenvector in $\mathcal{S}^\perp$. This eigenvector is clearly also an eigenvector of T, and therefore a contradiction is obtained. This concludes the proof of the spectral theorem.

Some comments about Theorem 6.2 are in order. If in its statement we drop either of the two assumptions (the compactness or symmetry of T), then T may have no eigenvectors. (See Exercises 32 and 33.) However, when T is a general bounded linear transformation which is symmetric, there is an appropriate extension of the spectral theorem that holds for it. Its formulation and proof require further ideas that are deferred to Chapter 6.

7 Exercises

1. Show that properties (i) and (ii) in the definition of a Hilbert space (Section 2) imply property (iii): the Cauchy-Schwarz inequality $|(f, g)| \leq \|f\| \cdot \|g\|$ and the triangle inequality $\|f + g\| \leq \|f\| + \|g\|$.

[Hint: For the first inequality, consider $(f + \lambda g, f + \lambda g)$ as a positive quadratic function of λ. For the second, write $\|f + g\|^2$ as $(f + g, f + g)$.]

2. In the case of equality in the Cauchy-Schwarz inequality we have the following. If $|(f, g)| = \|f\| \|g\|$ and $g \neq 0$, then $f = cg$ for some scalar c.

[Hint: Assume $\|f\| = \|g\| = 1$ and $(f, g) = 1$. Then $f - g$ and g are orthogonal, while $f = f - g + g$. Thus $\|f\|^2 = \|f - g\|^2 + \|g\|^2$.]

3. Note that $\|f + g\|^2 = \|f\|^2 + \|g\|^2 + 2\mathrm{Re}(f, g)$ for any pair of elements in a Hilbert space $\mathcal{H}$. As a result, verify the identity $\|f + g\|^2 + \|f - g\|^2 = 2(\|f\|^2 + \|g\|^2)$.

4. Prove from the definition that $\ell^2(\mathbb{Z})$ is complete and separable.

5. Establish the following relations between $L^2(\mathbb{R}^d)$ and $L^1(\mathbb{R}^d)$:

(a) Neither the inclusion $L^2(\mathbb{R}^d) \subset L^1(\mathbb{R}^d)$ nor the inclusion $L^1(\mathbb{R}^d) \subset L^2(\mathbb{R}^d)$ is valid.

(b) Note, however, that if f is supported on a set E of finite measure and if $f \in L^2(\mathbb{R}^d)$, applying the Cauchy-Schwarz inequality to $f\chi_E$ gives $f \in L^1(\mathbb{R}^d)$, and
$$\|f\|_{L^1(\mathbb{R}^d)} \leq m(E)^{1/2}\|f\|_{L^2(\mathbb{R}^d)}.$$

(c) If f is bounded $(|f(x)| \leq M)$, and $f \in L^1(\mathbb{R}^d)$, then $f \in L^2(\mathbb{R}^d)$ with
$$\|f\|_{L^2(\mathbb{R}^d)} \leq M^{1/2}\|f\|_{L^1(\mathbb{R}^d)}^{1/2}.$$

[Hint: For (a) consider $f(x) = |x|^{-\alpha}$, when $|x| \leq 1$ or when $|x| > 1$.]

6. Prove that the following are dense subspaces of $L^2(\mathbb{R}^d)$.

(a) The simple functions.

(b) The continuous functions of compact support.

7. Suppose $\{\varphi_k\}_{k=1}^{\infty}$ is an orthonormal basis for $L^2(\mathbb{R}^d)$. Prove that the collection $\{\varphi_{k,j}\}_{1 \leq k,j < \infty}$ with $\varphi_{k,j}(x, y) = \varphi_k(x)\varphi_j(y)$ is an orthonormal basis of $L^2(\mathbb{R}^d \times \mathbb{R}^d)$.

[Hint: First verify that the $\{\varphi_{k,j}\}$ are orthonormal, by Fubini's theorem. Next, for each j consider $F_j(x) = \int_{\mathbb{R}^d} F(x, y)\overline{\varphi_j(y)}\, dy$. If one assumes that $(F, \varphi_{k,j}) = 0$ for all j, then $\int F_j(x)\overline{\varphi_k(x)}\, dx = 0$.]

8. Let $\eta(t)$ be a fixed strictly positive continuous function on $[a, b]$. Define $\mathcal{H}_\eta = L^2([a, b], \eta)$ to be the space of all measurable functions f on $[a, b]$ such that
$$\int_a^b |f(t)|^2 \eta(t)\, dt < \infty.$$

Define the inner product on $\mathcal{H}_\eta$ by
$$(f, g)_\eta = \int_a^b f(t)\overline{g(t)}\eta(t)\, dt.$$

(a) Show that $\mathcal{H}_\eta$ is a Hilbert space, and that the mapping $U : f \mapsto \eta^{1/2} f$ gives a unitary correspondence between $\mathcal{H}_\eta$ and the usual space $L^2([a, b])$.

(b) Generalize this to the case when η is not necessarily continuous.

9. Let $\mathcal{H}_1 = L^2([-\pi, \pi])$ be the Hilbert space of functions $F(e^{i\theta})$ on the unit circle with inner product $(F, G) = \frac{1}{2\pi} \int_{-\pi}^{\pi} F(e^{i\theta}) \overline{G(e^{i\theta})} \, d\theta$. Let $\mathcal{H}_2$ be the space $L^2(\mathbb{R})$. Using the mapping

$$x \mapsto \frac{i - x}{i + x}$$

of $\mathbb{R}$ to the unit circle, show that:

(a) The correspondence $U : F \to f$, with

$$f(x) = \frac{1}{\pi^{1/2}(i + x)} F\left(\frac{i - x}{i + x}\right)$$

gives a unitary mapping of $\mathcal{H}_1$ to $\mathcal{H}_2$.

(b) As a result,

$$\left\{ \frac{1}{\pi^{1/2}} \left(\frac{i - x}{i + x}\right)^n \frac{1}{i + x} \right\}_{n = -\infty}^{\infty}$$

is an orthonormal basis of $L^2(\mathbb{R})$.

10. Let $\mathcal{S}$ denote a subspace of a Hilbert space $\mathcal{H}$. Prove that $(\mathcal{S}^\perp)^\perp$ is the smallest closed subspace of $\mathcal{H}$ that contains $\mathcal{S}$.

11. Let P be the orthogonal projection associated with a closed subspace $\mathcal{S}$ in a Hilbert space $\mathcal{H}$, that is,

$$P(f) = f \text{ if } f \in \mathcal{S} \quad \text{and} \quad P(f) = 0 \text{ if } f \in \mathcal{S}^\perp.$$

(a) Show that $P^2 = P$ and $P^* = P$.

(b) Conversely, if P is any bounded operator satisfying $P^2 = P$ and $P^* = P$, prove that P is the orthogonal projection for some closed subspace of $\mathcal{H}$.

(c) Using P, prove that if $\mathcal{S}$ is a closed subspace of a separable Hilbert space, then $\mathcal{S}$ is also a separable Hilbert space.

12. Let E be a measurable subset of $\mathbb{R}^d$, and suppose $\mathcal{S}$ is the subspace of $L^2(\mathbb{R}^d)$ of functions that vanish for a.e. $x \notin E$. Show that the orthogonal projection P on $\mathcal{S}$ is given by $P(f) = \chi_E \cdot f$, where χ_E is the characteristic function of E.

13. Suppose P_1 and P_2 are a pair of orthogonal projections on S_1 and S_2, respectively. Then $P_1 P_2$ is an orthogonal projection if and only if P_1 and P_2 commute, that is, $P_1 P_2 = P_2 P_1$. In this case, $P_1 P_2$ projects onto $S_1 \cap S_2$.

14. Suppose $\mathcal{H}$ and $\mathcal{H}'$ are two completions of a pre-Hilbert space $\mathcal{H}_0$. Show that there is a unitary mapping from $\mathcal{H}$ to $\mathcal{H}'$ that is the identity on $\mathcal{H}_0$.

[Hint: If $f \in \mathcal{H}$, pick a Cauchy sequence $\{f_n\}$ in $\mathcal{H}_0$ that converges to f in $\mathcal{H}$. This sequence will also converge to an element f' in $\mathcal{H}'$. The mapping $f \mapsto f'$ gives the required unitary mapping.]

15. Let T be any linear transformation from $\mathcal{H}_1$ to $\mathcal{H}_2$. If we suppose that $\mathcal{H}_1$ is finite-dimensional, then T is automatically bounded. (If $\mathcal{H}_1$ is not assumed to be finite-dimensional this may fail; see Problem 1 below.)

16. Let $F_0(z) = 1/(1-z)^i$.

(a) Verify that $|F_0(z)| \leq e^{\pi/2}$ in the unit disc, but that $\lim_{r \to 1} F_0(r)$ does not exist.

[Hint: Note that $|F_0(r)| = 1$ and $F_0(r)$ oscillates between ± 1 infinitely often as $r \to 1$.]

(b) Let $\{\alpha_n\}_{n=1}^{\infty}$ be an enumeration of the rationals, and let

$$F(z) = \sum_{j=1}^{\infty} \delta^j F_0(ze^{-i\alpha_j}),$$

where δ is sufficiently small. Show that $\lim_{r \to 1} F(re^{i\theta})$ fails to exist whenever $\theta = \alpha_j$, and hence F fails to have a radial limit for a dense set of points on the unit circle.

17. Fatou's theorem can be generalized by allowing a point to approach the boundary in larger regions, as follows.

For each $0 < s < 1$ and point z on the unit circle, consider the region $\Gamma_s(z)$ defined as the smallest closed convex set that contains z and the closed disc $D_s(0)$. In other words, $\Gamma_s(z)$ consists of all lines joining z with points in $D_s(0)$. Near the point z, the region $\Gamma_s(z)$ looks like a triangle. See Figure 2.

We say that a function F defined in the open unit disc has a **non-tangential limit** at a point z on the circle, if for every $0 < s < 1$, the limit

$$\lim_{\substack{w \to z \\ w \in \Gamma_s(z)}} F(w)$$

exists.

Prove that if F is holomorphic and bounded on the open unit disc, then F has a non-tangential limit for almost every point on the unit circle.

[Hint: Show that the Poisson integral of a function f has non-tangential limits at every point of the Lebesgue set of f.]

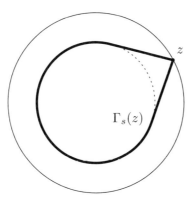

Figure 2. The region $\Gamma_s(z)$

18. Let $\mathcal{H}$ denote a Hilbert space, and $\mathcal{L}(\mathcal{H})$ the vector space of all bounded linear operators on $\mathcal{H}$. Given $T \in \mathcal{L}(\mathcal{H})$, we define the operator norm

$$\|T\| = \inf\{B : \|Tv\| \le B\|v\|, \quad \text{for all } v \in \mathcal{H}\}.$$

(a) Show that $\|T_1 + T_2\| \le \|T_1\| + \|T_2\|$ whenever $T_1, T_2 \in \mathcal{L}(\mathcal{H})$.

(b) Prove that

$$d(T_1, T_2) = \|T_1 - T_2\|$$

defines a metric on $\mathcal{L}(\mathcal{H})$.

(c) Show that $\mathcal{L}(\mathcal{H})$ is complete in the metric d.

19. If T is a bounded linear operator on a Hilbert space, prove that

$$\|TT^*\| = \|T^*T\| = \|T\|^2 = \|T^*\|^2.$$

20. Suppose $\mathcal{H}$ is an infinite-dimensional Hilbert space. We have seen an example of a sequence $\{f_n\}$ in $\mathcal{H}$ with $\|f_n\| = 1$ for all n, but for which no subsequence of $\{f_n\}$ converges in $\mathcal{H}$. However, show that for any sequence $\{f_n\}$ in $\mathcal{H}$ with $\|f_n\| = 1$ for all n, there exist $f \in \mathcal{H}$ and a subsequence $\{f_{n_k}\}$ such that for all $g \in \mathcal{H}$, one has

$$\lim_{k \to \infty} (f_{n_k}, g) = (f, g).$$

One says that $\{f_{n_k}\}$ **converges weakly** to f.

[Hint: Let g run through a basis for $\mathcal{H}$, and use a diagonalization argument. One can then define f by giving its series expansion with respect to the chosen basis.]

21. There are several senses in which a sequence of bounded operators $\{T_n\}$ can converge to a bounded operator T (in a Hilbert space $\mathcal{H}$). First, there is convergence in the norm, that is, $\|T_n - T\| \to 0$, as $n \to \infty$. Next, there is a weaker convergence, which happens to be called **strong convergence**, that requires that $T_n f \to T f$, as $n \to \infty$, for every vector $f \in \mathcal{H}$. Finally, there is **weak convergence** (see also Exercise 20) that requires $(T_n f, g) \to (T f, g)$ for every pair of vectors $f, g \in \mathcal{H}$.

   (a) Show by examples that weak convergence does not imply strong convergence, nor does strong convergence imply convergence in the norm.

   (b) Show that for any bounded operator T there is a sequence $\{T_n\}$ of bounded operators of finite rank so that $T_n \to T$ strongly as $n \to \infty$.

22. An operator T is an **isometry** if $\|Tf\| = \|f\|$ for all $f \in \mathcal{H}$.

   (a) Show that if T is an isometry, then $(Tf, Tg) = (f, g)$ for every $f, g \in \mathcal{H}$. Prove as a result that $T^*T = I$.

   (b) If T is an isometry and T is surjective, then T is unitary and $TT^* = I$.

   (c) Give an example of an isometry that is not unitary.

   (d) Show that if T^*T is unitary then T is an isometry.

[Hint: Use the fact that $(Tf, Tf) = (f, f)$ for f replaced by $f \pm g$ and $f \pm ig$.]

23. Suppose $\{T_k\}$ is a collection of bounded operators on a Hilbert space $\mathcal{H}$, with $\|T_k\| \leq 1$ for all k. Suppose also that

$$T_k T_j^* = T_k^* T_j = 0 \qquad \text{for all } k \neq j.$$

Let $S_N = \sum_{k=-N}^{N} T_k$.
   Show that $S_N(f)$ converges as $N \to \infty$, for every $f \in \mathcal{H}$. If $T(f)$ denotes the limit, prove that $\|T\| \leq 1$.
   A generalization is given in Problem 8* below.

[Hint: Consider first the case when only finitely many of the T_k are non-zero, and note that the ranges of the T_k are mutually orthogonal.]

24. Let $\{e_k\}_{k=1}^{\infty}$ denote an orthonormal set in a Hilbert space $\mathcal{H}$. If $\{c_k\}_{k=1}^{\infty}$ is a sequence of positive real numbers such that $\sum c_k^2 < \infty$, then the set

$$A = \{\sum_{k=1}^{\infty} a_k e_k : |a_k| \leq c_k\}$$

is compact in $\mathcal{H}$.

25. Suppose T is a bounded operator that is diagonal with respect to a basis $\{\varphi_k\}$, with $T\varphi_k = \lambda_k \varphi_k$. Then T is compact if and only if $\lambda_k \to 0$.

[Hint: If $\lambda_k \to 0$, then note that $\|P_n T - T\| \to 0$, where P_n is the orthogonal projection on the subspace spanned by $\varphi_1, \varphi_2, \ldots, \varphi_n$.]

26. Suppose w is a measurable function on $\mathbb{R}^d$ with $0 < w(x) < \infty$ for a.e. x, and K is a measurable function on $\mathbb{R}^{2d}$ that satisfies:

(i) $\displaystyle\int_{\mathbb{R}^d} |K(x,y)|w(y)\,dy \leq Aw(x)$ for almost every $x \in \mathbb{R}^d$, and

(ii) $\displaystyle\int_{\mathbb{R}^d} |K(x,y)|w(x)\,dx \leq Aw(y)$ for almost every $y \in \mathbb{R}^d$.

Prove that the integral operator defined by

$$Tf(x) = \int_{\mathbb{R}^d} K(x,y)f(y)\,dy, \qquad x \in \mathbb{R}^d$$

is bounded on $L^2(\mathbb{R}^d)$ with $\|T\| \leq A$.

Note as a special case that if $\int |K(x,y)|\,dy \leq A$ for all x, and $\int |K(x,y)|\,dx \leq A$ for all y, then $\|T\| \leq A$.

[Hint: Show that if $f \in L^2(\mathbb{R}^d)$, then

$$\int |K(x,y)|\,|f(y)|\,dy \leq A^{1/2}w(x)^{1/2}\left[\int |K(x,y)|\,|f(y)|^2 w(y)^{-1}\,dy\right]^{1/2}.]$$

27. Prove that the operator

$$Tf(x) = \frac{1}{\pi}\int_0^\infty \frac{f(y)}{x+y}\,dy$$

is bounded on $L^2(0,\infty)$ with norm $\|T\| \leq 1$.

[Hint: Use Exercise 26 with an appropriate w.]

28. Suppose $\mathcal{H} = L^2(B)$, where B is the unit ball in $\mathbb{R}^d$. Let $K(x,y)$ be a measurable function on $B \times B$ that satisfies $|K(x,y)| \leq A|x-y|^{-d+\alpha}$ for some $\alpha > 0$, whenever $x, y \in B$. Define

$$Tf(x) = \int_B K(x,y)f(y)dy.$$

(a) Prove that T is a bounded operator on $\mathcal{H}$.

(b) Prove that T is compact.

(c) Note that T is a Hilbert-Schmidt operator if and only if $\alpha > d/2$.

[Hint: For (b), consider the operators T_n associated with the truncated kernels $K_n(x,y) = K(x,y)$ if $|x-y| \geq 1/n$ and 0 otherwise. Show that each T_n is compact, and that $\|T_n - T\| \to 0$ as $n \to \infty$.]

29. Let T be a compact operator on a Hilbert space $\mathcal{H}$, and assume $\lambda \neq 0$.

(a) Show that the range of $\lambda I - T$ defined by

$$\{g \in \mathcal{H} : g = (\lambda I - T)f, \text{ for some } f \in \mathcal{H}\}$$

is closed. [Hint: Suppose $g_j \to g$, where $g_j = (\lambda I - T)f_j$. Let V_λ denote the eigenspace of T corresponding to λ, that is, the kernel of $\lambda I - T$. Why can one assume that $f_j \in V_\lambda^\perp$? Under this assumption prove that $\{f_j\}$ is a bounded sequence.]

(b) Show by example that this may fail when $\lambda = 0$.

(c) Show that the range of $\lambda I - T$ is all of $\mathcal{H}$ if and only if the null-space of $\overline{\lambda} I - T^*$ is trivial.

30. Let $\mathcal{H} = L^2([-\pi, \pi])$ with $[-\pi, \pi]$ identified as the unit circle. Fix a bounded sequence $\{\lambda_n\}_{n=-\infty}^{\infty}$ of complex numbers, and define an operator Tf by

$$Tf(x) \sim \sum_{n=-\infty}^{\infty} \lambda_n a_n e^{inx} \quad \text{whenever} \quad f(x) \sim \sum_{n=-\infty}^{\infty} a_n e^{inx}.$$

Such an operator is called a **Fourier multiplier operator**, and the sequence $\{\lambda_n\}$ is called the **multiplier sequence**.

(a) Show that T is a bounded operator on $\mathcal{H}$ and $\|T\| = \sup_n |\lambda_n|$.

(b) Verify that T commutes with translations, that is, if we define $\tau_h(x) = f(x - h)$ then

$$T \circ \tau_h = \tau_h \circ T \quad \text{for every } h \in \mathbb{R}.$$

(c) Conversely, prove that if T is any bounded operator on $\mathcal{H}$ that commutes with translations, then T is a Fourier multiplier operator. [Hint: Consider $T(e^{inx})$.]

31. Consider a version of the sawtooth function defined on $[-\pi, \pi)$ by[5]

$$K(x) = i(sgn(x)\pi - x),$$

and extended to $\mathbb{R}$ with period 2π. Suppose $f \in L^1([-\pi, \pi])$ is extended to $\mathbb{R}$ with period 2π, and define

$$Tf(x) = \frac{1}{2\pi} \int_{-\pi}^{\pi} K(x - y)f(y) \, dy$$

$$= \frac{1}{2\pi} \int_{-\pi}^{\pi} K(y)f(x - y) \, dy.$$

---

[5]The symbol $sgn(x)$ denotes the sign function: it equals 1 or -1 if x is positive or negative respectively, and 0 if $x = 0$.

(a) Show that $F(x) = Tf(x)$ is absolutely continuous, and if $\int_{-\pi}^{\pi} f(y)dy = 0$, then $F'(x) = if(x)$ a.e. x.

(b) Show that the mapping $f \mapsto Tf$ is compact and symmetric on $L^2([-\pi, \pi])$.

(c) Prove that $\varphi(x) \in L^2([-\pi, \pi])$ is an eigenfunction for T if and only if $\varphi(x)$ is (up to a constant multiple) equal to e^{inx} for some integer $n \neq 0$ with eigenvalue $1/n$, or $\varphi(x) = 1$ with eigenvalue 0.

(d) Show as a result that $\{e^{inx}\}_{n \in \mathbb{Z}}$ is an orthonormal basis of $L^2([-\pi, \pi])$.

Note that in Book I, Chapter 2, Exercise 8, it is shown that the Fourier series of K is

$$K(x) \sim \sum_{n \neq 0} \frac{e^{inx}}{n}.$$

32. Consider the operator $T : L^2([0, 1]) \to L^2([0, 1])$ defined by

$$T(f)(t) = tf(t).$$

(a) Prove that T is a bounded linear operator with $T = T^*$, but that T is not compact.

(b) However, show that T has no eigenvectors.

33. Let $\mathcal{H}$ be a Hilbert space with basis $\{\varphi_k\}_{k=1}^{\infty}$. Verify that the operator T defined by

$$T(\varphi_k) = \frac{1}{k} \varphi_{k+1}$$

is compact, but has no eigenvectors.

34. Let K be a Hilbert-Schmidt kernel which is real and symmetric. Then, as we saw, the operator T whose kernel is K is compact and symmetric. Let $\{\varphi_k(x)\}$ be the eigenvectors (with eigenvalues λ_k) that diagonalize T. Then:

(a) $\sum_k |\lambda_k|^2 < \infty$.

(b) $K(x, y) \sim \sum \lambda_k \varphi_k(x) \varphi_k(y)$ is the expansion of K in the basis $\{\varphi_k(x) \varphi_k(y)\}$.

(c) Suppose T is a compact operator which is symmetric. Then T is of Hilbert-Schmidt type if and only if $\sum_n |\lambda_n|^2 < \infty$, where $\{\lambda_n\}$ are the eigenvalues of T counted according to their multiplicities.

35. Let $\mathcal{H}$ be a Hilbert space. Prove the following variants of the spectral theorem.

(a) If T_1 and T_2 are two linear symmetric and compact operators on $\mathcal{H}$ that commute (that is, $T_1T_2 = T_2T_1$), show that they can be diagonalized simultaneously. In other words, there exists an orthonormal basis for $\mathcal{H}$ which consists of eigenvectors for both T_1 and T_2.

(b) A linear operator on $\mathcal{H}$ is **normal** if $TT^* = T^*T$. Prove that if T is normal and compact, then T can be diagonalized.

[Hint: Write $T = T_1 + iT_2$ where T_1 and T_2 are symmetric, compact and commute.]

(c) If U is unitary, and $U = \lambda I - T$, where T is compact, then U can be diagonalized.

8 Problems

1. Let $\mathcal{H}$ be an infinite-dimensional Hilbert space. There exists a linear functional ℓ defined on $\mathcal{H}$ that is not bounded (and hence not continuous).

[Hint: Using the axiom of choice (or one of its equivalent forms), construct an **algebraic basis** of $\mathcal{H}$, $\{e_\alpha\}$; it has the property that every element of $\mathcal{H}$ is uniquely a finite linear combination of the $\{e_\alpha\}$. Select a denumerable collection $\{e_n\}_{n=1}^\infty$, and define ℓ to satisfy the requirement that $\ell(e_n) = n\|e_n\|$ for all $n \in \mathbb{N}$.]

2.[*] The following is an example of a non-separable Hilbert space. We consider the collection of exponentials $\{e^{i\lambda x}\}$ on $\mathbb{R}$, where λ ranges over the real numbers. Let $\mathcal{H}_0$ denote the space of finite linear combinations of these exponentials. For $f, g \in \mathcal{H}_0$, we define the inner product as

$$(f, g) = \lim_{T \to \infty} \frac{1}{2T} \int_{-T}^{T} f(x)\overline{g(x)}\, dx.$$

(a) Show that this limit exists, and

$$(f, g) = \sum_{k=1}^{N} a_{\lambda_k}\overline{b_{\lambda_k}}$$

if $f(x) = \sum_{k=1}^{N} a_{\lambda_k} e^{i\lambda_k x}$ and $g(x) = \sum_{k=1}^{N} b_{\lambda_k} e^{i\lambda_k x}$.

(b) With this inner product $\mathcal{H}_0$ is a pre-Hilbert space. Notice that $\|f\| \leq \sup_x |f(x)|$, if $f \in \mathcal{H}_0$, where $\|f\|$ denotes the norm $\langle f, f \rangle^{1/2}$. Let $\mathcal{H}$ be the completion of $\mathcal{H}_0$. Then $\mathcal{H}$ is not separable because $e^{i\lambda x}$ and $e^{i\lambda' x}$ are orthonormal if $\lambda \neq \lambda'$.

A continuous function F defined on $\mathbb{R}$ is called **almost periodic** if it is the uniform limit (on $\mathbb{R}$) of elements in $\mathcal{H}_0$. Such functions can be identified with (certain) elements in the completion $\mathcal{H}$: We have $\mathcal{H}_0 \subset AP \subset \mathcal{H}$, where AP denotes the almost periodic functions.

(c) A continuous function F is in AP if for every $\epsilon > 0$ we can find a length $L = L_\epsilon$ such that any interval $I \subset \mathbb{R}$ of length L contains an "almost period" τ satisfying

$$\sup_x |F(x + \tau) - F(x)| < \epsilon.$$

(d) An equivalent characterization is that F is in AP if and only if every sequence $F(x + h_n)$ of translates of F contains a subsequence that converges uniformly.

3. The following is a direct generalization of Fatou's theorem: if $u(re^{i\theta})$ is harmonic in the unit disc and bounded there, then $\lim_{r \to 1} u(re^{i\theta})$ exists for a.e. θ.

[Hint: Let $a_n(r) = \frac{1}{2\pi} \int_0^{2\pi} u(re^{i\theta}) e^{-in\theta}\, d\theta$. Then $a_n''(r) + \frac{1}{r} a_n'(r) - \frac{n^2}{r^2} a_n(r) = 0$, hence $a_n(r) = A_n r^n + B_n r^{-n}$, $n \neq 0$, and as a result[6] $u(re^{i\theta}) = \sum_{-\infty}^{\infty} a_n r^{|n|} e^{in\theta}$. From this one can proceed as in the proof of Theorem 3.3.]

4.[*] This problem provides some examples of functions that fail to have radial limits almost everywhere.

(a) At almost every point of the boundary unit circle, the function $\sum_{n=0}^{\infty} z^{2^n}$ fails to have a radial limit.

(b) More generally, suppose $F(z) = \sum_{n=0}^{\infty} a_n z^{2^n}$. Then, if $\sum |a_n|^2 = \infty$ the function F fails to have radial limits at almost every boundary point. However, if $\sum |a_n|^2 < \infty$, then $F \in H^2(\mathbb{D})$, and we know by the proof of Theorem 3.3 that F does have radial limits almost everywhere.

5.[*] Suppose F is holomorphic in the unit disc, and

$$\sup_{0 \leq r < 1} \frac{1}{2\pi} \int_{-\pi}^{\pi} \log^+ |F(re^{i\theta})|\, d\theta < \infty,$$

where $\log^+ u = \log u$ if $u \geq 1$, and $\log^+ u = 0$ if $u < 1$.
Then $\lim_{r \to 1} F(re^{i\theta})$ exists for almost every θ.
The above condition is satisfied whenever (say)

$$\sup_{0 \leq r < 1} \frac{1}{2\pi} \int_{-\pi}^{\pi} |F(re^{i\theta})|^p\, d\theta < \infty, \qquad \text{for some } p > 0,$$

(since $e^{pu} \geq pu$, $u \geq 0$).
Functions that satisfy the latter condition are said to belong to the Hardy space $H^p(\mathbb{D})$.

6.[*] If T is compact, and $\lambda \neq 0$, show that

---

[6] See also Section 5, Chapter 2 in Book I.

(a) $\lambda I - T$ is injective if and only if $\overline{\lambda} I - T^*$ is injective.

(b) $\lambda I - T$ is injective if and only if $\lambda I - T$ is surjective.

This result, known as the **Fredholm alternative**, is often combined with that in Exercise 29.

7. Show that the identity operator on $L^2(\mathbb{R}^d)$ cannot be given as an (absolutely) convergent integral operator. More precisely, if $K(x, y)$ is a measurable function on $\mathbb{R}^d \times \mathbb{R}^d$ with the property that for each $f \in L^2(\mathbb{R}^d)$, the integral $T(f)(x) = \int_{\mathbb{R}^d} K(x, y) f(y) \, dy$ converges for almost every x, then $T(f) \neq f$ for some f.

[Hint: Prove that otherwise for any pair of disjoint balls B_1 and B_2 in $\mathbb{R}^d$, we would have that $K(x, y) = 0$ for a.e. $(x, y) \in B_1 \times B_2$.]

8.[*] Suppose $\{T_k\}$ is a collection of bounded opeartors on a Hilbert space $\mathcal{H}$. Assume that

$$\|T_k T_j^*\| \leq a_{k-j} \quad \text{and} \quad \|T_k^* T_j\| \leq a_{k-j},$$

for positive constants $\{a_n\}$ with the property that $\sum_{-\infty}^{\infty} a_n = A < \infty$. Then $S_N(f)$ converges as $N \to \infty$, for every $f \in \mathcal{H}$, with $S_N = \sum_{-N}^{N} T_k$. Moreover, $T = \lim_{N \to \infty} S_N$ satisfies $\|T\| \leq A$.

9. A discussion of a class of regular Sturm-Liouville operators follows. Other special examples are given in the problems below.

Suppose $[a, b]$ is a bounded interval, and L is defined on functions f that are twice continuously differentiable in $[a, b]$ (we write, $f \in C^2([a, b])$) by

$$L(f)(x) = \frac{d^2 f}{dx^2} - q(x) f(x).$$

Here the function q is continuous and real-valued on $[a, b]$, and we assume for simplicity that q is non-negative. We say that $\varphi \in C^2([a, b])$ is an **eigenfunction** of L with eigenvalue μ if $L(\varphi) = \mu\varphi$, under the assumption that φ satisfies the boundary conditions $\varphi(a) = \varphi(b) = 0$. Then one can show:

(a) The eigenvalues μ are strictly negative, and the eigenspace corresponding to each eigenvalue is one-dimensional.

(b) Eigenvectors corresponding to distinct eigenvalues are orthogonal in $L^2([a, b])$.

(c) Let $K(x, y)$ be the "Green's kernel" defined as follows. Choose $\varphi_-(x)$ to be a solution of $L(\varphi_-) = 0$, with $\varphi_-(a) = 0$ but $\varphi'_-(a) \neq 0$. Similarly, choose $\varphi_+(x)$ to be a solution of $L(\varphi_+) = 0$ with $\varphi_+(b) = 0$, but $\varphi'_+(b) \neq 0$. Let $w = \varphi'_+(x)\varphi_-(x) - \varphi'_-(x)\varphi_+(x)$, be the "Wronskian" of these solutions, and note that w is a non-zero constant.

Set

$$K(x, y) = \begin{cases} \frac{\varphi_-(x)\varphi_+(y)}{w} & \text{if } a \leq x \leq y \leq b, \\ \frac{\varphi_+(x)\varphi_-(y)}{w} & \text{if } a \leq y \leq x \leq b. \end{cases}$$

Then the operator T defined by

$$T(f)(x) = \int_a^b K(x,y)f(y)\,dy$$

is a Hilbert-Schmidt operator, and hence compact. It is also symmetric. Moreover, whenever f is continuous on $[a,b]$, Tf is of class $C^2([a,b])$ and

$$L(Tf) = f.$$

(d) As a result, each eigenvector of T (with eigenvalue λ) is an eigenvector of L (with eigenvalue $\mu = 1/\lambda$). Hence Theorem 6.2 proves the completeness of the orthonormal set arising from normalizing the eigenvectors of L.

10.* Let L be defined on $C^2([-1,1])$ by

$$L(f)(x) = (1-x^2)\frac{d^2 f}{dx^2} - 2x\frac{df}{dx}.$$

If φ_n is the nth Legendre polynomial, given by

$$\varphi_n(x) = \left(\frac{d}{dx}\right)^n (1-x^2)^n, \qquad n = 0, 1, 2, \ldots,$$

then $L\varphi_n = -n(n+1)\varphi_n$.

When normalized the φ_n form an orthonormal basis of $L^2([-1,1])$ (see also Problem 2, Chapter 3 in Book I, where φ_n is denoted by L_n.)

11.* The **Hermite functions** $h_k(x)$ are defined by the generating identity

$$\sum_{k=0}^{\infty} h_k(x)\frac{t^k}{k!} = e^{-(x^2/2 - 2tx + t^2)}.$$

(a) They satisfy the "creation" and "annihilation" identities $\left(x - \frac{d}{dx}\right)h_k(x) = h_{k+1}(x)$ and $\left(x + \frac{d}{dx}\right)h_k(x) = h_{k-1}(x)$ for $k \geq 0$ where $h_{-1}(x) = 0$. Note that $h_0(x) = e^{-x^2/2}$, $h_1(x) = 2xe^{-x^2/2}$, and more generally $h_k(x) = P_k(x)e^{-x^2/2}$, where P_k is a polynomial of degree k.

(b) Using (a) one sees that the h_k are eigenvectors of the operator $L = -d^2/dx^2 + x^2$, with $L(h_k) = \lambda_k h_k$, where $\lambda_k = 2k + 1$. One observes that these functions are mutually orthogonal. Since

$$\int_{\mathbb{R}} [h_k(x)]^2\,dx = \pi^{1/2} 2^k k! = c_k,$$

we can normalize them obtaining a orthonormal sequence $\{H_k\}$, with $H_k = c_k^{-1/2} h_k$. This sequence is complete in $L^2(\mathbb{R}^d)$ since $\int_{\mathbb{R}} f H_k\,dx = 0$ for all k implies $\int_{-\infty}^{\infty} f(x)e^{-\frac{x^2}{2} + 2tx}\,dx = 0$ for all $t \in \mathbb{C}$.

(c) Suppose that $K(x, y) = \sum_{k=0}^{\infty} \frac{H_k(x) H_k(y)}{\lambda_k}$, and also $F(x) = T(f)(x) = \int_{\mathbb{R}} K(x, y) f(y) \, dy$. Then T is a symmetric Hilbert-Schmidt operator, and if $f \sim \sum_{k=0}^{\infty} a_k H_k$, then $F \sim \sum_{k=0}^{\infty} \frac{a_k}{\lambda_k} H_k$.

One can show on the basis of (a) and (b) that whenever $f \in L^2(\mathbb{R})$, not only is $F \in L^2(\mathbb{R})$, but also $x^2 F(x) \in L^2(\mathbb{R})$. Moreover, F can be corrected on a set of measure zero, so it is continuously differentiable, F' is absolutely continuous, and $F'' \in L^2(\mathbb{R})$. Finally, the operator T is the inverse of L in the sense that

$$LT(f) = LF = -F'' + x^2 F = f \quad \text{for every } f \in L^2(\mathbb{R}).$$

(See also Problem 7* in Chapter 5 of Book I.)

5 Hilbert Spaces: Several Examples

> What is the difference between a mathematician and a physicist? It is this: To a mathematician all Hilbert spaces are the same; for a physicist, however, it is their different realizations that really matter.
>
> *Attributed to E. Wigner,* ca. 1960

Hilbert spaces arise in a large number of different contexts in analysis. Although it is a truism that all (infinite-dimensional) Hilbert spaces are the same, it is in fact their varied and distinct realizations and separate applications that make them of such interest in mathematics. We shall illustrate this via several examples.

To begin with, we consider the Plancherel formula and the resulting unitary character of the Fourier transform. The relevance of these ideas to complex analysis is then highlighted by the study of holomorphic functions in a half-space that belong to the Hardy space H^2. That function space itself is another interesting realization of a Hilbert space. The considerations here are analogous to the ideas that led us to Fatou's theorem for the unit disc, but are of a more involved character.

We next see how complex analysis and the Fourier transform combine to guarantee the existence of solutions to linear partial differential equations with constant coefficients. The proof relies on a basic L^2 estimate, which once established can be exploited by simple Hilbert space techniques.

Our final example is Dirichlet's principle and its applications to the boundary value problem for harmonic functions. Here the Hilbert space that arises is given by Dirichlet's integral, and the solution is expressed by aid of an appropriate orthogonal projection operator.

1 The Fourier transform on L^2

The Fourier transform of a function f on $\mathbb{R}^d$ is defined by

$$(1) \qquad \hat{f}(\xi) = \int_{\mathbb{R}^d} f(x)\, e^{-2\pi i x \cdot \xi}\, dx,$$

and its attached inversion is given by

$$(2) \qquad f(x) = \int_{\mathbb{R}^d} \hat{f}(\xi)\, e^{2\pi i x \cdot \xi}\, d\xi.$$

These formulas have already appeared in several different contexts. We considered first (in Book I) the properties of the Fourier transform in the elementary setting by restricting to functions in the Schwartz class $\mathcal{S}(\mathbb{R}^d)$. The class $\mathcal{S}$ consists of functions f that are smooth (indefinitely differentiable) and such that for each multi-index α and β, the function $x^\alpha (\frac{\partial}{\partial x})^\beta f$ is bounded on $\mathbb{R}^d$.[1] We saw that on this class the Fourier transform is a bijection, that the inversion formula (2) holds, and moreover we have the Plancherel identity

$$(3) \qquad \int_{\mathbb{R}^d} |\hat{f}(\xi)|^2\, d\xi = \int_{\mathbb{R}^d} |f(x)|^2\, dx.$$

Turning now to more general (in particular, non-continuous) functions, we note that the largest class for which the integral defining $\hat{f}(\xi)$ converges (absolutely) is the space $L^1(\mathbb{R}^d)$. For it, we saw in Chapter 2 that a (relative) inversion formula is valid.

Beyond these particular facts, what we would like here is to reestablish in the general context the symmetry between f and $\hat{f}$ that holds for $\mathcal{S}$. This is where the special role of the Hilbert space $L^2(\mathbb{R}^d)$ enters.

We shall define the Fourier transform on $L^2(\mathbb{R}^d)$ as an extension of its definition on $\mathcal{S}$. For this purpose, we temporarily adopt the notational device of denoting by $\mathcal{F}_0$ and $\mathcal{F}$ the Fourier transform on $\mathcal{S}$ and its extension to L^2, respectively.

The main results we prove are the following.

Theorem 1.1 *The Fourier transform $\mathcal{F}_0$, initially defined on $\mathcal{S}(\mathbb{R}^d)$, has a (unique) extension $\mathcal{F}$ to a unitary mapping of $L^2(\mathbb{R}^d)$ to itself. In particular,*

$$\|\mathcal{F}(f)\|_{L^2(\mathbb{R}^d)} = \|f\|_{L^2(\mathbb{R}^d)}$$

for all $f \in L^2(\mathbb{R}^d)$.

The extension $\mathcal{F}$ will be given by a limiting process: if $\{f_n\}$ is a sequence in the Schwartz space that converges to f in $L^2(\mathbb{R}^d)$, then $\{\mathcal{F}_0(f_n)\}$ will

---

[1]Recall that $x^\alpha = x_1^{\alpha_1} x_2^{\alpha_2} \cdots x_d^{\alpha_d}$ and $(\frac{\partial}{\partial x})^\beta = (\frac{\partial}{\partial x_1})^{\beta_1} \cdots (\frac{\partial}{\partial x_d})^{\beta_d}$, where $\alpha = (\alpha_1, \ldots, \alpha_d)$ and $\beta = (\beta_1, \ldots, \beta_d)$, with α_j and β_j positive integers. The order of α is denoted by $|\alpha|$ and defined to be $\alpha_1 + \cdots + \alpha_d$.

converge to an element in $L^2(\mathbb{R}^d)$ which we will *define* as the Fourier transform of f. To implement this approach we have to see that every L^2 function can be approximated by elements in the Schwartz space.

Lemma 1.2 *The space $\mathcal{S}(\mathbb{R}^d)$ is dense in $L^2(\mathbb{R}^d)$. In other words, given any $f \in L^2(\mathbb{R}^d)$, there exists a sequence $\{f_n\} \subset \mathcal{S}(\mathbb{R}^d)$ such that*

$$\|f - f_n\|_{L^2(\mathbb{R}^d)} \to 0 \quad \text{as } n \to \infty.$$

For the proof of the lemma, we fix $f \in L^2(\mathbb{R}^d)$ and $\epsilon > 0$. Then, for each $M > 0$, we define

$$g_M(x) = \begin{cases} f(x) & \text{if } |x| \le M \text{ and } |f(x)| \le M, \\ 0 & \text{otherwise.} \end{cases}$$

Then, $|f(x) - g_M(x)| \le 2|f(x)|$, hence $|f(x) - g_M(x)|^2 \le 4|f(x)|^2$, and since $g_M(x) \to f(x)$ as $M \to \infty$ for almost every x, the dominated convergence theorem guarantees that for some M, we have

$$\|f - g_M\|_{L^2(\mathbb{R}^d)} < \epsilon.$$

We write $g = g_M$, note that this function is bounded and supported on a bounded set, and observe that it now suffices to approximate g by functions in the Schwartz space. To achieve this goal, we use a method called **regularization**, which consists of "smoothing" g by convolving it with an approximation of the identity. Consider a function $\varphi(x)$ on $\mathbb{R}^d$ with the following properties:

(a) φ is smooth (indefinitely differentiable).

(b) φ is supported in the unit ball.

(c) $\varphi \ge 0$.

(d) $\displaystyle\int_{\mathbb{R}^d} \varphi(x)\, dx = 1.$

For instance, one can take

$$\varphi(x) = \begin{cases} c\, e^{-\frac{1}{1-|x|^2}} & \text{if } |x| < 1, \\ 0 & \text{if } |x| \ge 1, \end{cases}$$

where the constant c is chosen so that (d) holds.

Next, we consider the approximation to the identity defined by

$$K_\delta(x) = \delta^{-d} \varphi(x/\delta).$$

The key observation is that $g * K_\delta$ belongs to $\mathcal{S}(\mathbb{R}^d)$, with this convolution in fact bounded and supported on a fixed bounded set, uniformly in δ (assuming for example that $\delta \leq 1$). Indeed, we may write

$$(g * K_\delta)(x) = \int g(y) K_\delta(x - y)\, dy = \int g(x - y) K_\delta(y)\, dy,$$

in view of the identity (6) in Chapter 2. We note that since g is supported on some bounded set and K_δ vanishes outside the ball of radius δ, the function $g * K_\delta$ is supported in some fixed bounded set independent of δ. Also, the function g is bounded by construction, hence

$$|(g * K_\delta)(x)| \leq \int |g(x - y)| K_\delta(y)\, dy$$

$$\leq \sup_{z \in \mathbb{R}^d} |g(z)| \int K_\delta(y)\, dy = \sup_{z \in \mathbb{R}^d} |g(z)|,$$

which shows that $g * K_\delta$ is also uniformly bounded in δ. Moreover, from the first integral expression for $g * K_\delta$ above, one may differentiate under the integral sign to see that $g * K_\delta$ is smooth and all of its derivatives have support in some fixed bounded set.

The proof of the lemma will be complete if we can show that $g * K_\delta$ converges to g in $L^2(\mathbb{R}^d)$. Now Theorem 2.1 in Chapter 3 guarantees that for almost every x, the quantity $|(g * K_\delta)(x) - g(x)|^2$ converges to 0 as δ tends to 0. An application of the bounded convergence theorem (Theorem 1.4 in Chapter 2) yields

$$\|(g * K_\delta) - g\|^2_{L^2(\mathbb{R}^d)} \to 0 \quad \text{as } \delta \to 0.$$

In particular, $\|(g * K_\delta) - g\|_{L^2(\mathbb{R}^d)} < \epsilon$ for an appropriate δ and hence $\|f - g * K_\delta\|_{L^2(\mathbb{R}^d)} < 2\epsilon$, and choosing a sequence of ϵ tending to zero gives the construction of the desired sequence $\{f_n\}$.

For later purposes it is useful to observe that the proof of the above lemma establishes the following assertion: if f belongs to both $L^1(\mathbb{R}^d)$ and $L^2(\mathbb{R}^d)$, then there is a sequence $\{f_n\}$, $f_n \in \mathcal{S}(\mathbb{R}^d)$, that converges to f in both the L^1-norm and the L^2-norm.

Our definition of the Fourier transform on $L^2(\mathbb{R}^d)$ combines the above density of $\mathcal{S}$ with a general "extension principle."

Lemma 1.3 *Let $\mathcal{H}_1$ and $\mathcal{H}_2$ denote Hilbert spaces with norms $\|\cdot\|_1$ and $\|\cdot\|_2$, respectively. Suppose $\mathcal{S}$ is a dense subspace of $\mathcal{H}_1$ and $T_0 : \mathcal{S} \to \mathcal{H}_2$ a linear transformation that satisfies $\|T_0(f)\|_2 \leq c\|f\|_1$ whenever $f \in \mathcal{S}$.*

Then T_0 extends to a (unique) linear transformation $T : \mathcal{H}_1 \to \mathcal{H}_2$ that satisfies $\|T(f)\|_2 \le c\|f\|_1$ for all $f \in \mathcal{H}_1$.

Proof. Given $f \in \mathcal{H}_1$, let $\{f_n\}$ be a sequence in $\mathcal{S}$ that converges to f, and define

$$T(f) = \lim_{n \to \infty} T_0(f_n),$$

where the limit is taken in $\mathcal{H}_2$. To see that T is well-defined we must verify that the limit exists, and that it is independent of the sequence $\{f_n\}$ used to approximate f. Indeed, for the first point, we note that $\{T(f_n)\}$ is a Cauchy sequence in $\mathcal{H}_2$ because by construction $\{f_n\}$ is Cauchy in $\mathcal{H}_1$, and the inequality verified by T_0 yields

$$\|T_0(f_n) - T_0(f_m)\|_2 \le c\|f_n - f_m\|_1 \to 0 \quad \text{as } n, m \to \infty;$$

thus $\{T_0(f_n)\}$ is Cauchy, hence converges in $\mathcal{H}_2$.

Second, to justify that the limit is independent of the approximating sequence, let $\{g_n\}$ be another sequence in $\mathcal{S}$ that converges to f in $\mathcal{H}_1$. Then

$$\|T_0(f_n) - T_0(g_n)\|_2 \le c\|f_n - g_n\|_1,$$

and since $\|f_n - g_n\|_1 \le \|f_n - f\|_1 + \|f - g_n\|_1$, we conclude that $\{T_0(g_n)\}$ converges to a limit in $\mathcal{H}_2$ that equals the limit of $\{T_0(f_n)\}$.

Finally, we recall that if $f_n \to f$ and $T_0(f_n) \to T(f)$, then $\|f_n\|_1 \to \|f\|_1$ and $\|T_0(f_n)\|_2 \to \|T(f)\|_2$, so in the limit as $n \to \infty$, the inequality $\|T(f)\|_2 \le c\|f\|_1$ holds for all $f \in \mathcal{H}_1$.

In the present case of the Fourier transform, we apply this lemma with $\mathcal{H}_1 = \mathcal{H}_2 = L^2(\mathbb{R}^d)$ (equipped with the L^2-norm), $\mathcal{S} = \mathcal{S}(\mathbb{R}^d)$, and $T_0 = \mathcal{F}_0$ the Fourier transform defined on the Schwartz space. The Fourier transform on $L^2(\mathbb{R}^d)$ is by definition the unique (bounded) extension of $\mathcal{F}_0$ to L^2 guaranteed by Lemma 1.3. Thus if $f \in L^2(\mathbb{R}^d)$ and $\{f_n\}$ is any sequence in $\mathcal{S}(\mathbb{R}^d)$ that converges to f (that is, $\|f - f_n\|_{L^2(\mathbb{R}^d)} \to 0$ as $n \to \infty$), we define the **Fourier transform** of f by

(4) $$\mathcal{F}(f) = \lim_{n \to \infty} \mathcal{F}_0(f_n),$$

where the limit is taken in the L^2 sense. Clearly, the argument in the proof of the lemma shows that in our special case the extension $\mathcal{F}$ continues to satisfy the identity (3):

$$\|\mathcal{F}(f)\|_{L^2(\mathbb{R}^d)} = \|f\|_{L^2(\mathbb{R}^d)} \quad \text{whenever } f \in L^2(\mathbb{R}^d).$$

The fact that $\mathcal{F}$ is invertible on L^2 (and thus $\mathcal{F}$ is a unitary mapping) is also a consequence of the analogous property on $\mathcal{S}(\mathbb{R}^d)$. Recall that on the Schwartz space, $\mathcal{F}_0^{-1}$ is given by formula (2), that is,

$$\mathcal{F}_0^{-1}(g)(x) = \int_{\mathbb{R}^d} g(\xi) e^{2\pi i x \cdot \xi} \, d\xi,$$

and satisfies again the identity $\|\mathcal{F}_0^{-1}(g)\|_{L^2} = \|g\|_{L^2}$. Therefore, arguing in the same fashion as above, we can extend $\mathcal{F}_0^{-1}$ to $L^2(\mathbb{R}^d)$ by a limiting argument. Then, given $f \in L^2(\mathbb{R}^d)$, we choose a sequence $\{f_n\}$ in the Schwartz space so that $\|f - f_n\|_{L^2} \to 0$. We have

$$f_n = \mathcal{F}_0^{-1}\mathcal{F}_0(f_n) = \mathcal{F}_0\mathcal{F}_0^{-1}(f_n),$$

and taking the limit as n tends to infinity, we see that

$$f = \mathcal{F}^{-1}\mathcal{F}(f) = \mathcal{F}\mathcal{F}^{-1}(f),$$

and hence $\mathcal{F}$ is invertible. This concludes the proof of Theorem 1.1.

Some remarks are in order.

(i) Suppose f belongs to both $L^1(\mathbb{R}^d)$ and $L^2(\mathbb{R}^d)$. Are the two definitions of the Fourier transform the same? That is, do we have $\mathcal{F}(f) = \hat{f}$, with $\mathcal{F}(f)$ defined by the limiting process in Theorem 1.1 and $\hat{f}$ defined by the convergent integral (1)? To prove that this is indeed the case we recall that we can approximate f by a sequence $\{f_n\}$ in $\mathcal{S}$ so that $f_n \to f$ both in the L^1-norm and the L^2-norm. Since $\mathcal{F}_0(f_n) = \hat{f}_n$, a passage to the limit gives the desired conclusion. In fact, $\mathcal{F}_0(f_n)$ converges to $\mathcal{F}(f)$ in the L^2-norm, so a subsequence converges to $\mathcal{F}(f)$ almost everywhere; see the analogous statement for L^1 in Corollary 2.3, Chapter 2. Moreover,

$$\sup_{\xi \in \mathbb{R}^d} |\hat{f}_n(\xi) - \hat{f}(\xi)| \leq \|f_n - f\|_{L^1(\mathbb{R}^d)},$$

hence $\hat{f}_n$ converges to $\hat{f}$ everywhere, and the assertion is established.

(ii) The theorem gives a rather abstract definition of the Fourier transform on L^2. In view of what we have just said, we can also define the Fourier transform more concretely as follows. If $f \in L^2(\mathbb{R}^d)$, then

$$\hat{f}(\xi) = \lim_{R \to \infty} \int_{|x| \leq R} f(x) e^{-2\pi i x \cdot \xi} \, dx,$$

where the limit is taken in the L^2-norm. Note in fact that if χ_R denotes the characteristic function of the ball $\{x \in \mathbb{R}^d : |x| \leq R\}$, then for each R the function $f\chi_R$ is in both L^1 and L^2, and $f\chi_R \to f$ in the L^2-norm.

(iii) The identity of the various definitions of the Fourier transform discussed above allows us to choose $\hat{f}$ as the preferred notation for the Fourier transform. We adopt this practice in what follows.

2 The Hardy space of the upper half-plane

We will apply the L^2 theory of the Fourier transform to holomorphic functions in the upper half-plane. This leads us to consider the relevant analogues of the Hardy space and Fatou's theorem discussed in the previous chapter.[2] It incidentally provides an answer to the following natural question: What are the functions $f \in L^2(\mathbb{R})$ whose Fourier transforms are supported on the half-line $(0, \infty)$?

Let $\mathbb{R}_+^2 = \{z = x + iy, \; x \in \mathbb{R}, \; y > 0\}$ be the upper half-plane. We define the **Hardy space** $H^2(\mathbb{R}_+^2)$ to consist of all functions F analytic in $\mathbb{R}_+^2$ with the property that

$$(5) \qquad \sup_{y>0} \int_{\mathbb{R}} |F(x+iy)|^2 \, dx < \infty.$$

We define the corresponding norm, $\|F\|_{H^2(\mathbb{R}_+^2)}$, to be the square root of the quantity (5).

Let us first describe a (typical) example of a function F in $H^2(\mathbb{R}_+^2)$. We start with a function $\hat{F}_0$ that belongs to $L^2(0, \infty)$, and write

$$(6) \qquad F(x+iy) = \int_0^\infty \hat{F}_0(\xi) e^{2\pi i \xi z} \, d\xi, \qquad z = x + iy, \; y > 0.$$

(The choice of the particular notation $\hat{F}_0$ will become clearer below.) We claim that for any $\delta > 0$ the integral (6) converges absolutely and uniformly as long as $y \geq \delta$. Indeed, $|\hat{F}_0(\xi) e^{2\pi i \xi z}| = |\hat{F}_0(\xi)| e^{-2\pi \xi y}$, hence by the Cauchy-Schwarz inequality

$$\int_0^\infty |\hat{F}_0(\xi) e^{2\pi i \xi z}| \, d\xi \leq \left(\int_0^\infty |\hat{F}_0(\xi)|^2 d\xi \right)^{1/2} \left(\int_0^\infty e^{-4\pi \xi \delta} d\xi \right)^{1/2},$$

from which the asserted convergence is established. From the uniform convergence it follows that $F(z)$ is holomorphic in the upper half-plane. Moreover, by Plancherel's theorem

$$\int_{\mathbb{R}} |F(x+iy)|^2 \, dx = \int_0^\infty |\hat{F}_0(\xi)|^2 \, e^{-4\pi \xi y} \, d\xi \leq \|\hat{F}_0\|_{L^2(0,\infty)}^2,$$

---

[2]Further motivation and some elementary background material may be found in Theorem 3.5 in Chapter 4 of Book II.

and in fact, by the monotone convergence theorem,

$$\sup_{y>0} \int_{\mathbb{R}} |F(x+iy)|^2 \, dx = \|\hat{F}_0\|_{L^2(0,\infty)}^2.$$

In particular, F belongs to $H^2(\mathbb{R}_+^2)$. The main result we prove next is the converse, that is, every element of the space $H^2(\mathbb{R}_+^2)$ is in fact of the form (6).

Theorem 2.1 *The elements F in $H^2(\mathbb{R}_+^2)$ are exactly the functions given by (6), with $\hat{F}_0 \in L^2(0,\infty)$. Moreover*

$$\|F\|_{H^2(\mathbb{R}_+^2)} = \|\hat{F}_0\|_{L^2(0,\infty)}.$$

This shows incidentally that $H^2(\mathbb{R}_+^2)$ is a Hilbert space that is isomorphic to $L^2(0,\infty)$ via the correspondence (6).

The crucial point in the proof of the theorem is the following fact. For any fixed strictly positive y, we let $\hat{F}_y(\xi)$ denote the Fourier transform of the L^2 function $F(x+iy)$, $x \in \mathbb{R}$. Then for any pair of choices of y, y_1 and y_2, we have that

$$(7) \qquad \hat{F}_{y_1}(\xi)e^{2\pi y_1 \xi} = \hat{F}_{y_2}(\xi)e^{2\pi y_2 \xi} \qquad \text{for a.e. } \xi.$$

To establish this assertion we rely on a useful technical observation.

Lemma 2.2 *If F belongs to $H^2(\mathbb{R}_+^2)$, then F is bounded in any proper half-plane $\{z = x+iy, \ y \geq \delta\}$, where $\delta > 0$.*

To prove this we exploit the mean-value property of holomorphic functions. This property may be stated in two alternative ways. First, in terms of averages over circles,

$$(8) \qquad F(\zeta) = \frac{1}{2\pi} \int_0^{2\pi} F(\zeta + re^{i\theta}) \, d\theta \qquad \text{if } 0 < r \leq \delta.$$

(Note that if ζ lies in the upper half-plane, $\text{Im}(\zeta) > \delta$, then the disc centered at ζ of radius r belongs to $\mathbb{R}_+^2$.) Alternatively, integrating over r, we have the mean-value property in terms of discs,

$$(9) \qquad F(\zeta) = \frac{1}{\pi\delta^2} \int_{|z|<\delta} F(\zeta + z) \, dx \, dy, \qquad z = x + iy.$$

These assertions actually hold for harmonic functions in $\mathbb{R}^2$ (see Corollary 7.2, Chapter 3 in Book II for the result about holomorphic functions, and Lemma 2.8, Chapter 5 in Book I for the case of harmonic functions); later in this chapter we in fact prove the extension of (9) to $\mathbb{R}^d$.

From (9) we see from the Cauchy-Schwarz inequality that

$$|F(\zeta)|^2 \leq \frac{1}{\pi\delta^2} \int_{|z|<\delta} |F(\zeta + z)|^2 \, dx \, dy.$$

Writing $z = x + iy$ and $\zeta = \xi + i\eta$, with $\eta > \delta$, we see that the disc $B_\delta(\zeta)$ of center ζ and radius δ is contained in the strip $\{z + \zeta : z = x + iy, \; -\delta < y < \delta\}$, and moreover this strip lies in the half-plane $\mathbb{R}^2_+$. See Figure 1.

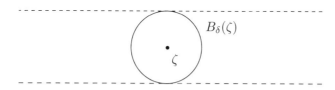

Figure 1. Disc contained in a strip

This gives the following majorization:

$$\int_{|z|<\delta} |F(\zeta + z)|^2 \, dx \, dy \leq \int_{|y|<\delta} \int_{\mathbb{R}} |F(\zeta + x + iy)|^2 \, dx \, dy$$

$$\leq 2\delta \sup_{-\delta < y < \delta} \int_{\mathbb{R}} |F(x + i(\eta + y))|^2 \, dx.$$

Recalling that $\eta > \delta$, we see that the last expression is in fact majorized by

$$2\delta \sup_{y>0} \int_{\mathbb{R}} |F(x + iy)|^2 \, dx = 2\delta \, \|F\|^2_{H^2(\mathbb{R}^2_+)}.$$

In all $|F(\zeta)|^2 \leq \frac{2}{\pi\delta} \|F\|^2_{H^2}$ in the half-plane $\mathrm{Im}(\zeta) > 0$, which proves the lemma.

We now turn to the proof of the identity (7). Starting with F in $H^2(\mathbb{R}^2_+)$, we improve it by replacing it with the function F^ϵ defined by

$$F^\epsilon(z) = F(z) \frac{1}{(1 - i\epsilon z)^2}, \quad \text{with } \epsilon > 0.$$

Observe that $|F^\epsilon(z)| \leq |F(z)|$ when $\mathrm{Im}(z) > 0$; also $F^\epsilon(z) \to F(z)$ for each such z, as $\epsilon \to 0$. This shows that for each $y > 0$, $F^\epsilon(x + iy) \to$

$F(x + iy)$ in the L^2-norm. Moreover, the lemma guarantees that each F^ϵ satisfies the decay estimate

$$F^\epsilon(z) = O\left(\frac{1}{1 + x^2}\right) \quad \text{whenever } \mathrm{Im}(z) > \delta, \text{ for some } \delta > 0.$$

We assert first that (7) holds with F replaced by F^ϵ. This is a simple consequence of contour integration applied to the function

$$G(z) = F^\epsilon(z)e^{-2\pi i z \xi}.$$

In fact we integrate $G(z)$ over the rectangle with vertices $-R + iy_1$, $R + iy_1$, $R + iy_2$, $-R + iy_2$, and let $R \to \infty$. If we take into account that $G(z) = O(1/(1 + x^2))$ in this rectangle, then we find that

$$\int_{L_1} G(z)\, dz = \int_{L_2} G(z)\, dz,$$

where L_j is the line $\{x + iy_j : x \in \mathbb{R}\}$, $j = 1, 2$. Since

$$\int_{L_j} G(z)\, dz = \int_{\mathbb{R}} F^\epsilon(x + iy_j)e^{-2\pi i(x + iy_j)\xi}\, dx,$$

This means that

$$\hat{F}^\epsilon_{y_1}(\xi)e^{2\pi y_1 \xi} = \hat{F}^\epsilon_{y_2}(\xi)e^{2\pi y_2 \xi}.$$

Since $F^\epsilon(x + iy_j) \to F(x + iy_j)$ in the L^2-norm as $\epsilon \to 0$, we then obtain (7).

The identity we have just proved states that $\hat{F}_y(\xi)e^{2\pi y \xi}$ is independent of y, $y > 0$, and thus there is a function $\hat{F}_0(\xi)$ so that $\hat{F}_y(\xi)e^{2\pi \xi y} = \hat{F}_0(\xi)$; as a result

$$\hat{F}_y(\xi) = \hat{F}_0(\xi)e^{-2\pi \xi y} \quad \text{for all } y > 0.$$

Therefore by Plancherel's identity

$$\int_{\mathbb{R}} |F(x + iy)|^2\, dx = \int_{\mathbb{R}} |\hat{F}_0(\xi)|^2 e^{-4\pi \xi y}\, d\xi,$$

and hence

$$\sup_{y > 0} \int_{\mathbb{R}} |\hat{F}_0(\xi)|^2 e^{-4\pi \xi y}\, d\xi = \|F\|^2_{H^2(\mathbb{R}^2_+)} < \infty.$$

Finally this in turn implies that $\hat{F}_0(\xi) = 0$ for almost every $\xi \in (-\infty, 0)$. For if this were not the case, then for appropriate positive numbers a, b, and c we could have that $|\hat{F}_0(\xi)| \geq a$ for ξ in a set E in $(-\infty, -b)$, with $m(E) \geq c$. This would give $\int |\hat{F}_0(\xi)|^2 e^{-4\pi\xi y} \, d\xi \geq a^2 c e^{4\pi b y}$, which grows indefinitely as $y \to \infty$. The contradiction thus obtained shows that $\hat{F}_0(\xi)$ vanishes almost everywhere when $\xi \in (-\infty, 0)$.

To summarize, for each $y > 0$ the function $\hat{F}_y(\xi)$ equals $\hat{F}_0(\xi)e^{-2\pi\xi y}$, with $\hat{F}_0 \in L^2(0, \infty)$. The Fourier inversion formula then yields the representation (6) for an arbitrary element of H^2, and the proof of the theorem is concluded.

The second result we deal with may be viewed as the half-plane analogue of Fatou's theorem in the previous chapter.

Theorem 2.3 *Suppose F belongs to $H^2(\mathbb{R}_+^2)$. Then $\lim_{y \to 0} F(x + iy) = F_0(x)$ exists in the following two senses:*

(i) *As a limit in the $L^2(\mathbb{R})$-norm.*

(ii) *As a limit for almost every x.*

Thus F has boundary values (denoted by F_0) in either of the two senses above. The function F_0 is sometimes referred to as the **boundary-value function** of f. The proof of (i) is immediate from what we already know. Indeed, if F_0 is the L^2 function whose Fourier transform is $\hat{F}_0$, then

$$\|F(x + iy) - F_0(x)\|_{L^2(\mathbb{R})}^2 = \int_0^\infty |\hat{F}_0(\xi)|^2 |e^{-2\pi\xi y} - 1|^2 \, dy,$$

and this tends to zero as $y \to 0$ by the dominated convergence theorem.

To prove the almost everywhere convergence, we establish the **Poisson integral representation**

$$(10) \qquad \int_{\mathbb{R}} \hat{f}(\xi)e^{-2\pi|\xi|y}e^{2\pi i x\xi} \, d\xi = \int_{\mathbb{R}} f(x - t)\mathcal{P}_y(t) \, dt,$$

with

$$\mathcal{P}_y(x) = \frac{1}{\pi} \frac{y}{y^2 + x^2}$$

the Poisson kernel.[3] This identity holds for every $(x, y) \in \mathbb{R}_+^2$ and any function f in $L^2(\mathbb{R})$. To see this, we begin by noting the following elementary integration formulas:

$$(11) \qquad \int_0^\infty e^{2\pi i \xi z} \, d\xi = \frac{i}{2\pi z} \qquad \text{if } \mathrm{Im}(z) > 0,$$

---

[3]This is the analogue in $\mathbb{R}$ of the identity (3) for the circle, given in Chapter 4.

and

$$\text{(12)} \qquad \int_{\mathbb{R}} e^{-2\pi|\xi|y} e^{2\pi i \xi x} \, d\xi = \frac{1}{\pi} \frac{y}{y^2 + x^2} \qquad \text{if } y > 0.$$

The first is an immediate consequence of the fact that

$$\int_0^N e^{2\pi i \xi z} \, d\xi = \frac{1}{2\pi i z} [e^{2\pi i N z} - 1]$$

if we let $N \to \infty$. To prove the second formula, we write the integral as

$$\int_0^\infty e^{-2\pi \xi y} e^{2\pi i \xi x} \, d\xi + \int_0^\infty e^{-2\pi \xi y} e^{-2\pi i \xi x} \, d\xi,$$

which equals

$$\frac{i}{2\pi} \left[\frac{1}{x + iy} + \frac{1}{-x + iy} \right] = \frac{1}{\pi} \frac{y}{y^2 + x^2}$$

by (11).

Next we establish (10) when f belongs to (say) the space $\mathcal{S}$. Indeed, for fixed $(x, y) \in \mathbb{R}_+^2$ consider the function $\Phi(t, \xi) = f(t) e^{-2\pi i \xi t} e^{-2\pi |\xi| y} e^{2\pi i \xi x}$ on $\mathbb{R}^2 = \{(\xi, t)\}$. Since $|\Phi(t, \xi)| = |f(t)| e^{-2\pi |\xi| y}$, then (because f is rapidly decreasing) Φ is integrable over $\mathbb{R}^2$. Applying Fubini's theorem yields

$$\int_{\mathbb{R}} \left(\int_{\mathbb{R}} \Phi(t, \xi) \, d\xi \right) dt = \int_{\mathbb{R}} \left(\int_{\mathbb{R}} \Phi(t, \xi) \, dt \right) d\xi.$$

The right-hand side obviously gives $\int_{\mathbb{R}} \hat{f}(\xi) e^{-2\pi |\xi| y} e^{2\pi i x \xi} \, d\xi$, while the left-hand side yields $\int_{\mathbb{R}} f(t) \mathcal{P}_y(x - y) \, dt$ in view of (12) above. However, if we use the relation (6) in Chapter 2 we see that

$$\int_{\mathbb{R}} f(t) \mathcal{P}_y(x - y) \, dt = \int_{\mathbb{R}} f(x - t) \mathcal{P}_y(t) \, dt.$$

Thus the Poisson integral representation (10) holds for every $f \in \mathcal{S}$. For a general $f \in L^2(\mathbb{R})$ we consider a sequence $\{f_n\}$ of elements in $\mathcal{S}$, so that $f_n \to f$ (and also $\hat{f}_n \to \hat{f}$) in the L^2-norm. A passage to the limit then yields the formula for f from the corresponding formula for each f_n. Indeed, by the Cauchy-Schwarz inequality we have

$$\left| \int_{\mathbb{R}} [\hat{f}(\xi) - \hat{f}_n(\xi)] e^{-2\pi |\xi| y} e^{2\pi i x \xi} \, d\xi \right| \leq \|\hat{f} - \hat{f}_n\|_{L^2} \left(\int_{\mathbb{R}} e^{-4\pi |\xi| y} \, d\xi \right)^{1/2},$$

and also

$$\left| \int_{\mathbb{R}} [f(x-t) - f_n(x-t)] \mathcal{P}_y(t)\, dt \right| \leq \| f - f_n \|_{L^2} \left(\int_{\mathbb{R}} |\mathcal{P}_y(t)|^2\, dt \right)^{1/2},$$

and the right-hand sides tend to 0 because for each fixed $(x, y) \in \mathbb{R}^2_+$ the functions $e^{-2\pi|\xi|y}$, $\xi \in \mathbb{R}$, and $\mathcal{P}_y(t)$, $t \in \mathbb{R}$, belong to $L^2(\mathbb{R})$.

Having established the Poisson integral representation (10), we return to our given element $F \in H^2(\mathbb{R}^2_+)$. We know that there is an L^2 function $\hat{F}_0(\xi)$ (which vanishes when $\xi < 0$) such that (6) holds. With F_0 the $L^2(\mathbb{R})$ function whose Fourier transform is $\hat{F}_0(\xi)$, we see from (10), with $f = F_0$, that

$$F(x + iy) = \int_{\mathbb{R}} F_0(x - t) \mathcal{P}_y(t)\, dt.$$

From this we deduce the fact that $F(x + iy) \to F_0(x)$ a.e in x as $y \to 0$, since the family $\{\mathcal{P}_y\}$ is an approximation of the identity for which Theorem 2.1 in Chapter 3 applies. There is, however, one small obstacle that has to be overcome: the theorem as stated applied to L^1 functions and not to functions in L^2. Nevertheless, given the nature of the approximation to the identity, a simple "localization" argument will succeed. We proceed as follows.

It will suffice to see that for any large N, which is fixed, $F(x + iy) \to F_0(x)$, for a.e x with $|x| < N$. To do this, decompose F_0 as $G + H$, where $G(x) = F_0(x)$ when $|x| > 2N$, $G(x) = 0$ when $|x| \geq 2N$; thus $H(x) = 0$ if $|x| \leq 2N$ but $|H(x)| \leq |F_0(x)|$. Note that now $G \in L^1$ and

$$\int_{\mathbb{R}} F_0(x - t) \mathcal{P}_y(t)\, dt = \int_{\mathbb{R}} G(x - t) \mathcal{P}_y(t)\, dt + \int_{\mathbb{R}} H(x - t) \mathcal{P}_y(t)\, dt.$$

Therefore, by the above mentioned theorem in Chapter 3, the first integral on the right-hand side converges for a.e x to $G(x) = F_0(x)$ when $|x| < N$. While when $|x| < N$ the integrand of the second integral vanishes when $|t| < N$ (since then $|x - t| < 2N$). That integral is therefore majorized by

$$\left(\int_{\mathbb{R}} |H(x - t)|^2\, dt \right)^{1/2} \left(\int_{|t| \geq N} |\mathcal{P}_y(t)|^2\, dt \right)^{1/2}.$$

However $\left(\int_{\mathbb{R}} |H(x - t)|^2\, dt \right)^{1/2} \leq \| F_0 \|_{L^2}$, while (as is easily seen) $\int_{|t| \geq N} |\mathcal{P}_y(t)|^2\, dt \to 0$ as $y \to 0$. Hence $F(x + iy) \to F_0(x)$ for a.e x with

$|x| < N$, as $y \to 0$, and since N is arbitrary, the proof of Theorem 2.3 is now complete.

The following comments may help clarify the thrust of the above theorems.

(i) Let S be the subspace of $L^2(\mathbb{R})$ consisting of all functions F_0 arising in Theorem 2.3. Then, since the functions F_0 are exactly those functions in L^2 whose Fourier transform is supported on the half-line $(0, \infty)$, we see that S is a closed subspace. We might be tempted to say that S consists of those functions in L^2 that arise as boundary values of holomorphic functions in the upper half-plane; but this heuristic assertion is not exact if we do not add a quantitative restriction such as in the definition (5) of the Hardy space. See Exercise 4.

(ii) Suppose we defined P to be the orthogonal projection on the subspace S of L^2. Then, as is easily seen, $\widehat{(Pf)}(\xi) = \chi(\xi)\hat{f}(\xi)$ for any $f \in L^2(\mathbb{R})$; here χ is the characteristic function of $(0, \infty)$. The operator P is also closely related to the **Cauchy integral**. Indeed, if F is the (unique) element in $H^2(\mathbb{R}^2_+)$ whose boundary function (according to Theorem 2.3) is $P(f)$, then

$$F(z) = \frac{1}{2\pi i} \int_{\mathbb{R}} \frac{f(t)}{t - z}\, dt, \quad z \in \mathbb{R}^2_+.$$

To prove this it suffices to verify that for any $f \in L^2(\mathbb{R})$ and any fixed $z = x + iy \in \mathbb{R}^2_+$, we have

$$\int_0^\infty \hat{f}(\xi)e^{2\pi i \xi z}\, d\xi = \frac{1}{2\pi i} \int_{\mathbb{R}} \frac{f(t)}{t - z}\, dt.$$

This is proved in the same way as the Poisson integral representation (10) except here we use the identity (11) instead of (12). The details may be left to the interested reader. Also, the reader might note the close analogy between this version of the Cauchy integral for the upper-half plane, and a corresponding version for the unit disc, as given in Example 2, Section 4 of Chapter 4.

(iii) In analogy with the periodic case discussed in Exercise 30 of Chapter 4, we define a **Fourier multiplier operator** T on $\mathbb{R}$ to be a linear operator on $L^2(\mathbb{R})$ determined by a bounded function m (the **multiplier**), such that T is defined by the formula $\widehat{(Tf)}(\xi) = m(\xi)\hat{f}(\xi)$ for any $f \in L^2(\mathbb{R})$. The orthogonal projection P above is such an operator and its multiplier is the characteristic function $\chi(\xi)$. Another closely related operator of this type is the **Hilbert transform** H defined by

$P = \frac{I+iH}{2}$. Then H is a Fourier multiplier operator corresponding to the multiplier $\frac{1}{i}\text{sign}(\xi)$. Among the many important properties of H is its connection to conjugate harmonic functions. Indeed, for f a real-valued function in $L^2(\mathbb{R})$, f and $H(f)$ are, respectively, the real and imaginary parts of the boundary values of a function in the Hardy space. More about the Hilbert transform can be found in Exercises 9 and 10 and Problem 5 below.

3 Constant coefficient partial differential equations

We turn our attention to solving the linear partial differential equation

$$(13) \qquad\qquad L(u) = f,$$

where the operator L takes the form

$$L = \sum_{|\alpha|\leq n} a_\alpha \left(\frac{\partial}{\partial x}\right)^\alpha$$

with $a_\alpha \in \mathbb{C}$ constants.

In the study of the classical examples of L, such as the wave equation, the heat equation, and Laplace's equation, one already sees the Fourier transform entering in an important way.[4] For general L, this key role is further indicated by the following simple observation. If, for example, we try to solve this equation with both u and f elements in $\mathcal{S}$, then this is equivalent to the algebraic equation

$$P(\xi)\hat{u}(\xi) = \hat{f}(\xi),$$

where $P(\xi)$ is the **characteristic polynomial** of f defined by

$$P(\xi) = \sum_{|\alpha|\leq n} a_\alpha (2\pi i\xi)^\alpha.$$

This is because one has the Fourier transform identity

$$\widehat{\left(\frac{\partial^\alpha f}{\partial x^\alpha}\right)}(\xi) = (2\pi i\xi)^\alpha \hat{f}(\xi).$$

Thus a solution u in the space $\mathcal{S}$ (if it exists) would be uniquely determined by

$$\hat{u}(\xi) = \frac{\hat{f}(\xi)}{P(\xi)}.$$

---

[4]See for example Chapters 5 and 6 in Book I.

In a more general setting, matters are not so easy: aside from the question of defining (13), the Fourier transform is not directly applicable; also, solutions that we prove to exist (but are not unique!) have to be understood in a wider sense.

3.1 Weak solutions

As the reader may have guessed, it will not suffice to restrict our attention to those functions for which $L(u)$ is defined in the usual way, but instead a broader notion is needed, one involving the idea of "weak solutions." To describe this concept, we start with a given open set Ω in $\mathbb{R}^d$ and consider the space $C_0^\infty(\Omega)$, which consists of the indefinitely differentiable functions[5] having compact support in Ω.[6] We have the following fact.

Lemma 3.1 *The space $C_0^\infty(\Omega)$ is dense in $L^2(\Omega)$ in the norm $\|\cdot\|_{L^2(\Omega)}$.*

The proof is essentially a repetition of that of Lemma 1.2. We take the precaution of modifying the definition of g_M given there to be: $g_M(x) = f(x)$ if $|x| \leq M$, $d(x, \Omega^c) \geq 1/M$ and $|f(x)| \leq M$, and $g_M(x) = 0$ otherwise. Also, when we regularize g_M, we replace it with $g_M * \varphi_\delta$, with $\delta < 1/2M$. Then the support of $g_M * \varphi_\delta$ is still compact and at a distance $\geq 1/2M$ from Ω^c.

We next consider the **adjoint operator** of L defined by

$$L^* = \sum_{|\alpha| \leq n} (-1)^{|\alpha|} \overline{a_\alpha} \left(\frac{\partial}{\partial x}\right)^\alpha.$$

The operator L^* is called the adjoint of L because, in analogy with the definition of the adjoint of a bounded linear transformation given in Section 5.2 of the previous chapter, we have

(14) $(L\varphi, \psi) = (\varphi, L^*\psi)$ whenever $\varphi, \psi \in C_0^\infty(\Omega)$,

where $(\cdot, \cdot)$ denotes the inner product on $L^2(\Omega)$ (which is the restriction of the usual inner product on $L^2(\mathbb{R}^d)$). The identity (14) is proved by successive integration by parts. Indeed, consider first the special case when $L = \partial/\partial x_j$, and then $L^* = -\partial/\partial x_j$. If we use Fubini's theorem, integrating first in the x_j variable, then in this case (14) reduces to the

---

[5]Indefinitely differentiable functions are also referred to as C^∞ functions, or **smooth** functions.

[6]This means that the closure of the support of f, as defined in Section 1 of Chapter 2, is compact and contained in Ω.

familiar one-dimensional formula

$$\int_{-\infty}^{\infty} \left(\frac{d\varphi}{dx} \right) \overline{\psi} \, dx = - \int_{-\infty}^{\infty} \varphi \left(\frac{\overline{d\psi}}{dx} \right) \, dx,$$

with the integrated boundary terms vanishing because of the assumed support properties of ψ (or φ). Once established for $L = \partial/\partial x_j$, $1 \leq j \leq n$, then (14) follows for $L = (\partial/\partial x)^\alpha$ by iteration, and hence for general L by linearity.

At this point we digress momentarily to consider besides $C_0^\infty(\Omega)$ some other spaces of differentiable functions on Ω that will be useful later. The space $C^n(\Omega)$ consists of all functions f on Ω that have continuous partial derivatives of order $\leq n$. Also, the space $C^n(\overline{\Omega})$ consists of those functions on $\overline{\Omega}$ that can be extended to functions in $\mathbb{R}^d$ that belong to $C^n(\mathbb{R}^d)$. Thus, in an obvious sense, we have the inclusion relation

$$C_0^\infty(\Omega) \subset C^n(\overline{\Omega}) \subset C^n(\Omega), \quad \text{for each positive integer } n.$$

Returning to our partial differential operator L, it is useful to observe that the formula

$$(Lu, \psi) = (u, L^*\psi)$$

continues to hold (with the same proof) if we merely assume that $u \in C^n(\Omega)$ without assuming it has compact support, while still supposing $\psi \in C_0^\infty(\Omega)$.

In particular, if we have $L(u) = f$ in the ordinary sense (sometimes called the "strong" sense), which requires the assumption that $u \in C^n(\Omega)$ in order to define the partial derivatives entering in Lu, then we would also have

(15) $$(f, \psi) = (u, L^*\psi) \quad \text{for all } \psi \in C_0^\infty(\Omega).$$

This leads to the following important definition: if $f \in L^2(\Omega)$, a function $u \in L^2(\Omega)$ is a **weak solution** of the equation $Lu = f$ in Ω if (15) holds. Of course an ordinary solution is always a weak solution.

Significant instances of weak solutions that are not ordinary solutions already arise in elementary situations such as in the study of the one-dimensional wave equation. Here $L(u) = (\partial^2 u/\partial x^2) - (\partial^2 u/\partial t^2)$, so the underlying space is $\mathbb{R}^2 = \{(x_1, x_2) : \text{with } x_1 = x, x_2 = t\}$. Suppose, for example, we consider the case of the "plucked string."[7] We are then

---

[7]See Chapter 1 in Book I.

looking at the solution of $L(u) = 0$ subject to the boundary conditions $u(x,0) = f(x)$ and $(\partial u/\partial t)(x,0) = 0$ for $0 \leq x \leq \pi$, where the graph of f is piecewise linear and is illustrated in Figure 2.

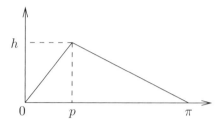

Figure 2. Initial position of a plucked string

If one extends f to $[-\pi, \pi]$ by making it odd, and then to all of $\mathbb{R}$ by periodicity (of period 2π), then the solution is given by d'Alembert's formula

$$u(x,t) = \frac{f(x+t) + f(x-t)}{2}.$$

In the present case u is *not* twice continuously differentiable, and it is therefore not an ordinary solution. Nevertheless it is a weak solution. To see this, approximate f by a sequence of functions f_n that are C^∞ and such that $f_n \to f$ uniformly on every compact subset of $\mathbb{R}$.[8] If we define $u_n(x,t)$ as $[f_n(x+t) + f_n(x-t)]/2$, we can check directly that $L(u_n) = 0$ and hence $(u_n, L^*\psi) = 0$ for all $\psi \in C_0^\infty(\mathbb{R}^2)$, and thus by uniform convergence we obtain that $(u, L^*\psi) = 0$ as desired.

A different example illustrating the nature of weak solutions arises for the operator $L = d/dx$ on $\mathbb{R}$. If we suppose $\Omega = (0,1)$, then with u and f in $L^2(\Omega)$, we have that $Lu = f$ in the weak sense if and only if there is an absolutely continuous function F on $[0,1]$ such that $F(x) = u(x)$ and $F'(x) = f(x)$ almost everywhere. For more about this, see Exercise 14.

3.2 The main theorem and key estimate

We now turn to the general theorem guaranteeing the existence of solutions of partial differential equations with constant coefficients

Theorem 3.2 *Suppose Ω is a bounded open subset of $\mathbb{R}^d$. Given a linear partial differential operator L with constant coefficients, there exists a*

---

[8] One may write, for example, $f_n = f * \varphi_{1/n}$, where $\{\varphi_\epsilon\}$ is the approximation to the identity, as in the proof of Lemma 1.2.

bounded linear operator K on $L^2(\Omega)$ such that whenever $f \in L^2(\Omega)$, then

$$L(Kf) = f \quad \text{in the weak sense.}$$

In other words, $u = K(f)$ is a weak solution to $L(u) = f$.

The heart of the matter lies in an inequality that we state next, but whose proof (which uses the Fourier transform) is postponed until the next section.

Lemma 3.3 *There exists a constant c such that*

$$\|\psi\|_{L^2(\Omega)} \le c\|L^*\psi\|_{L^2(\Omega)} \quad \text{whenever } \psi \in C_0^\infty(\Omega).$$

The usefulness of this lemma comes about for the following reason. If L is a finite-dimensional linear transformation, the solvability of L (the fact that it is surjective) is of course equivalent with the fact that its adjoint L^* is injective. In effect, the lemma provides the analytic substitute for this reasoning in an infinite-dimensional setting.

We first prove the theorem assuming the validity of the inequality in the lemma.

Consider the pre-Hilbert space $\mathcal{H}_0 = C_0^\infty(\Omega)$ equipped with the inner product and norm

$$\langle \varphi, \psi \rangle = (L^*\varphi, L^*\psi), \qquad \|\psi\|_0^2 = \|L^*\psi\|_{L^2(\Omega)}.$$

Following the results in Section 2.3 of Chapter 4, we let $\mathcal{H}$ denote the completion of $\mathcal{H}_0$. By Lemma 3.3, a Cauchy sequence in the $\|\cdot\|_0$-norm is also Cauchy in the $L^2(\Omega)$-norm; hence we may identify $\mathcal{H}$ with a subspace of $L^2(\Omega)$. Also, L^*, initially defined as a bounded operator from $\mathcal{H}_0$ to $L^2(\Omega)$, extends to a bounded operator L^* from $\mathcal{H}$ to $L^2(\Omega)$ (by Lemma 1.3). For a fixed $f \in L^2(\Omega)$, consider the linear map $\ell_0 : C_0^\infty(\Omega) \to \mathbb{C}$ defined by

$$\ell_0(\psi) = (\psi, f) \quad \text{for } \psi \in C_0^\infty(\Omega).$$

The Cauchy-Schwarz inequality together with another application of Lemma 3.3 yields

$$\begin{aligned}
|\ell_0(\psi)| = |(\psi, f)| &\le \|\psi\|_{L^2(\Omega)}\|f\|_{L^2(\Omega)} \\
&\le c\|L^*\psi\|_{L^2(\Omega)}\|f\|_{L^2(\Omega)} \\
&\le c'\|\psi\|_0,
\end{aligned}$$

with $c' = c\|f\|_{L^2(\Omega)}$. Hence ℓ_0 is bounded on the pre-Hilbert space $\mathcal{H}_0$. Therefore, ℓ extends to a bounded linear functional on $\mathcal{H}$ (see Section 5.1, Chapter 4), and the above inequalities show that $\|\ell\| \leq c\|f\|_{L^2(\Omega)}$. By the Riesz representation theorem applied to ℓ on the Hilbert space $\mathcal{H}$ (Theorem 5.3 in Chapter 4), there exists $U \in \mathcal{H}$ such that

$$\ell(\psi) = \langle \psi, U \rangle = (L^*\psi, L^*U) \quad \text{for all } \psi \in C_0^\infty(\Omega).$$

Here $\langle \cdot, \cdot \rangle$ denotes the extension to $\mathcal{H}$ of the initial inner product on $\mathcal{H}_0$, and L^* also denotes the extension of L^* originally given on $\mathcal{H}_0$.

If we let $u = L^*U$, then $u \in L^2(\Omega)$, and we find that

$$\ell(\psi) = (\psi, f) = (L^*\psi, u) \quad \text{for all } \psi \in C_0^\infty(\mathbb{R}^d).$$

Hence

$$(f, \psi) = (u, L^*\psi) \quad \text{for all } \psi \in C_0^\infty(\mathbb{R}^d),$$

and by definition, u is a weak solution to the equation $Lu = f$ in Ω. If we let $Kf = u$, we see that once f is given, Kf is uniquely determined by the above steps. Since $\|U\|_0 = \|\ell\| \leq c\|f\|_{L^2(\Omega)}$ we see that

$$\|Kf\|_{L^2(\Omega)} = \|u\|_{L^2(\Omega)} = \|L^*U\|_{L^2(\Omega)} = \|U\|_0 \leq c\|f\|_{L^2(\Omega)},$$

whence $K : L^2(\Omega) \to L^2(\Omega)$ is bounded.

Proof of the main estimate

To complete the proof of the theorem, we must still prove the estimate in Lemma 3.3, that is,

$$\|\psi\|_{L^2(\Omega)} \leq c\|L^*\psi\|_{L^2(\Omega)} \quad \text{whenever } \psi \in C_0^\infty(\Omega).$$

The reasoning below relies on an important fact: if f has compact support in $\mathbb{R}$, then $\hat{f}(\xi)$ initially defined for $\xi \in \mathbb{R}$ extends to an entire function for $\zeta = \xi + i\eta \in \mathbb{C}$. This observation reduces the problem to an inequality about holomorphic functions and polynomials.

Lemma 3.4 *Suppose* $P(z) = z^m + \cdots + a_1 z + a_0$ *is a polynomial of degree* m *with leading coefficient* 1. *If* F *is a holomorphic function on* $\mathbb{C}$, *then*

$$|F(0)|^2 \leq \frac{1}{2\pi} \int_0^{2\pi} |P(e^{i\theta})F(e^{i\theta})|^2 d\theta.$$

Proof. The lemma is a consequence of the special case when $P = 1$

$$(16) \qquad |F(0)|^2 \le \frac{1}{2\pi} \int_0^{2\pi} \int_0^{2\pi} |F(e^{i\theta})|^2 \, d\theta.$$

This assertion follows directly from the mean-value identity (8) in Section 2 with $\zeta = 0$ and $r = 1$, via the Cauchy-Schwarz inequality. With it we begin by factoring P:

$$P(z) = \prod_{|\alpha| \ge 1} (z - \alpha) \prod_{|\beta| < 1} (z - \beta) = P_1(z) P_2(z),$$

where each product is finite and taken over the roots of P whose absolute values are ≥ 1 and < 1, respectively.

Note that $|P_1(0)| = \prod_{|\alpha| \ge 1} |\alpha| \ge 1$.

For P_2 we write

$$(z - \beta) = -(1 - \bar{\beta} z) \psi_\beta(z),$$

where $\psi_\beta(z) = \frac{\beta - z}{1 - \bar{\beta} z}$ are the "Blaschke factors" that have the obvious property that they are holomorphic in a region containing the closed unit disc and $|\psi_\beta(e^{i\theta})| = 1$; see also Chapter 8 in Book II. We write $\tilde{P}_2 = \prod_{|\beta| < 1} (1 - \bar{\beta} z)$ and $\tilde{P} = P_1 \tilde{P}_2$. Thus $|\tilde{P}(0)| \ge 1$, while $|\tilde{P}(e^{i\theta})| = |P(e^{i\theta})|$ for every θ. We now apply (16) to the function $\tilde{P} F$ in place of F and find that

$$|F(0)|^2 \le |\tilde{P}(0) F(0)|^2 \le \frac{1}{2\pi} \int_0^{2\pi} |\tilde{P}(e^{i\theta}) F(e^{i\theta})|^2 \, d\theta$$

$$= \frac{1}{2\pi} \int_0^{2\pi} |P(e^{i\theta}) F(e^{i\theta})|^2 \, d\theta,$$

which gives the desired conclusion.

We turn to the proof of the inequality $\|\psi\| \le c\|L^*\psi\|$ for all $\psi \in C_0^\infty(\Omega)$ in the special case of one dimension, that is, $\Omega \subset \mathbb{R}$.

Suppose f is an L^2 function supported on the interval $[-M, M]$. Then

$$\hat{f}(\xi) = \int_{-M}^{M} f(x) e^{-2\pi i x \xi} \, dx$$

whenever $\xi \in \mathbb{R}$. In fact, the above integral converges whenever ξ is replaced by $\zeta = \xi + i\eta \in \mathbb{C}$, and we may extend $\hat{f}$ to a holomorphic function of ζ in the whole complex plane. An application of the Plancherel formula (for fixed η) yields

$$\int_{-\infty}^{\infty} |\hat{f}(\xi + i\eta)|^2 \, d\xi \leq e^{4\pi M|\eta|} \int_{-\infty}^{\infty} |f(x)|^2 \, dx.$$

We use this observation in the following context. We may assume (upon multiplying L by a suitable constant) that

$$L^* = \sum_{0 \leq k \leq n} (-1)^k \overline{a_k} \left(\frac{\partial}{\partial x} \right)^k,$$

where $a_n = (2\pi i)^{-n}$. If we let $Q(\xi) = \sum_{0 \leq k \leq n} (-1)^k \overline{a_k} (2\pi i\xi)^k$ be its characteristic polynomial, then we note that

$$\widehat{L^* \psi}(\xi) = Q(\xi)\hat{\psi}(\xi) \quad \text{whenever } \psi \in C_0^\infty(\mathbb{R}).$$

If M is chosen so large that $\Omega \subset [-M, M]$, then our previous observation gives

$$(17) \qquad \int_{-\infty}^{\infty} |Q(\xi + i\eta)\hat{\psi}(\xi + i\eta)|^2 d\xi \leq e^{4\pi M|\eta|} \int_{-\infty}^{\infty} |L^* \psi(x)|^2 dx.$$

Picking $\eta = i \sin \theta$, and making a translation by $\cos \theta$ yields

$$\int_{-\infty}^{\infty} |Q(\xi + \cos\theta + i\sin\theta)\hat{\psi}(\xi + \cos\theta + i\sin\theta)|^2 d\xi \leq$$

$$\leq e^{4\pi M} \int_{-\infty}^{\infty} |L^* \psi(x)|^2 dx.$$

An application of Lemma 3.4 with $F(z) = \hat{\psi}(\xi + z)$ and $Q(\xi + z)$ in place of $P(z)$ then gives

$$|\hat{\psi}(\xi)|^2 \leq \frac{1}{2\pi} \int_0^{2\pi} |Q(\xi + \cos\theta + i\sin\theta)\hat{\psi}(\xi + \cos\theta + i\sin\theta)|^2 d\theta.$$

We now integrate in ξ over $\mathbb{R}$, and on the right-hand side interchange the order of the ξ and θ integrations; also by translation invariance we replace the integration in the ξ variable by that in the variable $\xi + \cos\theta$. Using (17) the result is

$$\|\hat{\psi}\|_{L^2(\mathbb{R})}^2 \leq \frac{1}{2\pi} \int_0^{2\pi} \int_{\mathbb{R}} |Q(\xi + i\sin\theta)\hat{\psi}(\xi + i\sin\theta)|^2 \, d\xi \, d\theta$$

$$\leq e^{4\pi M} \int_{\mathbb{R}} |L^* \psi(x)|^2 \, dx,$$

which by Plancherel's identity proves the main lemma in the one-dimensional case.

The higher dimensional case is a modification of the argument above. Let $Q = \sum_{|\alpha| \le n} (-1)^\alpha a_\alpha (2\pi i \xi)^\alpha$ be the characteristic polynomial of L^*. Then we can choose a new set of orthogonal axes (whose coordinates we denote by $(\xi_1, \ldots, \xi_d)$) so that if $\xi = (\xi_1, \xi')$ with $\xi' = (\xi_2, \ldots, \xi_d)$, then after multiplying by a suitable constant

$$(18) \qquad Q(\xi) = (2\pi i)^{-n} \xi_1^n + \sum_{j=0}^{n-1} \xi_1^j q_j(\xi'),$$

where $q_j(\xi')$ are polynomials of ξ' (of degrees $\le n - j$).

To see that such a choice is possible, write $Q = Q_n + Q'$, where Q_n is homogeneous of degree n and Q' has degree $< n$. Then since we may assume $Q_n \ne 0$ there is (after multiplying Q by a suitable constant), a unit vector γ so that $Q_n(\gamma) = (2\pi i)^{-n}$. Then $Q_n(\xi) = (2\pi i)^{-n} r^n$ if $\xi = \gamma r$, $r \in \mathbb{R}$. We can then take the ξ_1-axis to lie along γ, and the $\xi_2, \ldots, \xi_d$-axes to be in mutually orthogonal directions, from which the form (18) is clear.

Proceeding now as before we obtain

$$|\hat{\psi}(\xi_1, \xi')|^2 \le \frac{1}{2\pi} \int_0^{2\pi} |Q(\xi_1 + e^{i\theta}, \xi') \hat{\psi}(\xi_1 + e^{i\theta}, \xi')|^2 \, d\theta$$

for each $(\xi_1, \xi') \in \mathbb{R}^d$. An integration[9] then gives

$$\|\hat{\psi}\|_{L^2(\mathbb{R}^d)}^2 \le \frac{1}{2\pi} \int_0^{2\pi} \int_{\mathbb{R}^d} |Q(\xi_1 + i \sin\theta, \xi') \hat{\psi}(\xi_1 + i \sin\theta, \xi')|^2 \, d\xi \, d\theta.$$

If we suppose that the projection of the (bounded) set Ω on the x_1-axis is contained in $[-M, M]$, we see as before that the right-hand side above is majorized by $e^{4\pi M} \int_{\mathbb{R}^d} |L^* \psi(x)|^2 \, dx$, finishing the proof of Lemma 3.3 and hence that of the theorem.

4* The Dirichlet principle

Dirichlet's principle arose in the study of the boundary-value problem for Laplace's equation. Stated in the case of two dimensions it refers to the classical problem of finding the steady-state temperature of a plate

---

[9]We note that by the rotational invariance of Lebesgue measure (Problem 4 in Chapter 2 and Exercise 26 in Chapter 3), integration in ξ can be carried out in the new coordinates as well.

whose boundary is exposed to a given temperature distribution. The issue raised is the following question, called the **Dirichlet problem**: If Ω is a bounded open set in $\mathbb{R}^2$ and f a continuous function on the boundary $\partial\Omega$, we wish to find a function $u(x_1, x_2)$ such that

(19)
$$\begin{cases} \triangle u = 0 & \text{in } \Omega, \\ \quad u = f & \text{on } \partial\Omega. \end{cases}$$

Thus we need to determine a function that is C^2 (twice continuously differentiable) in Ω, whose Laplacian[10] is zero, and which is continuous on the closure of Ω, with $u|_{\partial\Omega} = f$.

With either Ω or f satisfying special symmetry conditions, the solution to this problem can sometimes be written out explicitly. For instance, if Ω is the unit disc, then

$$u(re^{i\theta}) = \frac{1}{2\pi} \int_{-\pi}^{\pi} f(\varphi) P_r(\theta - \varphi) \, d\varphi,$$

where P_r is the Poisson kernel (for the disc). We also obtained (in Books I and II) explicit formulas for the solution of the Dirichlet problem for some unbounded domains. For example, when Ω is the upper half-plane the solution is

$$u(x, y) = \int_{\mathbb{R}} \mathcal{P}_y(x - t) f(t) \, dt,$$

where $\mathcal{P}_y(x)$ is the analogous Poisson kernel for the upper half-plane. A somewhat similar convolution formula was obtained when Ω is a strip. Also, the Dirichlet problem can be solved explicitly for certain Ω by using conformal mappings.[11]

In general, however, there are no explicit solutions, and other methods must be found. An idea that was used intially was based on an approach of wide utility in mathematics and physics: to find the equilibrium state of a system one seeks to minimize an appropriate "energy" or "action." In the present case the role of this energy is played by the **Dirichlet integral**, which is defined for appropriate functions U by

$$\mathcal{D}(U) = \int_{\Omega} |\nabla U|^2 = \int_{\Omega} \left| \frac{\partial U}{\partial x_1} \right|^2 + \left| \frac{\partial U}{\partial x_2} \right|^2 dx_1 dx_2.$$

(Note the similarity with the expression of the "potential energy" in the case of the vibrating string in Chapters 3 and 6 of Book I.) In fact,

---

[10]The **Laplacian** of a function u in $\mathbb{R}^d$ is defined by $\triangle u = \sum_{k=1}^{d} \partial^2 u / \partial x_k^2$.
[11]The close relation between conformal maps and the Dirichlet problem is discussed in the last part of Section 1 of Chapter 8, in Book II.

that approach underlies the proof Riemann proposed for his well-known mapping theorem. About this early history R. Courant has written:

> Already some years before the rise of Riemann's genius, C.F. Gauss and W. Thompson had observed that the boundary value problem of the harmonic differential equation $\triangle u = u_{xx} + u_{yy} = 0$ for a domain G in the x, y-plane can be reduced to the problem of minimizing the integral $\mathcal{D}[\phi]$ for the domain G, under the condition that the functions ϕ admitted to competition have the prescribed boundary values. Because of the positive character of $\mathcal{D}[\phi]$ the existence of a solution for the latter problem was considered obvious and hence the existence for the former assured. As a student in Dirichlet's lectures, Riemann had been fascinated by this convincing argument: soon afterwards he used it, under the name "Dirichlet's Principle," in a more varied and spectacular manner as the very foundation of his new geometric function theory.

The application of Dirichlet's principle was thought to have been justified by the following simple observation:

Proposition 4.1 *Suppose there exists a function $u \in C^2(\overline{\Omega})$ that minimizes $\mathcal{D}(U)$ among all $U \in C^2(\overline{\Omega})$ with $U|_{\partial\Omega} = f$. Then u is harmonic in Ω.*

Proof. For functions F and G in $C^2(\overline{\Omega})$ define the following inner-product

$$\langle F, G \rangle = \int_\Omega \left(\frac{\partial F}{\partial x_1} \frac{\overline{\partial G}}{\partial x_1} + \frac{\partial F}{\partial x_2} \frac{\overline{\partial G}}{\partial x_2} \right) dx_1 dx_2.$$

We then note that $\mathcal{D}(u) = \langle u, u \rangle$. If v is any function in $C^2(\overline{\Omega})$ with $v|_{\partial\Omega} = 0$, then for all ϵ we have

$$\mathcal{D}(u + \epsilon v) \geq \mathcal{D}(u),$$

since $u + \epsilon v$ and u have the same boundary values, and u minimizes the Dirichlet integral. We note, however, that

$$\mathcal{D}(u + \epsilon v) = \mathcal{D}(u) + \epsilon^2 \mathcal{D}(v) + \epsilon \langle u, v \rangle + \epsilon \langle v, u \rangle.$$

Hence

$$\epsilon^2 \mathcal{D}(v) + \epsilon \langle u, v \rangle + \epsilon \langle v, u \rangle \geq 0,$$

and since ϵ can be both positive or negative, this can happen only if $\mathrm{Re}\langle u, v \rangle = 0$. Similarly, considering the perturbation $u + i\epsilon v$, we find $\mathrm{Im}\langle u, v \rangle = 0$, and therefore $\langle u, v \rangle = 0$. An integration by parts then provides

$$0 = \langle u, v \rangle = - \int_\Omega (\triangle u)\overline{v}$$

for all $v \in C^2(\overline{\Omega})$ with $v|_{\partial\Omega} = 0$. This implies that $\triangle u = 0$ in Ω, and of course u equals f on the boundary.

Nevertheless, several serious objections were later raised to Dirichlet's principle. The first was by Weierstrass, who pointed out that it was not clear (and had not been proved) that a minimizing function for the Dirichlet integral exists, so there might simply be no winner to the implied competition in Proposition 4.1. He argued by analogy with a simpler one-dimensional problem: that of minimizing the integral

$$D(\varphi) = \int_{-1}^{1} |x\varphi'(x)|^2 dx$$

among all C^1 functions on $[-1, 1]$ that satisfy $\varphi(-1) = -1$ and $\varphi(1) = 1$. The minimum value achieved by this integral is zero. To verify this, let ψ be a smooth non-decreasing function on $\mathbb{R}$ that satisfies $\psi(x) = 1$ for $x \geq 1$, and $\psi(x) = -1$ if $x \leq -1$. For each $0 < \epsilon < 1$, we consider the function

$$\varphi_\epsilon(x) = \begin{cases} 1 & \text{if } \epsilon \leq x, \\ \psi(x/\epsilon) & \text{if } -\epsilon < x < \epsilon, \\ -1 & \text{if } x \leq -\epsilon. \end{cases}$$

Then φ_ϵ satisfies the desired constraints, and if M denotes a bound for the derivative of ψ, we find

$$D(\varphi_\epsilon) = \int_{-\epsilon}^{\epsilon} |x|^2 |\epsilon^{-1}\psi'(x/\epsilon)|^2 dx$$
$$\leq \int_{-\epsilon}^{\epsilon} |\psi'(x/\epsilon)|^2 dx$$
$$\leq 2\epsilon M^2.$$

In the limit as ϵ tends to 0, we find that the minimum value of the integral $D(\varphi)$ is zero. This minimum value cannot be reached by a C^1 function satisfying the boundary conditions, since $D(\varphi) = 0$ implies $\varphi'(x) = 0$ and thus φ is constant.

A further objection was raised by Hadamard, who remarked that $\mathcal{D}(u)$ may be infinite even for a solution u of the boundary value problem: thus, in effect, there may simply be no competitors who qualify for the competition!

To illustrate this point, we return to the disc, and consider the function

$$f(\theta) = f_\alpha(\theta) = \sum_{n=0}^{\infty} 2^{-n\alpha} e^{i2^n\theta}$$

for $\alpha > 0$. This function first appeared in Chapter 4 of Book I, where it is shown that f_α is continuous but nowhere differentiable if $\alpha \leq 1$. The solution of the Dirichlet problem on the unit disc with boundary value f_α is given by the Poisson integral

$$u(r,\theta) = \sum_{n=0}^{\infty} r^{2^n} 2^{-n\alpha} e^{i2^n\theta}.$$

However, the use of polar coordinates gives

$$\left| \frac{\partial u}{\partial x_1} \right|^2 + \left| \frac{\partial u}{\partial x_2} \right|^2 = \left| \frac{\partial u}{\partial r} \right|^2 + \frac{1}{r^2} \left| \frac{\partial u}{\partial \theta} \right|^2.$$

Thus

$$\iint_{D_\rho} \left(\left| \frac{\partial u}{\partial x_1} \right|^2 + \left| \frac{\partial u}{\partial x_2} \right|^2 \right) dx_1 dx_2 = \int_0^\rho \int_0^{2\pi} \left(\left| \frac{\partial u}{\partial r} \right|^2 + \frac{1}{r^2} \left| \frac{\partial u}{\partial \theta} \right|^2 \right) d\theta r \, dr$$

where D_ρ is the disc of radius $0 < \rho < 1$ centered at the origin. Since

$$\frac{\partial u}{\partial r} \sim \sum 2^n 2^{-n\alpha} r^{2^n-1} e^{i2^n\theta} \quad \text{and} \quad \frac{\partial u}{\partial \theta} \sim \sum r^{2^n} 2^{-n\alpha} i 2^n e^{i2^n\theta},$$

applications of Parseval's identity lead to

$$\iint_{D_\rho} \left(\left| \frac{\partial u}{\partial x_1} \right|^2 + \left| \frac{\partial u}{\partial x_2} \right|^2 \right) dx_1 dx_2 \approx \int_0^\rho \sum_{n=0}^{\infty} 2^{2n+1} 2^{-2n\alpha} r^{2^{n+1}-1} dr$$

$$= \sum_{n=0}^{\infty} \rho^{2^{n+1}} 2^n 2^{-2n\alpha},$$

which tends to infinity as $\rho \to 1$ if $\alpha \leq 1/2$.

One can formulate this objection in a more precise way by appealing to the result in Exercise 20.

Despite these significant difficulties, Dirichlet's principle can indeed be validated, if applied in the appropriate way. A key insight is that the space of competing functions arising in the proof of the above proposition is itself a pre-Hilbert space, with inner product $\langle \cdot, \cdot \rangle$ given there. The desired solution lies in the completion of this pre-Hilbert space, and this requires the L^2 theory for its analysis. These ideas were clearly not available at the time Dirichlet's principle was first formulated and used.

In what follows we shall describe how these additional concepts can be exploited. We will begin our presentation in the more general d-dimensional setting, but conclude with the application of these techniques to the solution of the two-dimensional problem (19). As an important preliminary matter we start with the study of some basic properties of harmonic functions.

4.1 Harmonic functions

Throughout this section Ω will denote an open subset of $\mathbb{R}^d$. A function u is **harmonic** in Ω if it is twice continuously differentiable[12] and u solves

$$\triangle u = \sum_{j=1}^{d} \frac{\partial^2 u}{\partial x_j^2} = 0.$$

We shall see that harmonic functions can be characterized by a number of equivalent properties.[13] Adapting the terminology used in Section 3, we say that u is **weakly harmonic** in Ω if

(20) $(u, \triangle \psi) = 0$ for every $\psi \in C_0^\infty(\Omega)$.

Note that the left-hand side of (20) is well-defined for any u that is integrable on compact subsets of Ω. Thus, in particular, a weakly harmonic function needs to be defined only almost everywhere. Clearly, however, any harmonic function is weakly harmonic.

Another notion is the **mean-value property** generalizing the identity (9) in Section 2 for holomorphic functions. A continuous function u defined in Ω satisfies this property if

(21) $u(x_0) = \frac{1}{m(B)} \int_B u(x) \, dx$

for each ball B whose center is x_0 and whose closure $\overline{B}$ is contained in Ω.

---

[12]In other words, u is in $C^2(\Omega)$ in the notation of Section 3.1.
[13]Note that in the case of one dimension, harmonic functions are linear and so their theory is essentially trivial.

The following two theorems give alternative characterizations of harmonic functions. Their proofs are closely intertwined.

Theorem 4.2 *If u is harmonic in Ω, then u satisfies the mean-value property (21). Conversely, a continuous function satisfying the mean-value property is harmonic.*

Theorem 4.3 *Any weakly harmonic function u in Ω can be corrected on a set of measure zero so that the resulting function is harmonic in Ω.*

The above statement says that for a given weakly harmonic function u there exists a harmonic function $\tilde{u}$, so that $\tilde{u}(x) = u(x)$ for a.e. $x \in \Omega$. Notice since $\tilde{u}$ is necessarily continuous it is uniquely determined by u.

Before we prove the theorems, we deduce a noteworthy corollary. It is a version of the **maximum principle**.

Corollary 4.4 *Suppose Ω is a bounded open set, and let $\partial\Omega = \overline{\Omega} - \Omega$ denote its boundary. Assume that u is continuous in $\overline{\Omega}$ and is harmonic in Ω. Then*

$$\max_{x \in \overline{\Omega}} |u(x)| = \max_{x \in \partial\Omega} |u(x)|.$$

Proof. Since the sets $\overline{\Omega}$ and $\partial\Omega$ are compact and u is continuous, the two maxima above are clearly attained. We suppose that $\max_{x \in \overline{\Omega}} |u(x)|$ is attained at an interior point $x_0 \in \Omega$, for otherwise there is nothing to prove.

Now by the mean-value property, $|u(x_0)| \leq \frac{1}{m(B)} \int_B |u(x)| \, dx$. If for some point $x' \in B$ we had $|u(x')| < |u(x_0)|$, then a similar inequality would hold in a small neighborhood of x', and since $|u(x)| \leq |u(x_0)|$ throughout B, the result would be that $\frac{1}{m(B)} \int_B |u(x)| \, dx < |u(x_0)|$, which is a contradiction. Hence $|u(x)| = |u(x_0)|$ for each $x \in B$. Now this is true for each ball B_r of radius r, centered at x_0, such that $B_r \subset \Omega$. Let r_0 be the least upper bound of such r; then $\overline{B}_{r_0}$ intersects the boundary Ω at some point $\tilde{x}$. Since $|u(x)| = |u(x_0)|$ for all $x \in \overline{B}_r$, $r < r_0$, it follows by continuity that $|u(\tilde{x})| = |u(x_0)|$, proving the corollary.

Turning to the proofs of the theorems, we first establish a variant of Green's formula (for the unit ball) that does not explicitly involve boundary terms.[14] Here u, v, and η are assumed to be twice continuously differentiable functions in a neighborhood of the closure of B, but η is also supposed to be supported in a compact subset of B.

---

[14]The more usual version requires integration over the (boundary) sphere, a topic deferred to the next chapter. See also Exercises 6 and 7 in that chapter.

Lemma 4.5 *We have the identity*

$$\int_B (v\triangle u - u\triangle v)\eta\, dx = \int_B u(\nabla v \cdot \nabla \eta) - v(\nabla u \cdot \nabla \eta)\, dx.$$

Here ∇u is the **gradient** of u, that is, $\nabla u = \left(\frac{\partial u}{\partial x_1}, \frac{\partial u}{\partial x_2}, \ldots, \frac{\partial u}{\partial x_d}\right)$ and

$$\nabla v \cdot \nabla \eta = \sum_{j=1}^{d} \frac{\partial v}{\partial x_j} \frac{\partial \eta}{\partial x_j},$$

with $\nabla u \cdot \nabla \eta$ defined similarly.

In fact, by integrating by parts as in the proof of (14) we have

$$\int_B \frac{\partial u}{\partial x_j} v\eta\, dx = -\int_B u\frac{\partial v}{\partial x_j}\eta\, dx - \int_B uv\frac{\partial \eta}{\partial x_j}\, dx.$$

We then repeat this with u replaced by $\partial u/\partial x_j$, and sum in j to obtain

$$\int_B (\triangle u)v\eta\, dx = -\int_B (\nabla u \cdot \nabla v)\eta\, dx - \int_B (\nabla u \cdot \nabla \eta)v\, dx.$$

This yields the lemma if we subtract from this the symmetric formula with u and v interchanged.

We shall apply the lemma when u is a given harmonic function, while v is one of the three following "test" functions: first, $v(x) = 1$; second, $v(x) = |x|^2$; and third, $v(x) = |x|^{-d+2}$ if $d \geq 3$, while $v(x) = \log|x|$ if $d = 2$. The relevance of these choices arises because $\triangle v = 0$ in the first case, while $\triangle v$ is a non-zero constant in the second case; also v in the third case is a constant multiple of a "fundamental solution," and in particular $v(x)$ is harmonic for $x \neq 0$.

When $v(x) = 1$, we take $\eta = \eta_\epsilon^+$, where $\eta_\epsilon^+(x) = 1$ for $|x| \leq 1 - \epsilon$, $\eta_\epsilon^+(x) = 0$ for $|x| \geq 1$, and $|\nabla \eta_\epsilon^+(x)| \leq c/\epsilon$. We accomplish this by setting $\eta_\epsilon^+(x) = \chi\left(\frac{|x|-1+\epsilon}{\epsilon}\right)$ for $1 - \epsilon \leq |x| \leq 1$, where χ is a fixed C^2 function on $[0,1]$ that equals 1 in $[0, 1/4]$ and equals 0 in $[3/4, 1]$. A picture of η_ϵ^+ is given in Figure 3.

Since u is harmonic, we see that with $v = 1$, Lemma 4.5 implies

$$(22) \qquad \int_B \nabla u \cdot \nabla \eta_\epsilon^+\, dx = 0.$$

Next we take $v(x) = |x|^2$; then clearly $\triangle v = 2d$, and with $\eta = \eta_\epsilon^+$ the lemma yields:

$$2d\int_B u\eta_\epsilon^+\, dx = \int_B |x|^2(\nabla u \cdot \nabla \eta_\epsilon^+)\, dx - 2\int_B u(x \cdot \nabla \eta_\epsilon^+)\, dx.$$

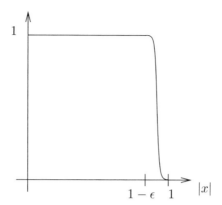

Figure 3. The function η_ϵ^+

However, since $\nabla\eta_\epsilon^+$ is supported in the spherical shell $S_\epsilon^+ = \{x : 1 - \epsilon \leq |x| \leq 1\}$, we see that

$$\int_B |x|^2 (\nabla u \cdot \nabla\eta_\epsilon^+)\, dx = \int_B (\nabla u \cdot \nabla\eta_\epsilon^+)\, dx + O(\epsilon),$$

and hence by (22) we get

$$(23) \qquad d\int_B u\, dx = -\lim_{\epsilon \to 0} \int_B u(x \cdot \nabla\eta_\epsilon^+)\, dx.$$

We finally turn to $v(x) = |x|^{-d+2}$, when $d \geq 3$, and calculate $(\triangle v)(x)$ for $x \neq 0$ to see that it vanishes there. In fact, since $\partial|x|/\partial x_j = x_j/|x|$, we note that

$$\frac{\partial |x|^a}{\partial x_j} = ax_j|x|^{a-2} \quad \text{and} \quad \frac{\partial^2 |x|^a}{\partial x_j^2} = a|x|^{a-2} + a(a-2)x_j^2|x|^{a-4}.$$

Upon adding in j, we obtain that $\triangle(|x|^a) = [da + a(a-2)]|x|^{a-2}$, and this is zero if $a = -d + 2$ (or $a = 0$). A similar argument shows that $\triangle(\log|x|) = 0$ when $d = 2$ and $x \neq 0$.

We now apply the lemma with this v and $\eta = \eta_\epsilon$ defined as follows:

$$\begin{aligned}
\eta_\epsilon(x) &= 1 - \chi(|x|/\epsilon) & &\text{for } |x| \leq \epsilon, \\
\eta_\epsilon(x) &= 1 & &\text{for } \epsilon \leq |x| \leq 1 - \epsilon, \\
\eta_\epsilon(x) &= \eta_\epsilon^+(x) = \chi\left(\frac{|x|-1+\epsilon}{\epsilon}\right) & &\text{for } 1 - \epsilon \leq |x| \leq 1.
\end{aligned}$$

The picture for η_ϵ is as follows (Figure 4):

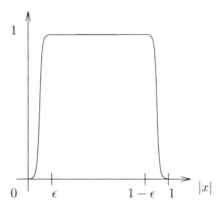

Figure 4. The function η_ϵ

We note that $|\nabla\eta_\epsilon|$ is $O(1/\epsilon)$ throughout. Now both u and v are harmonic in the support of η_ϵ, and in this case $\nabla\eta_\epsilon$ is supported only near the unit sphere (in the shell S_ϵ^+) or near the origin (in the ball $B_\epsilon = \{|x| < \epsilon\}$). Thus the right-hand side of the identity of the lemma gives two contributions, one over S_ϵ^+ and the other over B_ϵ. We consider the first contribution (when $d \geq 3$); it is

$$\int_{S_\epsilon^+} u\nabla(|x|^{-d+2}) \cdot \nabla\eta_\epsilon \, dx - \int_{S_\epsilon^+} |x|^{-d+2}(\nabla u \cdot \nabla\eta_\epsilon^+) \, dx.$$

Now the first integral is $(-d + 2)\int_{S_\epsilon^+} u|x|^{-d}(x \cdot \nabla\eta_\epsilon^+) \, dx$, which by (23) tends to $c\int_B u \, dx$ as $\epsilon \to 0$, where c is the constant $(2 - d)d$, since $|x|^{-d} - 1 = O(\epsilon)$ over S_ϵ^+. The second term tends to zero as $\epsilon \to 0$ because of (22) and the fact that the integrand there is supported in the shell S_ϵ^+. A similar argument for $d = 2$, with $v(x) = \log|x|$, yields the result with $c = 1$.

To consider the contribution near the origin, that is, over B_ϵ, we temporarily make the additional assumption that $u(0) = 0$. Then because of the differentiability assumption satisfied by a harmonic function, we have $u(x) = O(|x|)$ as $|x| \to 0$. Now over B_ϵ we have two terms, the first being $\int_{B_\epsilon} u\nabla(|x|^{-d+2})\nabla\eta_\epsilon \, dx$, which is majorized by

$$\int_{B_\epsilon} O(\epsilon)|x|^{-d+1}O(1/\epsilon) \, dx \leq O\left(\int_{|x|\leq\epsilon} |x|^{-d+1} \, dx\right) \leq O(\epsilon),$$

because of (8) in Section 2 of Chapter 2. This term tends to 0 with ϵ.

The second term is $\int_{B_\epsilon} |x|^{-d+2} (\nabla u \cdot \nabla \eta_\epsilon) \, dx$, which is majorized by

$$\frac{c_1}{\epsilon} \int_{|x| \le \epsilon} |x|^{-d+2} = c_2 \epsilon,$$

using the result just cited. We have used the fact that ∇u is bounded and $\nabla \eta_\epsilon$ is $O(1/\epsilon)$ throughout B. Letting $\epsilon \to 0$ we see that this term tends to zero also. A similar argument works when $d = 2$.

Thus we have proved that if u is harmonic in a neighborhood of the closure of the unit ball B, and $u(0) = 0$, then $\int_B u \, dx = 0$. We can drop the assumption $u(0) = 0$ by applying the conclusion we have just reached to $u(x) - u(0)$ in place of $u(x)$. Therefore we have achieved the mean-value property (21) for the unit ball.

Now suppose $B_r(x_0) = \{x : |x - x_0| < r\}$ is the ball of radius r centered at x_0, and consider $U(x) = u(x_0 + rx)$. If we suppose that u is harmonic in $B_r(x_0)$, then clearly U is harmonic in the unit ball (indeed, the property of being harmonic is unchanged under translations $x \to x + x_0$ and dilations $x \to rx$, as is easily verified). Thus if u were supported in Ω, and $B_r(x_0) \subset \Omega$, then by the result just proved $U(0) = \frac{1}{m(B)} \int_B U(x) \, dx$, which means that

$$u(x_0) = \frac{1}{m(B)} \int_{|x| \le 1} u(x_0 + rx) \, dx = \frac{1}{r^d m(B)} \int_{|x| \le r} u(x_0 + x) \, dx$$

$$= \frac{1}{m(B_r(x_0))} \int_{B_r} u(x) \, dx,$$

by the relative invariance of Lebesgue measure under dilations and translations. This establishes (21) in general.

The converse property

To prove this, we first show that the mean-value property allows a useful extension of itself. For this purpose, we fix a function $\varphi(y)$ that is continuous in the closed unit ball $\{|y| \le 1\}$ and is radial (that is, $\varphi(y) = \Phi(|y|)$ for an appropriate Φ), and extend φ to be zero when $|y| > 1$. Suppose in addition that $\int \varphi(y) \, dy = 1$. We then claim the following:

Lemma 4.6 *Whenever u satisfies the mean-value property (21) in Ω, and the closure of the ball $\{x : |x - x_0| < r\}$ lies in Ω, then*
(24)
$$u(x_0) = \int_{\mathbb{R}^d} u(x_0 - ry)\varphi(y) \, dy = \int_{\mathbb{R}^d} u(x_0 - y)\varphi_r(y) \, dy = (u * \varphi_r)(x_0),$$

where $\varphi_r(y) = r^{-d}\varphi(y/r)$.

That the second of the two identities holds is an immediate consequence of the change of variables $y \to y/r$; the rightmost equality is merely the definition of $u * \varphi_r$.

We can prove (24) as a consequence of a simple observation about integration. Let $\psi(y)$ be another function on the ball $\{|y| \le 1\}$, which we assume is bounded. For each N, a large positive integer, denote by $B(j)$ the ball $\{|y| \le j/N\}$. Recall that $\varphi(y) = \Phi(|y|)$. Then

$$(25) \qquad \int \varphi(y)\psi(y)\, dy = \lim_{N\to\infty} \sum_{j=1}^{N} \Phi\left(\frac{j}{N}\right) \int_{B(j)-B(j-1)} \psi(y)\, dy.$$

To verify this, note that the left-hand side of (25) equals

$$\sum_{j=1}^{N} \int_{B(j)-B(j-1)} \varphi(y)\psi(y)\, dy.$$

However, $\sup_{1\le j \le N} \sup_{y \in B(j)-B(j-1)} |\varphi(y) - \Phi(j/N)| = \epsilon_N$, which tends to zero as $N \to \infty$, since φ is radial, continuous, and $\varphi(y) = \Phi(|y|)$. Thus the left-hand side of (25) differs from $\sum_{j=1}^{N} \Phi(j/N) \int_{B(j)-B(j-1)} \psi(y)\, dy$ by at most $\epsilon_N \int_{|y|\le 1} |\psi(y)|\, dy$, proving (25).

We now use this in the case where $\psi(y) = u(x_0 - ry)$ and φ is as before. Then

$$\int u(x_0 - ry)\varphi(y)\, dy = \lim_{N\to\infty} \sum_{j=1}^{N} \Phi\left(\frac{j}{N}\right) \int_{B(j)-B(j-1)} u(x_0 - ry)\, dy.$$

However, it follows from the mean-value property assumed for u that

$$\int_{B(j)-B(j-1)} u(x_0 - ry)\, dy = u(x_0)[m(B(j)) - m(B(j-1))].$$

Therefore, the right-hand side above equals

$$u(x_0) \lim_{N\to\infty} \sum_{j=1}^{N} \Phi\left(\frac{j}{N}\right) \int_{B(j)-B(j-1)} dy,$$

and this is $u(x_0)$ if we use (25) again, this time with $\psi = 1$, and recall that $\int \varphi(y)\, dy = 1$. We have therefore proved the lemma.

We see from this that every continuous function which satisfies the mean-value property is its own regularization! To be precise, we have

$$(26) \qquad\qquad u(x) = (u * \varphi_r)(x)$$

whenever $x \in \Omega$ and the distance from x to the boundary of Ω is larger than r. If we now require in addition that $\varphi \in C_0^\infty\{|y| < 1\}$, then by the discussion in Section 1 we conclude that u is smooth throughout Ω.

Let us now establish that such functions are harmonic. Indeed, by Taylor's theorem, for every $x_0 \in \Omega$

$$(27) \qquad u(x_0 + x) - u(x_0) = \sum_{j=1}^{d} a_j x_j + \frac{1}{2} \sum_{j,k=1}^{d} a_{jk} x_j x_k + \epsilon(x),$$

where $\epsilon(x) = O(|x|^3)$ as $|x| \to 0$. We note next that $\int_{|x| \leq r} x_j \, dx = 0$ and $\int_{|x| \leq r} x_j x_k \, dx = 0$ for all j and k with $k \neq j$. This follows by carrying out the integrations first in the x_j variable and noting that the integral vanishes because x_j is an odd function. Also by an obvious symmetry $\int_{|x| \leq r} x_j^2 \, dx = \int_{|x| \leq r} x_k^2 \, dx$, and by the relative dilation-invariance (see Section 3, Chapter 1) these are equal to $r^2 \int_{|x| \leq r} (x_1/r)^2 \, dx = r^{d+2} \int_{|x| \leq 1} x_1^2 \, dx = cr^{d+2}$, with $c > 0$. We now integrate both sides of (27) over the ball $\{|x| \leq r\}$, divide by r^d, and use the mean-value property. The result is that

$$\frac{c}{2} r^2 \sum_{j=1}^{d} a_{jj} = \frac{cr^2}{2} (\triangle u)(x_0) = O\left(\frac{1}{r^d} \int_{|x| \leq r} |\epsilon(x)| \, dx\right) = O(r^3).$$

Letting $r \to 0$ then gives $\triangle u(x_0) = 0$. Since x_0 was an arbitrary point of Ω, the proof of Theorem 4.2 is concluded.

Theorem 4.3 and some corollaries

We come now to the proof of Theorem 4.3. Let us assume that u is weakly harmonic in Ω. For each $\epsilon > 0$ we define Ω_ϵ to be the set of points in Ω that are at a distance greater than ϵ from its boundary:

$$\Omega_\epsilon = \{x \in \Omega : d(x, \partial\Omega) > \epsilon\}.$$

Notice that Ω_ϵ is open, and that every point of Ω belongs to Ω_ϵ if ϵ is small enough. Then the regularization $u * \varphi_r = u_r$ considered in the previous theorem is defined in Ω_ϵ, for $r < \epsilon$, and as we have noted is a smooth function there. We next observe that it is weakly harmonic in

Ω_ϵ. In fact, for $\psi \in C_0^\infty(\Omega_\epsilon)$ we have

$$(u_r, \triangle\psi) = \int_{\mathbb{R}^d} \left(\int_{\mathbb{R}^d} u(x - ry)\varphi(y)\, dy \right) (\triangle\psi)(x)\, dx$$

$$= \int_{\mathbb{R}^d} \varphi(y) \left(\int_{\mathbb{R}^d} u(x - ry)(\triangle\psi)(x)\, dx \right) dy,$$

by Fubini's theorem, and the inner integral vanishes for y, $|y| \leq 1$, because it equals $(u, \triangle\psi_r)$, with $\psi_r = \psi(x + ry)$. Thus we have

$$(u * \varphi_r, \triangle\psi) = 0,$$

and hence $u * \varphi_r$ is weakly harmonic. Next, since this regularization is automatically smooth it is then also harmonic. Moreover, we claim that

(28) $(u * \varphi_{r_1})(x) = (u * \varphi_{r_2})(x)$

whenever $x \in \Omega_\epsilon$ and $r_1 + r_2 < \epsilon$. Indeed, $(u * \varphi_{r_1}) * \varphi_{r_2} = u * \varphi_{r_1}$ as we have shown in (26) above. However convolutions are commutative (see Remark (6) in Chapter 2); thus $(u * \varphi_{r_1}) * \varphi_{r_2} = (u * \varphi_{r_2}) * \varphi_{r_1} = u * \varphi_{r_2}$, and (28) is proved.

Now we can let r_1 tend to zero, while keeping r_2 fixed. We know by the properties of approximations to the identity that $u * \varphi_{r_1}(x) \to u(x)$ for almost every x in Ω_ϵ; hence $u(x)$ equals $u_{r_2}(x)$ for almost every $x \in \Omega_\epsilon$. Thus u can be corrected on Ω_ϵ (setting it equal to u_{r_2}), so that it becomes harmonic there. Now since ϵ can be taken arbitrarily small, the proof of the theorem is complete.

We state several further corollaries arising out of the above theorems.

Corollary 4.7 *Every harmonic function is indefinitely differentiable.*

Corollary 4.8 *Suppose $\{u_n\}$ is a sequence of harmonic functions in Ω that converges to a function u uniformly on compact subsets of Ω as $n \to \infty$. Then u is also harmonic.*

The first of these corollaries was already proved as a consequence of (26). For the second, we use the fact that each u_n satisfies the mean-value property

$$u_n(x_0) = \frac{1}{m(B)} \int_B u_n(x)\, dx$$

whenever B is a ball with center at x_0, and $\overline{B} \subset \Omega$. Thus by the uniform convergence it follows that u also satisfies this property, and hence u is harmonic.

We should point out that these properties of harmonic functions on $\mathbb{R}^d$ are reminiscent of similar properties of holomorphic functions. But this should not be surprising, given the close connection between these two classes of functions in the special case $d = 2$.

4.2 The boundary value problem and Dirichlet's principle

The d-dimensional Dirichlet boundary value problem we are concerned with may be stated as follows. Let Ω be an open bounded set in $\mathbb{R}^d$. Given a continuous function f defined on the boundary $\partial\Omega$, we wish to find a function u that is continuous in $\overline{\Omega}$, harmonic in Ω, and such that $u = f$ on $\partial\Omega$.

An important preliminary observation is that the solution to the problem, if it exists, is unique. Indeed, if u_1 and u_2 are two solutions then $u_1 - u_2$ is harmonic in Ω and vanishes on the boundary. Thus by the maximum principle (Corollary 4.4) we have $u_1 - u_2 = 0$, and hence $u_1 = u_2$.

Turning to the existence of a solution, we shall now pursue the approach of Dirichlet's principle outlined earlier.

We consider the class of functions $C^1(\overline{\Omega})$, and equip this space with the inner product

$$\langle u, v \rangle = \int_\Omega (\nabla u \cdot \overline{\nabla v}) \, dx,$$

where of course

$$\nabla u \cdot \overline{\nabla v} = \sum_{j=1}^{d} \frac{\partial u}{\partial x_j} \frac{\overline{\partial v}}{\partial x_j}.$$

With this inner product, we have a corresponding norm given by $\|u\|^2 = \langle u, u \rangle$. We note that $\|u\| = 0$ is the same as $\nabla u = 0$ throughout Ω, which means that u is constant on each connected component of Ω. Thus we are led to consider equivalence classes in $C^1(\overline{\Omega})$ of elements modulo functions that are constant on components of Ω. These then form a pre-Hilbert space with inner product and norm given as above. We call this pre-Hilbert space $\mathcal{H}_0$.

In studying the completion $\mathcal{H}$ of $\mathcal{H}_0$ and its applications to the boundary value problem, the following lemma is needed.

Lemma 4.9 *Let Ω be an open bounded set in $\mathbb{R}^d$. Suppose v belongs to $C^1(\overline{\Omega})$ and v vanishes on $\partial\Omega$. Then*

(29)
$$\int_\Omega |v(x)|^2 \, dx \le c_\Omega \int_\Omega |\nabla v(x)|^2 \, dx.$$

Proof. This conclusion could in fact be deduced from the considerations given in Lemma 3.3. We prefer to prove this easy version separately to highlight a simple idea that we shall also use later. It should be noted that the argument yields the estimate $c_\Omega \leq d(\Omega)^2$, where $d(\Omega)$ is the diameter of Ω.

We proceed on the basis of the following observation. Suppose f is a function in $C^1(\overline{I})$, where $I = (a, b)$ is an interval in $\mathbb{R}$. Assume that f vanishes at *one* of the end-points of I. Then

$$(30) \qquad \int_I |f(t)|^2 \, dt \leq |I|^2 \int_I |f'(t)|^2 \, dt,$$

where $|I|$ denotes the length of I.

Indeed, suppose $f(a) = 0$. Then $f(s) = \int_a^s f'(t) \, dt$, and by the Cauchy-Schwarz inequality

$$|f(s)|^2 \leq |I| \int_a^s |f'(t)|^2 \, dt \leq |I| \int_I |f'(t)|^2 \, dt.$$

Integrating this in s over I then yields (30).

To prove (29), write $x = (x_1, x')$ with $x_1 \in \mathbb{R}$ and $x' \in \mathbb{R}^{d-1}$ and apply (30) to f defined by $f(x_1) = v(x_1, x')$, with x' fixed. Let $J(x')$ be the open set in $\mathbb{R}$ that is the corresponding slice of Ω given by $\{x_1 \in \mathbb{R} : (x_1, x') \in \Omega\}$. The set $J(x')$ can be written as a disjoint union of open intervals I_j. (Note that in fact $f(x_1)$ vanishes at both end-points of each I_j.) For each j, on applying (30) we obtain

$$\int_{I_j} |v(x_1, x')|^2 \, dx_1 \leq |I_j|^2 \int_{I_j} |\nabla v(x_1, x')|^2 \, dx_1.$$

Now since $|I_j| \leq d(\Omega)$, summing over the disjoint intervals I_j gives

$$\int_{J(x')} |v(x_1, x')|^2 \, dx_1 \leq d(\Omega)^2 \int_{J(x')} |\nabla v(x_1, x')|^2 \, dx_1,$$

and an integration over $x' \in \mathbb{R}^d$ then leads to (29).

Now let S_0 denote the linear subspace of $C^1(\overline{\Omega})$ consisting of functions that vanish on the boundary of Ω. We note that distinct elements of S_0 remain distinct under the equivalence relation defining $\mathcal{H}_0$ (since constants on each component that vanish on the boundary are zero), and so S_0 may be identified with a subspace of $\mathcal{H}_0$. Denote by S the closure in $\mathcal{H}$ of this subspace, and let P_S be the orthogonal projection of $\mathcal{H}$ onto S.

With these preliminaries out of the way, we first try to solve the boundary value problem with f given on $\partial\Omega$ under the additional assumption that f is the restriction to $\partial\Omega$ of a function F in $C^1(\overline{\Omega})$. (How this additional hypothesis can be removed will be explained below.) Following the prescription of Dirichlet's principle, we seek a sequence $\{u_n\}$ with $u_n \in C^1(\overline{\Omega})$ and $u_n|_{\partial\Omega} = F|_{\partial\Omega}$, such that the Dirichlet integrals $\|u_n\|^2$ converge to a minimum value. This means that $u_n = F - v_n$, with $v_n \in S_0$, and that $\lim_{n\to\infty} \|u_n\|$ minimizes the distance from F to S_0. Since $S = \overline{S_0}$, this sequence also minimizes the distance from F to S in $\mathcal{H}$.

Now what do the elementary facts about orthogonal projections teach us? According to the proof of Lemma 4.1 in the previous chapter, we conclude that the sequence $\{v_n\}$, and hence also the sequence $\{u_n\}$, both converge in the norm of $\mathcal{H}$, the former having a limit $P_S(F)$. Now applying Lemma 4.9 to $v_n - v_m$ we deduce that $\{v_n\}$ and $\{u_n\}$ are also Cauchy in the $L^2(\Omega)$-norm, and thus converge also in the L^2-norm. Let $u = \lim_{n\to\infty} u_n$. Then

$$(31) \qquad\qquad u = F - P_S(F).$$

We see that u is weakly harmonic. Indeed, whenever $\psi \in C_0^\infty(\Omega)$, then $\psi \in S$, and hence by (31) $\langle u, \psi \rangle = 0$. Therefore $\langle u_n, \psi \rangle \to 0$, but by integration by parts, as we have seen,

$$\langle u_n, \psi \rangle = \int_\Omega (\nabla u_n \cdot \overline{\nabla \psi})\,dx = -\int_\Omega u_n \overline{\triangle \psi}\,dx = -(u_n, \triangle \psi).$$

As a result, $(u, \triangle \psi) = 0$, and so u is weakly harmonic and thus can be corrected on a set of measure zero to become harmonic.

This is the purported solution to our problem. However, two issues still remain to be resolved.

The first is that while u is the limit of a sequence $\{u_n\}$ of continuous functions in $\overline{\Omega}$ and $u_n|_{\partial\Omega} = f$, for each n, it is not clear that u itself is continuous in $\overline{\Omega}$ and $u|_{\partial\Omega} = f$.

The second issue is that we restricted our argument above to those f defined on the boundary of Ω that arise as restrictions of functions in $C^1(\overline{\Omega})$.

The second obstacle is the easier of the two to overcome, and this can be done by the use of the following lemma, applied to the set $\Gamma = \partial\Omega$.

Lemma 4.10 *Suppose Γ is a compact set in $\mathbb{R}^d$, and f is a continuous function on Γ. Then there exists a sequence $\{F_n\}$ of smooth functions on $\mathbb{R}^d$ so that $F_n \to f$ uniformly on Γ.*

In fact, supposing we can deal with the first issue raised, then with the lemma we proceed as follows. We find the functions U_n that are harmonic in Ω, continuous on $\overline{\Omega}$, and such that $U_n|_{\partial\Omega} = F_n|_{\partial\Omega}$. Now since the $\{F_n\}$ converges uniformly (to f) on $\partial\Omega$, it follows by the maximum principle that the sequence $\{U_n\}$ converges uniformly to a function u that is continuous on $\overline{\Omega}$, has the property that $u|_{\partial\Omega} = f$, and which is moreover harmonic (by Corollary 4.8 above). This achieves our goal.

The proof of Lemma 4.10 is based on the following extension principle.

Lemma 4.11 *Let f be a continuous function on a compact subset Γ of $\mathbb{R}^d$. Then there exists a function G on $\mathbb{R}^d$ that is continuous, and so that $G|_{\partial\Gamma} = f$.*

Proof. We begin with the observation that if K_0 and K_1 are two disjoint compact sets, there exists a continuous function $0 \leq g(x) \leq 1$ on $\mathbb{R}^d$ which takes the value 0 on K_0 and 1 on K_1. Indeed, if $d(x, \Omega)$ denotes the distance from x to Ω, we see that

$$g(x) = \frac{d(x, K_0)}{d(x, K_0) + d(x, K_1)}$$

has the required properties.

Now, we may assume without loss of generality that f is non-negative and bounded by 1 on Γ. Let

$$K_0 = \{x \in \Gamma : 2/3 \leq f(x) \leq 1\} \quad \text{and} \quad K_1 = \{x \in \Gamma : 0 \leq f(x) \leq 1/3\},$$

so that K_0 and K_1 are disjoint. Clearly, the observation before the lemma guarantees that there exists a function $0 \leq G_1(x) \leq 1/3$ on $\mathbb{R}^d$ which takes the value 1/3 on K_0 and 0 on K_1. Then we see that

$$0 \leq f(x) - G_1(x) \leq \frac{2}{3} \quad \text{for all } x \in \Gamma.$$

We now repeat the argument with f replaced by $f - G_1$. In the first step, we have gone from $0 \leq f \leq 1$ to $0 \leq f - G_1 \leq 2/3$. Consequently, we may find a continuous function G_2 on $\mathbb{R}^d$ so that

$$0 \leq f(x) - G_1(x) - G_2(x) \leq \left(\frac{2}{3}\right)^2 \quad \text{on } \Gamma,$$

and $0 \leq G_2 \leq \frac{1}{3}\frac{2}{3}$. Repeating this process, we find continuous functions G_n on $\mathbb{R}^d$ such that

$$0 \leq f(x) - G_1(x) - \cdots - G_N(x) \leq \left(\frac{2}{3}\right)^N \quad \text{on } \Gamma,$$

and $0 \leq G_N \leq \frac{1}{3} \left(\frac{2}{3}\right)^{N-1}$ on $\mathbb{R}^d$. If we define

$$G = \sum_{n=1}^{\infty} G_n,$$

then G is continuous and equals f on Γ.

To complete the proof of Lemma 4.10, we argue as follows. We regularize the function G obtained in Lemma 4.11 by defining

$$F_\epsilon(x) = \epsilon^{-d} \int_{\mathbb{R}^d} G(x-y)\varphi(y/\epsilon)\, dy = \int_{\mathbb{R}^d} G(y)\varphi_\epsilon(x-y)\, dy,$$

with $\varphi_\epsilon(y) = \epsilon^{-d}\varphi(y/\epsilon)$, where φ is a non-negative C_0^∞ function supported in the unit ball with $\int \varphi(y)\, dy = 1$. Then each F_ϵ is a C^∞ function. However,

$$F_\epsilon(x) - G(x) = \int (G(y) - G(x))\varphi_\epsilon(x-y)\, dy.$$

Since the integration above is restricted to $|x-y| \leq \epsilon$, then if $x \in \Gamma$, we see that

$$\begin{aligned}
|F_\epsilon(x) - G(x)| &\leq \sup_{|x-y|\leq\epsilon} |G(x) - G(y)| \int \varphi_\epsilon(x-y)\, dy \\
&\leq \sup_{|x-y|\leq\epsilon} |G(x) - G(y)|.
\end{aligned}$$

The last quantity tends to zero with ϵ by the uniform continuity of G near Γ, and if we choose $\epsilon = 1/n$ we obtain our desired sequence.

The two-dimensional theorem

We now take up the problem of whether the proposed solution u takes on the desired boundary values. Here we limit our discussion to the case of two dimensions for the reason that in the higher dimensional situation the problems that arise involve a number of questions that would take us beyond the scope of this book. In contrast, in two dimensions, while the proof of the result below is a little tricky, it is within the reach of the Hilbert space methods we have been illustrating.

The Dirichlet problem can be solved (in two dimensions as well as in higher dimensions) only if certain restrictions are made concerning the nature of the domain Ω. The regularity we shall assume, while not

optimal,[15] is broad enough to encompass many applications, and yet
has a simple geometric form. It can be described as follows. We fix an
initial triangle T_0 in $\mathbb{R}^2$. To be precise, we assume that T_0 is an isosceles
triangle whose two equal sides have length ℓ, and make an angle α at
their common vertex. The exact values of ℓ and α are unimportant;
they may both be taken as small as one wishes, but must be kept fixed
throughout our discussion. With the shape of T_0 thus determined, we
say that T is a **special triangle** if it is congruent to T_0, that is, T arises
from T_0 by a translation and rotation. The **vertex** of T is defined to be
the intersection of its two equal sides.

The regularity property of Ω we assume, the **outside-triangle condition**, is as follows: with ℓ and α fixed, for each x in the boundary of
Ω, there is a special triangle with vertex x whose interior lies outside Ω.
(See Figure 5.)

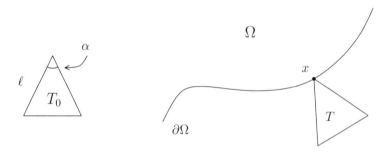

Figure 5. The triangle T_0 and the special triangle T

Theorem 4.12 *Let Ω be an open bounded set in $\mathbb{R}^2$ that satisfies the
outside-triangle condition. If f is a continuous function on $\partial\Omega$, then the
boundary value problem $\triangle u = 0$ with u continuous in $\overline{\Omega}$ and $u|_{\partial\Omega} = f$ is
always uniquely solvable.*

Some comments are in order.

(1) If Ω is bounded by a polygonal curve, it satisfies the conditions of
the theorem.

(2) More generally, if Ω is appropriately bounded by finitely many Lipschitz curves, or in particular C^1 curves, the conditions are also satisfied.

(3) There are simple examples where the problem is not solvable: for
instance, if Ω is the punctured disc. This example of course does not
satisfy the outside-triangle condition.

---

[15] The optimal conditions involve the notion of capacity of sets.

(4) The conditions on Ω in this theorem are not optimal: one can construct examples of Ω when the problem is solvable for which the above regularity fails.

For more details on the above, see Exercise 19 and Problem 4.

We turn to the proof of the theorem. It is based on the following proposition, which may be viewed as a refined version of Lemma 4.9 above.

Proposition 4.13 *For any bounded open set Ω in $\mathbb{R}^2$ that satisfies the outside-triangle condition there are two constants $c_1 < 1$ and $c_2 > 1$ such that the following holds. Suppose z is a point in Ω whose distance from $\partial\Omega$ is δ. Then whenever v belongs to $C^1(\overline{\Omega})$ and $v|_{\partial\Omega} = 0$, we have*

$$(32) \qquad \int_{B_{c_1\delta}(z)} |v(x)|^2 \, dx \leq C\delta^2 \int_{B_{c_2\delta}(z)\cap\Omega} |\nabla v(x)|^2 \, dx.$$

The bound C can be chosen to depend only on the diameter of Ω and the parameters ℓ and α which determine the triangles T.

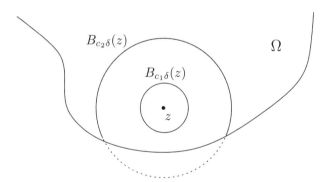

Figure 6. The situation in Proposition 4.13

Let us see how the proposition proves the theorem. We have already shown that it suffices to assume that f is the restriction to $\partial\Omega$ of an F that belongs to $C^1(\overline{\Omega})$. We recall we had the minimizing sequence $u_n = F - v_n$, with $v_n \in C^1(\overline{\Omega})$ and $v_n|_{\partial\Omega} = 0$. Moreover, this sequence converges in the norm of $\mathcal{H}$ and $L^2(\Omega)$ to a limit v, such that $u = F - v$ is harmonic in Ω. Then since (32) holds for each v_n, it also holds for $v = F - u$; that is,

$$(33) \qquad \int_{B_{c_1\delta}(z)} |(F - u)(x)|^2 \, dx \leq C\delta^2 \int_{B_{c_2\delta}(z)\cap\Omega} |\nabla(F - u)(x)|^2 \, dx.$$

To prove the theorem it suffices, in view of the continuity of u in Ω, to show that if y is any fixed point in $\partial\Omega$, and z is a variable point in Ω, then $u(z) \to f(y)$ as $z \to y$. Let $\delta = \delta(z)$ denote the distance of z from the boundary. Then $\delta(z) \leq |z - y|$ and therefore $\delta(z) \to 0$ as $z \to y$.

We now consider the averages of F and u taken over the discs centered at z of radius $c_1\delta(z)$ (recall that $c_1 < 1$). We denote these averages by $\mathrm{Av}(F)(z)$ and $\mathrm{Av}(u)(z)$, respectively. Then by the Cauchy-Schwarz inequality, we have

$$|\mathrm{Av}(F)(z) - \mathrm{Av}(u)(z)|^2 \leq \frac{1}{\pi(c_1\delta)^2} \int_{B_{c_1\delta}(z)\cap\Omega} |F - u|^2 \, dx,$$

which by (33) is then majorized by

$$C' \int_{B_{c_2\delta}(z)\cap\Omega} |\nabla(F - u)|^2 \, dx.$$

The absolute continuity of the integral guarantees that the last integral tends to zero with δ, since $m(B_{c_2\delta}) \to 0$. However, by the mean-value property, $\mathrm{Av}(u)(z) = u(z)$, while by the continuity of F in $\overline{\Omega}$,

$$\mathrm{Av}(F)(z) = \frac{1}{m(B_{c_1\delta}(z))} \int_{B_{c_1\delta}(z)} F(x) \, dx \to f(y),$$

because $F|_{\partial\Omega} = f$ and $z \to y$. Altogether this gives $u(z) \to f(y)$, and the theorem is proved, once the proposition is established.

To prove the proposition, we construct for each $z \in \Omega$ whose distance from $\partial\Omega$ is δ, and for δ sufficiently small, a rectangle R with the following properties:

(1) R has side lengths $2c_1\delta$ and $M\delta$ (with $c_1 \leq 1/2$, $M \leq 4$).

(2) $B_{c_1\delta}(z) \subset R$.

(3) Each segment in R, that is parallel to and of length equal to the length of the long side, intersects the boundary of Ω.

To obtain R we let y be a point in $\partial\Omega$ so that $\delta = |z - y|$, and we apply the outside-triangle condition at y. As a result, the line joining z with y and *one* of the sides of the special triangle whose vertex is at y must make an angle $\beta < \pi$. (In fact $\beta \leq \pi - \alpha/2$, as is easily seen.) Now after a suitable rotation and translation we may assume that $y = 0$ and that the angle going from the x_2-axis to the line joining z to 0 is equal to the

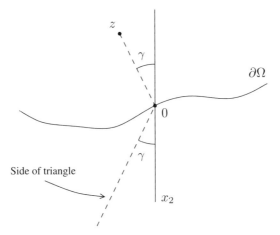

Figure 7. Placement of the rectangle R

angle of the side of the triangle to the x_2-axis. This angle can be taken to be γ, with $\gamma > \alpha/4$. (See Figure 7.)

There is an alternate possibility that occurs with this figure reflected through the x_2-axis.

With this picture in mind we construct the rectangle R as indicated in Figure 8.

It has its long side parallel to the x_2-axis, contains the disc $B_{c_1\delta}(z)$, and every segment R parallel to the x_2-axis intersects the (extension) of the side of the triangle.

Note that the coordinates of z are $(-\delta \sin \gamma, \delta \cos \gamma)$. We choose $c_1 < \sin \gamma$, then $B_{c_1\delta}(z)$ lies in the same (left) half-plane as z.

We next focus our attention on two points: P_1, which lies on the x_1-axis at the intersection of this axis with the far side of the rectangle; and P_2, which is at the corner of that side of the rectangle, that is, at the intersection of the (continuation) of the side of the outside triangle and the further side of the rectangle. The coordinates of P_1 are $(-a, 0)$, where $a = \delta c_1 + \delta \sin \gamma$. The coordinates of P_2 are $(-a, -a\frac{\cos \gamma}{\sin \gamma})$. Note that the distance of P_2 from the origin is $a/\sin \gamma$, which is $\delta + c_1\delta/\sin \gamma \leq 2\delta$, since $c_1 < \sin \gamma$.

Now we observe that the length of the larger side of the rectangle is the sum of the part that lies above the x_1-axis and the part that lies below. The upper part has length the sum of the radius of the disc plus the height of z, and this is $c_1\delta + \delta \cos \gamma \leq 2\delta$. The lower part has length equal to $a/\tan \gamma$, which is $\delta \cos \gamma + \delta c_1 \frac{\cos \gamma}{\sin \gamma} \leq 2\delta$, since $c_1 < \sin \gamma$. Thus

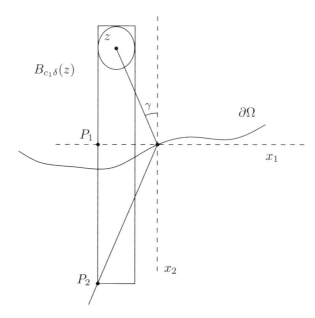

Figure 8. The disc $B_{c_1\delta}(z)$ and the rectangle R containing it

we find that the length of the side is $\leq 4\delta$.

Now it is clear from the construction that each vertical segment in R starting from the disc $B_{c_1\delta}(z)$ when continued downward and parallel to the x_2-axis intersects the line joining 0 to P_2, (which is a continuation of the side of the triangle). Moreover, if the length ℓ of this side of the triangle exceeds the distance of P_2 from the origin, then the segment intersects the triangle. When this intersection occurs the segment starting from $B_{c_2\delta}(z)$ must also intersect the boundary of Ω, since the triangle lies outside Ω. Therefore if $\ell \geq 2\delta$ the desired intersection occurs, and each of the conclusions (1), (2), and (3) are verified. (We shall lift the restriction $\delta \leq \ell/2$ momentarily.)

Now we integrate over each line segment parallel to the x_2-axis in R, including its portion in $B_{c_1\delta}(z)$, which is continued downward until it meets $\partial\Omega$. Call such a segment $I(x_1)$. Then, using (30) we see that

$$\int_{I(x_1)} |v(x_1, x_2)|^2 \, dx_2 \leq M^2\delta^2 \int_{I(x_1)} \left|\frac{\partial v}{\partial x_2}(x_1, x_2)\right|^2 \, dx_2,$$

and an integration in x_1 gives

$$\int_{R\cap\Omega} |v(x)|^2 \, dx \leq M\delta^2 \int_{R\cap\Omega} |\nabla v(x)|^2 \, dx.$$

However, we note that $B_{c_1\delta}(z) \subset R$, and $B_{c_2\delta}(z) \supset R$ when $c_2 \geq 2$. Thus the desired inequality (32) is established, still under the assumption that δ is small, that is, $\delta \leq \ell/2$. When $\delta > \ell/2$ it suffices merely to use the crude estimate (29) and the proposition is then proved. The proof of the theorem is therefore complete.

5 Exercises

1. Suppose $f \in L^2(\mathbb{R}^d)$ and $k \in L^1(\mathbb{R}^d)$.

(a) Show that $(f * k)(x) = \int f(x - y)k(y)\,dy$ converges for a.e. x.

(b) Prove that $\|f * k\|_{L^2(\mathbb{R}^d)} \leq \|f\|_{L^2(\mathbb{R}^d)}\|k\|_{L^1(\mathbb{R}^d)}$.

(c) Establish $\widehat{(f * k)}(\xi) = \hat{k}(\xi)\hat{f}(\xi)$ for a.e. ξ.

(d) The operator $Tf = f * k$ is a Fourier multiplier operator with multiplier $m(\xi) = \hat{k}(\xi)$.

[Hint: See Exercise 21 in Chapter 2.]

2. Consider the **Mellin transform** defined initially for continuous functions f of compact support in $\mathbb{R}^+ = \{t \in \mathbb{R} : t > 0\}$ and $x \in \mathbb{R}$ by

$$\mathcal{M}f(x) = \int_0^\infty f(t)t^{ix-1}dt.$$

Prove that $(2\pi)^{-1/2}\mathcal{M}$ extends to a unitary operator from $L^2(\mathbb{R}^+, dt/t)$ to $L^2(\mathbb{R})$. The Mellin transform serves on $\mathbb{R}^+$, with its multiplicative structure, the same purpose as the Fourier transform on $\mathbb{R}$, with its additive structure.

3. Let $F(z)$ be a bounded holomorphic function in the half-plane. Show in two ways that $\lim_{y\to 0} F(x + iy)$ exists for a.e. x.

(a) By using the fact that $F(z)/(z + i)$ is in $H^2(\mathbb{R}^2_+)$.

(b) By noting that $G(z) = F\left(i\frac{1-z}{1+z}\right)$ is a bounded holomorphic function in the unit disc, and using Exercise 17 in the previous chapter.

4. Consider $F(z) = e^{i/z}/(z + i)$ in the upper half-plane. Note that $F(x + iy) \in L^2(\mathbb{R})$, for each $y > 0$ and $y = 0$. Observe also that $F(z) \to 0$ as $|z| \to 0$. However, $F \notin H^2(\mathbb{R}^2_+)$. Why?

5. For $a < b$, let $S_{a,b}$ denote the strip $\{z = x + iy, \ a < y < b\}$. Define $H^2(S_{a,b})$ to consist of the holomorphic functions F in $S_{a,b}$ so that

$$\|F\|^2_{H^2(S_{a,b})} = \sup_{a<y<b} \int_{\mathbb{R}^2} |F(x + iy)|^2\,dx < \infty.$$

Define $H^2(S_{a,\infty})$ and $H^2(S_{-\infty,b})$ to be the obvious variants of the Hardy spaces for the half-planes $\{z = x + iy,\ y > a\}$ and $\{z = x + iy,\ y < b\}$, respectively.

(a) Show that $F \in H^2(S_{a,b})$ if and only if F can be written as

$$F(z) = \int_{\mathbb{R}} f(\xi) e^{-2\pi i z \xi}\, d\xi,$$

with $\int_{\mathbb{R}} |f(\xi)|^2 (e^{4\pi a \xi} + e^{4\pi b \xi})\, d\xi < \infty$.

(b) Prove that every $F \in H^2(S_{a,b})$ can be decomposed as $F = G_1 + G_2$, where $G \in H^2(S_{a,\infty})$ and $G_2 \in H^2(S_{-\infty,b})$.

(c) Show that $\lim_{a < y < b, y \to a} F(x + iy) = F_a(x)$ exists in the L^2-norm and also almost everywhere, with a similar result for $\lim_{a < y < b, y \to b} F(x + iy)$.

6. Suppose Ω is an open set in $\mathbb{C} = \mathbb{R}^2$, and let $\mathcal{H}$ be the subspace of $L^2(\Omega)$ consisting of holomorphic functions on Ω. Show that $\mathcal{H}$ is a closed subspace of $L^2(\Omega)$, and hence is a Hilbert space with inner product

$$(f, g) = \int_{\Omega} f(z) \overline{g}(z)\, dx\, dy, \qquad \text{where } z = x + iy.$$

[Hint: Prove that for $f \in \mathcal{H}$, we have $|f(z)| \le \frac{c}{d(z, \Omega^c)} \|f\|$ for $z \in \Omega$, where $c = \pi^{-1/2}$, using the mean-value property (9). Thus if $\{f_n\}$ is a Cauchy sequence in $\mathcal{H}$, it converges uniformly on compact subsets of Ω.]

7. Following up on the previous exercise, prove:

(a) If $\{\varphi_n\}_{n=0}^{\infty}$ is an orthonormal basis of $\mathcal{H}$, then

$$\sum_{n=0}^{\infty} |\varphi_n(z)|^2 \le \frac{c^2}{d(z, \Omega^c)} \qquad \text{for } z \in \Omega.$$

(b) The sum

$$B(z, w) = \sum_{n=0}^{\infty} \varphi_n(z) \overline{\varphi_n}(w)$$

converges absolutely for $(z, w) \in \Omega \times \Omega$, and is independent of the choice of the orthonormal basis $\{\varphi_n\}$ of $\mathcal{H}$.

(c) To prove (b) it is useful to characterize the function $B(z, w)$, called the **Bergman kernel**, by the following property. Let T be the linear transformation on $L^2(\Omega)$ defined by

$$Tf(z) = \int_{\Omega} B(z, w) f(w)\, du\, dv, \qquad w = u + iv.$$

Then T is the orthogonal projection of $L^2(\Omega)$ to $\mathcal{H}$.

(d) Suppose that Ω is the unit disc. Then $f \in \mathcal{H}$ exactly when $f(z) = \sum_{n=0}^{\infty} a_n z^n$, with

$$\sum_{n=0}^{\infty} |a_n|^2 (n+1)^{-1} < \infty.$$

Also, the sequence $\{\frac{z^n (n+1)}{\pi^{1/2}}\}_{n=0}^{\infty}$ is an orthonormal basis of $\mathcal{H}$. Moreover, in this case

$$B(z, w) = \frac{1}{\pi(1 - z\overline{w})^2}.$$

8. Continuing with Exercise 6, suppose Ω is the upper half-plane $\mathbb{R}_+^2$. Then every $f \in \mathcal{H}$ has a representation

(34) $$f(z) = \sqrt{4\pi} \int_0^{\infty} \hat{f}_0(\xi) e^{2\pi i \xi z} \, d\xi, \qquad z \in \mathbb{R}_+^2,$$

where $\int_0^{\infty} |\hat{f}_0(\xi)|^2 \frac{d\xi}{\xi} < \infty$. Moreover, the mapping $\hat{f}_0 \to f$ given by (34) is a unitary mapping from $L^2((0, \infty), \frac{d\xi}{\xi})$ to $\mathcal{H}$.

9. Let H be the Hilbert transform. Verify that

(a) $H^* = -H$, $H^2 = -I$, and H is unitary.

(b) If τ_h denotes the translation operator, $\tau_h(f)(x) = f(x - h)$, then H commutes with τ_h, $\tau_h H = H \tau_h$.

(c) If δ_a denotes the dilation operator, $\delta_a(f)(x) = f(ax)$ with $a > 0$, then H commutes with δ_a, $\delta_a H = H \delta_a$.

A converse is given in Problem 5 below.

10. Let $f \in L^2(\mathbb{R})$ and let $u(x, y)$ be the Poisson integral of f, that is $u = (f * \mathcal{P}_y)(x)$, as given in (10) above. Let $v(x, y) = (Hf * \mathcal{P}_y)(x)$, the Poisson integral of the Hilbert transform of f. Prove that:

(a) $F(x + iy) = u(x, y) + iv(x, y)$ is analytic in the half-plane $\mathbb{R}_+^2$, so that u and v are conjugate harmonic functions. We also have $f = \lim_{y \to 0} u(x, y)$ and $Hf = \lim_{y \to 0} v(x, y)$.

(b) $F(z) = \frac{1}{\pi i} \int_{\mathbb{R}} f(t) \frac{dt}{t - z}$.

(c) $v(x, y) = f * \mathcal{Q}_y$, where $\mathcal{Q}_y(x) = \frac{1}{\pi} \frac{x}{x^2 + y^2}$ is the **conjugate Poisson kernel**.

[Hint: Note that $\frac{i}{\pi z} = \mathcal{P}_y(x) + i \mathcal{Q}_y(x)$, $z = x + iy$.]

11. Show that

$$\left\{ \frac{1}{\pi^{1/2}(i + z)} \left(\frac{i - z}{i + z} \right)^n \right\}_{n=0}^{\infty}$$

is an orthonormal basis of $H^2(\mathbb{R}^2_+)$.

Note that $\left\{ \frac{1}{\pi^{1/2}(i+x)} \left(\frac{i-x}{i+x} \right)^n \right\}_{n=0}^\infty$ is an orthonormal basis of $L^2(\mathbb{R})$; see Exercise 9 in the previous chapter.

[Hint: It suffices to show that if $F \in H^2(\mathbb{R}^2_+)$ and

$$\int_{-\infty}^\infty F(x) \frac{(x+i)^n}{(x-i)^{n+1}} \, dx = 0 \quad \text{for } n = 0, 1, 2, \ldots,$$

then $F = 0$. Use the Cauchy integral formula to prove that

$$\left(\frac{d}{dz} \right)^n (F(z)(z+i)^n)|_{z=i} = 0,$$

and thus $F^{(n)}(i) = 0$ for $n = 0, 1, 2, \ldots$.]

12. We consider whether the inequality

$$\|u\|_{L^2(\Omega)} \leq c\|L(u)\|_{L^2(\Omega)}$$

can hold for open sets Ω that are unbounded.

(a) Assume $d \geq 2$. Show that for each constant coefficient partial differential operator L, there are unbounded connected open sets Ω for which the above holds for all $u \in C_0^\infty(\Omega)$.

(b) Show that $\|u\|_{L^2(\mathbb{R}^d)} \leq c\|L(u)\|_{L^2(\mathbb{R}^d)}$ for all $u \in C_0^\infty(\mathbb{R}^d)$ if and only if $|P(\xi)| \geq c > 0$ all ξ, where P is the characteristic polynomial of L.

[Hint: For (a) consider first $L = (\partial/\partial x_1)^n$ and a strip $\{x : -1 < x_1 < 1\}$.]

13. Suppose L is a linear partial differential operator with constant coefficients. Show that when $d \geq 2$, the linear space of solutions u of $L(u) = 0$ with $u \in C^\infty(\mathbb{R}^d)$ is not finite-dimensional.

[Hint: Consider the zeroes ζ of $P(\zeta)$, $\zeta \in \mathbb{C}^d$, where P is the characteristic polynomial of L.]

14. Suppose F and G are two integrable functions on a bounded interval $[a, b]$. Show that G is the weak derivative of F if and only if F can be corrected on a set of measure 0, such that F is absolutely continuous and $F'(x) = G(x)$ for almost every x.

[Hint: If G is the weak derivative of F, use an approximation to show that

$$\int_a^b G(x)\varphi(x)dx = -\int_a^b F(x)\varphi'(x)dx$$

holds for the function φ illustrated in Figure 9.]

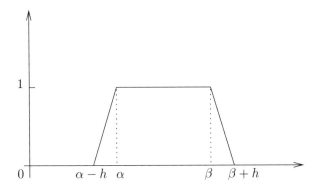

Figure 9. The function φ in Exercise 14

15. Suppose $f \in L^2(\mathbb{R}^d)$. Prove that there exists $g \in L^2(\mathbb{R}^d)$ such that

$$\left(\frac{\partial}{\partial x}\right)^\alpha f(x) = g(x)$$

in the weak sense, if and only if

$$(2\pi i \xi)^\alpha \hat{f}(\xi) = \hat{g}(\xi) \in L^2(\mathbb{R}^d).$$

16. Sobolev embedding theorem. Suppose n is the smallest integer $> d/2$. If

$$f \in L^2(\mathbb{R}^d) \quad \text{and} \quad \left(\frac{\partial}{\partial x}\right)^\alpha f \in L^2(\mathbb{R}^d)$$

in the weak sense, for all $1 \leq |\alpha| \leq n$, then f can be modified on a set of measure zero so that f is continuous and bounded.

[Hint: Express f in terms of $\hat{f}$, and show that $\hat{f} \in L^1(\mathbb{R}^d)$ by the Cauchy-Schwarz inequality.]

17. The conclusion of the Sobolev embedding theorem fails when $n = d/2$. Consider the case $d = 2$, and let $f(x) = (\log 1/|x|)^\alpha \eta(x)$, where η is a smooth cutoff function with $\eta = 1$ for x near the origin, but $\eta(x) = 0$ if $|x| \geq 1/2$. Let $0 < \alpha < 1/2$.

(a) Verify that $\partial f/\partial x_1$ and $\partial f/\partial x_2$ are in L^2 in the weak sense.

(b) Show that f cannot be corrected on a set of measure zero such that the resulting function is continuous at the origin.

18. Consider the linear partial differential operator

$$L = \sum_{|\alpha| \leq n} a_\alpha \left(\frac{\partial}{\partial x}\right)^\alpha.$$

Then

$$P(\xi) = \sum_{|\alpha|\le n} a_\alpha (2\pi i \xi)^\alpha$$

is called the **characteristic polynomial** of L. The differential operator L is said to be **elliptic** if

$$|P(\xi)| \ge c|\xi|^n \quad \text{for some } c > 0 \text{ and all } \xi \text{ sufficiently large.}$$

(a) Check that L is elliptic if and only if $\sum_{|\alpha|=n} a_\alpha (2\pi\xi)^\alpha$ vanishes only when $\xi = 0$.

(b) If L is elliptic, prove that for some $c > 0$ the inequality

$$\left\| \left(\frac{\partial}{\partial x}\right)^\alpha \varphi \right\|_{L^2(\mathbb{R}^d)} \le c \left(\|L\varphi\|_{L^2(\mathbb{R}^d)} + \|\varphi\|_{L^2(\mathbb{R}^d)} \right)$$

holds for all $\varphi \in C_0^\infty(\Omega)$ and $|\alpha| \le n$.

(c) Conversely, if (b) holds then L is elliptic.

19. Suppose u is harmonic in the punctured unit disc $\mathbb{D}^* = \{z \in \mathbb{C} : 0 < |z| < 1\}$.

(a) Show that if u is also continuous at the origin, then u is harmonic throughout the unit disc.

[Hint: Show that u is weakly harmonic.]

(b) Prove that the Dirichlet problem for the punctured unit disc is in general not solvable.

20. Let F be a continuous function on the closure $\overline{\mathbb{D}}$ of the unit disc. Assume that F is in C^1 on the (open) disc $\mathbb{D}$, and $\int_{\mathbb{D}} |\nabla F|^2 < \infty$.
   Let $f(e^{i\theta})$ denote the restriction of F to the unit circle, and write $f(e^{i\theta}) \sim \sum_{n=-\infty}^{\infty} a_n e^{in\theta}$. Prove that $\sum_{n=-\infty}^{\infty} |n| \, |a_n|^2 < \infty$.
[Hint: Write $F(re^{i\theta}) \sim \sum_{n=-\infty}^{\infty} F_n(r)e^{in\theta}$, with $F_n(1) = a_n$. Express $\int_{\mathbb{D}} |\nabla F|^2$ in polar coordinates, and use the fact that

$$\frac{1}{2}|F(1)|^2 \le L^{-1}\int_{1/2}^1 |F'(r)|^2 \, dr + L\int_{1/2}^1 |F(r)|^2 \, dr,$$

for $L \ge 2$; apply this to $F = F_n$, $L = |n|$.]

6 Problems

1. Suppose $F_0(x) \in L^2(\mathbb{R})$. Then a necessary and sufficient condition that there exists an entire analytic function F, such that $|F(z)| \leq Ae^{a|z|}$ for all $z \in \mathbb{C}$, and $F_0(x) = F(x)$ a.e. $x \in \mathbb{R}$, is that $\hat{F}_0(\xi) = 0$ whenever $|\xi| > a/2\pi$.

[Hint: Consider the regularization $F^\epsilon(z) = \int_{-\infty}^{\infty} F(z - t)\varphi_\epsilon(t)\, dt$ and apply to it the considerations in Theorem 3.3 of Chapter 4 in Book II.]

2. Suppose Ω is an open bounded subset of $\mathbb{R}^2$. A **boundary Lipschitz arc** γ is a portion of $\partial\Omega$ which after a rotation of the axes is represented as

$$\gamma = \{(x_1, x_2) : x_2 = \eta(x_1), \ a \leq x_1 \leq b\},$$

where $a < b$ and $\gamma \subset \partial\Omega$. It is also supposed that

$$(35) \qquad |\eta(x_1) - \eta(x_1')| \leq M|x_1 - x_1'|, \qquad \text{whenever } x_1, x_1' \in [a, b],$$

and moreover if $\gamma_\delta = \{(x_1, x_2) : x_2 - \delta \leq \eta(x_1) \leq x_2\}$, then $\gamma_\delta \cap \Omega = \emptyset$ for some $\delta > 0$. (Note that the condition (35) is satisfied if $\eta \in C^1([a, b])$.)

Suppose Ω satisfies the following condition. There are finitely many open discs $D_1, D_2, \ldots, D_N$ with the property that $\bigcup_j D_j$ contains $\partial\Omega$ and for each j, $\partial\Omega \cap D_j$ is a boundary Lipschitz arc (see Figure 10). Then Ω verifies the outside-triangle condition of Theorem 4.12, guaranteeing the solvability of the boundary value problem.

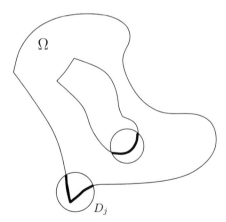

Figure 10. A domain with boundary Lipschitz arcs

3.[*] Suppose the bounded domain Ω has as its boundary a closed simple continuous curve. Then the boundary value problem is solvable for Ω. This is because there exists a conformal map Φ of the unit disc $\mathbb{D}$ to Ω that extends to a continuous bijection from $\overline{\mathbb{D}}$ to $\overline{\Omega}$. (See Section 1.3 and Problem 6* in Chapter 8 of Book II.)

4. Consider the two domains Ω in $\mathbb{R}^2$ given by Figure 11.

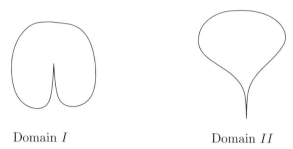

Domain I Domain II

Figure 11. Domains with a cusp

The set I has as its boundary a smooth curve, with the exception of an (inside) cusp. The set II is similar, except it has an outside cusp. Both I and II fall within the scope of the result of Problem 3, and hence the boundary value problem is solvable in each case. However, II satisfies the outside-triangle condition while I does not.

5. Let T be a Fourier multiplier operator on $L^2(\mathbb{R}^d)$. That is, suppose there is a bounded function m such that $\widehat{(Tf)}(\xi) = m(\xi)\hat{f}(\xi)$, all $f \in L^2(\mathbb{R}^d)$. Then T commutes with translations, $\tau_h T = T\tau_h$, where $\tau_h(f)(x) = f(x-h)$, for all $h \in \mathbb{R}^d$.

Conversely any bounded operator on $L^2(\mathbb{R}^d)$ that commutes with translations is a Fourier multiplier operator.

[Hint: It suffices to prove that if a bounded operator $\hat{T}$ commutes with multiplication by exponentials $e^{2\pi i \xi \cdot h}$, $h \in \mathbb{R}^d$, then there is an m so that $\hat{T}g(\xi) = m(\xi)g(\xi)$ for all $g \in L^2(\mathbb{R}^d)$. To do this, show first that

$$\hat{T}(\Phi g) = \Phi \hat{T}(g), \quad \text{all } g \in L^2(\mathbb{R}^d), \text{ whenever } \Phi \in C_0^\infty(\mathbb{R}^d).$$

Next, for large N, choose Φ so that it equals 1 in the ball $|\xi| \leq N$. Then $m(\xi) = \hat{T}(\Phi)(\xi)$ for $|\xi| \leq N$.]

As a consequence of this theorem show that if T is a bounded operator on $L^2(\mathbb{R})$ that commutes with translations and dilations (as in Exercise 9 above), then

(a) If $(Tf)(-x) = T(f(-x))$ it follows $T = cI$, where c is an appropriate constant and I the identity operator.

(b) If $(Tf)(-x) = -T(f(-x))$, then $T = cH$, where c is an appropriate constant and H the Hilbert transform.

6. This problem provides an example of the contrast between analysis on $L^1(\mathbb{R}^d)$ and $L^2(\mathbb{R}^d)$.

Recall that if f is locally integrable on $\mathbb{R}^d$, the maximal function f^* is defined by

$$f^*(x) = \sup_{x \in B} \frac{1}{m(B)} \int_B |f(y)|\, dy,$$

where the supremum is taken over all balls containing the point x.

Complete the following outline to prove that there exists a constant C so that

$$\|f^*\|_{L^2(\mathbb{R}^d)} \leq C\|f\|_{L^2(\mathbb{R}^d)}.$$

In other words, the map that takes f to f^* (although not linear) is bounded on $L^2(\mathbb{R}^d)$. This differs notably from the situation in $L^1(\mathbb{R}^d)$, as we observed in Chapter 3.

(a) For each $\alpha > 0$, prove that if $f \in L^2(\mathbb{R}^d)$, then

$$m(\{x: f^*(x) > \alpha\}) \leq \frac{2A}{\alpha} \int_{|f| > \alpha/2} |f(x)|\, dx.$$

Here, $A = 3^d$ will do.

[Hint: Consider $f_1(x) = f(x)$ if $|f(x)| \geq \alpha/2$ and 0 otherwise. Check that $f_1 \in L^1(\mathbb{R}^d)$, and

$$\{x: f^*(x) > \alpha\} \subset \{x: f_1^*(x) > \alpha/2\}.]$$

(b) Show that

$$\int_{\mathbb{R}^d} |f^*(x)|^2 dx = 2 \int_0^\infty \alpha m(E_\alpha) d\alpha,$$

where $E_\alpha = \{x: f^*(x) > \alpha\}$.

(c) Prove that $\|f^*\|_{L^2(\mathbb{R}^d)} \leq C\|f\|_{L^2(\mathbb{R}^d)}$.

6 Abstract Measure and Integration Theory

What immediately suggest itself, then, is that these characteristic properties themselves be treated as the main object of investigation, by defining and dealing with abstract objects which need satisfy no other conditions than those required by the very theory to be developed.

This procedure has been made use of − more or less consciously − by mathematicians of every era. The geometry of Euclid and the literal algebra of the sixteenth and seventeenth centuries arose in this way. But only in more recent times has this method, called the *axiomatic* method, been consistently developed and carried through to its logical conclusion.

It is our intention to treat the theories of measure and integration by means of the axiomatic method just described.

C. Carathéodory, 1918

In much of mathematics integration plays a significant role. It is used, in one form or another, when dealing with questions that arise in analysis on a variety of different spaces. While in some situations it suffices to integrate continuous or other simple functions on these spaces, the deeper study of a number of other problems requires integration based on the more refined ideas of measure theory. The development of these ideas, going beyond the setting of the Euclidean space $\mathbb{R}^d$, is the goal of this chapter.

The starting point is a fruitful insight of Carathéodory and the resulting theorems that lead to construction of measures in very general circumstances. Once this has been achieved, the deduction of the fundamental facts about integration in the general context then follows a familiar path.

We apply the abstract theory to obtain several useful results: the theory of product measures; the polar coordinate integration formula, which is a consequence of this; the construction of the Lebesgue-Stieltjes integral and its corresponding Borel measure on the real line; and the general notion of absolute continuity. Finally, we treat some of the basic limit theorems of ergodic theory. This not only gives an illustration of the abstract framework we have established, but also provides a link with the differentiation theorems studied in Chapter 3.

1 Abstract measure spaces

A **measure space** consists of a set X equipped with two fundamental objects:

(I) A σ-algebra $\mathcal{M}$ of "measurable" sets, which is a non-empty collection of subsets of X closed under complements and countable unions and intersections.

(II) A **measure** $\mu : \mathcal{M} \to [0, \infty]$ with the following defining property: if $E_1, E_2, \ldots$ is a countable family of disjoint sets in $\mathcal{M}$, then

$$\mu \left(\bigcup_{n=1}^{\infty} E_n \right) = \sum_{n=1}^{\infty} \mu(E_n).$$

A measure space is therefore often denoted by the triple $(X, \mathcal{M}, \mu)$ to emphasize its three main components. Sometimes, however, when there is no ambiguity we will abbreviate this notation by referring to the measure space as (X, μ), or simply X.

A feature that a measure space often enjoys is the property of being σ-**finite**. This means that X can be written as the union of countably many measurable sets of finite measure.

At this early stage we give only two simple examples of measure spaces:

(i) The first is the discrete example with X a countable set, $X = \{x_n\}_{n=1}^{\infty}$, $\mathcal{M}$ the collection of all subsets of X, and the measure μ determined by $\mu(x_n) = \mu_n$, with $\{\mu_n\}_{n=1}^{\infty}$ a given sequence of (extended) non-negative numbers. Note that $\mu(E) = \sum_{x_n \in E} \mu_n$. When $\mu_n = 1$ for all n, we call μ the **counting measure**, and also denote it by $\#$. In this case integration will amount to nothing but the summation of (absolutely) convergent series.

(ii) Here $X = \mathbb{R}^d$, $\mathcal{M}$ is the collection of Lebesgue measurable sets, and $\mu(E) = \int_E f \, dx$, where f is a given non-negative measurable function on $\mathbb{R}^d$. The case $f = 1$ corresponds to the Lebesgue measure. The countable additivity of μ follows from the usual additivity and limiting properties of integrals of non-negative functions proved in Chapter 2.

The construction of measure spaces relevant for most applications require further ideas, and to these we now turn.

1.1 Exterior measures and Carathéodory's theorem

To begin the construction of a measure and its corresponding measurable sets in the general setting requires, as in the special case of Lebesgue measure considered in Chapter 1, a prerequisite notion of "exterior" measure. This is defined as follows.

Let X be a set. An **exterior measure** (or **outer measure**) μ_* on X is a function μ_* from the collection of *all* subsets of X to $[0, \infty]$ that satisfies the following properties:

(i) $\mu_*(\emptyset) = 0$.

(ii) If $E_1 \subset E_2$, then $\mu_*(E_1) \leq \mu_*(E_2)$.

(iii) If $E_1, E_2, \ldots$ is a countable family of sets, then

$$\mu_* \left(\bigcup_{j=1}^{\infty} E_j \right) \leq \sum_{j=1}^{\infty} \mu_*(E_j).$$

For instance, the exterior Lebesgue measure m_* in $\mathbb{R}^d$ defined in Chapter 1 enjoys all these properties. In fact, this example belongs to a large class of exterior measures that can be obtained using "coverings" by a family of special sets whose measures are taken as known. This idea is systematized by the notion of a "premeasure" taken up below in Section 1.3. A different type of example is the exterior α-dimensional Hausdorff measure m_α^* defined in Chapter 7.

Given an exterior measure μ_*, the problem that one faces is how to define the corresponding notion of measurable sets. In the case of Lebesgue measure in $\mathbb{R}^d$ such sets were characterized by their difference from open (or closed) sets, when considered in terms of μ_*. For the general case, Carathéodory found an ingenious substitute condition. It is as follows.

A set E in X is **Carathéodory measurable** or simply **measurable** if one has

(1) $\mu_*(A) = \mu_*(E \cap A) + \mu_*(E^c \cap A)$ for every $A \subset X$.

In other words, E separates any set A in two parts that behave well in regard to the exterior measure μ_*. For this reason, (1) is sometimes referred to as the separation condition. One can show that in $\mathbb{R}^d$ with the Lebesgue exterior measure the notion of measurability (1) is equivalent to the definition of Lebesgue measurability given in Chapter 1. (See Exercise 3.)

A first observation we make is that to prove a set E is measurable, it suffices to verify

$$\mu_*(A) \geq \mu_*(E \cap A) + \mu_*(E^c \cap A) \quad \text{for all } A \subset X,$$

since the reverse inequality is automatically verified by the sub-additivity property (iii) of the exterior measure. We see immediately from the definition that sets of exterior measure zero are necessarily measurable.

The remarkable fact about the definition (1) is summarized in the next theorem.

Theorem 1.1 *Given an exterior measure μ_* on a set X, the collection $\mathcal{M}$ of Carathéodory measurable sets forms a σ-algebra. Moreover, μ_* restricted to $\mathcal{M}$ is a measure.*

Proof. Clearly, $\emptyset$ and X belong to $\mathcal{M}$ and the symmetry inherent in condition (1) shows that $E^c \in \mathcal{M}$ whenever $E \in \mathcal{M}$. Thus $\mathcal{M}$ is non-empty and closed under complements.

Next, we prove that $\mathcal{M}$ is closed under finite unions of disjoint sets, and μ_* is finitely additive on $\mathcal{M}$. Indeed, if $E_1, E_2 \in \mathcal{M}$, and A is any subset of X, then

$$\begin{aligned}
\mu_*(A) &= \mu_*(E_2 \cap A) + \mu_*(E_2^c \cap A) \\
&= \mu_*(E_1 \cap E_2 \cap A) + \mu_*(E_1^c \cap E_2 \cap A) + \\
&\quad\quad + \mu_*(E_1 \cap E_2^c \cap A) + \mu_*(E_1^c \cap E_2^c \cap A) \\
&\geq \mu_*((E_1 \cup E_2) \cap A) + \mu_*((E_1 \cup E_2)^c \cap A),
\end{aligned}$$

where in the first two lines we have used the measurability condition on E_2 and then E_1, and where the last inequality was obtained using the sub-additivity of μ_* and the fact that $E_1 \cup E_2 = (E_1 \cap E_2) \cup (E_1^c \cap E_2) \cup (E_1 \cap E_2^c)$. Therefore, we have $E_1 \cup E_2 \in \mathcal{M}$, and if E_1 and E_2 are disjoint, we find

$$\begin{aligned}
\mu_*(E_1 \cup E_2) &= \mu_*(E_1 \cap (E_1 \cup E_2)) + \mu_*(E_1^c \cap (E_1 \cup E_2)) \\
&= \mu_*(E_1) + \mu_*(E_2).
\end{aligned}$$

Finally, it suffices to show that $\mathcal{M}$ is closed under countable unions of disjoint sets, and that μ_* is countably additive on $\mathcal{M}$. Let $E_1, E_2, \ldots$ denote a countable collection of disjoint sets in $\mathcal{M}$, and define

$$G_n = \bigcup_{j=1}^{n} E_j \quad \text{and} \quad G = \bigcup_{j=1}^{\infty} E_j.$$

For each n, the set G_n is a finite union of sets in $\mathcal{M}$, hence $G_n \in \mathcal{M}$. Moreover, for any $A \subset X$ we have

$$\mu_*(G_n \cap A) = \mu_*(E_n \cap (G_n \cap A)) + \mu_*(E_n^c \cap (G_n \cap A))$$
$$= \mu_*(E_n \cap A) + \mu_*(G_{n-1} \cap A)$$
$$= \sum_{j=1}^{n} \mu_*(E_j \cap A),$$

where the last equality is obtained by induction. Since we know that $G_n \in \mathcal{M}$, and $G^c \subset G_n^c$, we find that

$$\mu_*(A) = \mu_*(G_n \cap A) + \mu_*(G_n^c \cap A) \geq \sum_{j=1}^{n} \mu_*(E_j \cap A) + \mu_*(G^c \cap A).$$

Letting n tend to infinity, we obtain

$$\mu_*(A) \geq \sum_{j=1}^{\infty} \mu_*(E_j \cap A) + \mu_*(G^c \cap A) \geq \mu_*(G \cap A) + \mu_*(G^c \cap A)$$
$$\geq \mu_*(A).$$

Therefore all the inequalities above are equalities, and we conclude that $G \in \mathcal{M}$, as desired. Moreover, by taking $A = G$ in the above, we find that μ_* is countably additive on $\mathcal{M}$, and the proof of the theorem is complete.

Our previous observation that sets of exterior measure 0 are Carathéodory measurable shows that the measure space $(X, \mathcal{M}, \mu)$ in the theorem is **complete**: whenever $F \in \mathcal{M}$ satisfies $\mu(F) = 0$ and $E \subset F$, then $E \in \mathcal{M}$.

1.2 Metric exterior measures

If the underlying set X is endowed with a "distance function" or "metric," there is a particular class of exterior measures that is of interest in practice. The importance of these exterior measures is that they induce measures on the natural σ-algebra generated by the open sets in X.

A **metric space** is a set X equipped with a function $d : X \times X \rightarrow [0, \infty)$ that satisfies:

(i) $d(x, y) = 0$ if and only if $x = y$.

(ii) $d(x, y) = d(y, x)$ for all $x, y \in X$.

(iii) $d(x, z) \leq d(x, y) + d(y, z)$, for all $x, y, z \in X$.

The last property is of course called the triangle inequality, and a function d that satisfies all these conditions is called a **metric** on X. For example, the set $\mathbb{R}^d$ with $d(x, y) = |x - y|$ is a metric space. Another example is provided by the space of continuous functions on a compact set K with $d(f, g) = \sup_{x \in K} |f(x) - g(x)|$.

A metric space (X, d) is naturally equipped with a family of open balls. Here

$$B_r(x) = \{y \in X : d(x, y) < r\}$$

defines the open ball of radius r centered at x. Together with this, we say that a set $\mathcal{O} \subset X$ is **open** if for any $x \in \mathcal{O}$ there exists $r > 0$ so that the open ball $B_r(x)$ is contained in $\mathcal{O}$. A set is **closed** if its complement is open. With these definitions, one checks easily that an (arbitrary) union of open sets is open, and a similar intersection of closed sets is closed.

Finally, on a metric space X we can define, as in Section 3 of Chapter 1, the **Borel σ-algebra**, $\mathcal{B}_X$, that is the smallest σ-algebra of sets in X that contains the open sets of X. In other words $\mathcal{B}_X$ is the intersection of all σ-algebras that contain the open sets. Elements in $\mathcal{B}_X$ are called **Borel sets**.

We now turn our attention to those exterior measures on X with the special property of being additive on sets that are "well separated." We show that this property guarantees that this exterior measure defines a measure on the Borel σ-algebra. This is achieved by proving that all Borel sets are Carathéodory measurable.

Given two sets A and B in a metric space (X, d), the **distance** between A and B is defined by

$$d(A, B) = \inf\{d(x, y) : x \in A \text{ and } y \in B\}.$$

Then an exterior measure μ_* on X is a **metric exterior measure** if it satisfies

$$\mu_*(A \cup B) = \mu_*(A) + \mu_*(B) \quad \text{whenever } d(A, B) > 0.$$

This property played a key role in the case of exterior Lebesgue measure.

Theorem 1.2 *If μ_* is a metric exterior measure on a metric space X, then the Borel sets in X are measurable. Hence μ_* restricted to $\mathcal{B}_X$ is a measure.*

Proof.   By the definition of $\mathcal{B}_X$ it suffices to prove that closed sets in X are Carathéodory measurable. Therefore, let F denote a closed set and A a subset of X with $\mu_*(A) < \infty$. For each $n > 0$, let

$$A_n = \{x \in F^c \cap A : \ d(x, F) \geq 1/n\}.$$

Then $A_n \subset A_{n+1}$, and since F is closed we have $F^c \cap A = \bigcup_{n=1}^{\infty} A_n$. Also, the distance between $F \cap A$ and A_n is $\geq 1/n$, and since μ_* is a metric exterior measure, we have

$$(2) \qquad \mu_*(A) \geq \mu_*((F \cap A) \cup A_n) = \mu_*(F \cap A) + \mu_*(A_n).$$

Next, we claim that

$$(3) \qquad\qquad \lim_{n \to \infty} \mu_*(A_n) = \mu_*(F^c \cap A).$$

To see this, let $B_n = A_{n+1} \cap A_n^c$ and note that

$$d(B_{n+1}, A_n) \geq \frac{1}{n(n+1)}.$$

Indeed, if $x \in B_{n+1}$ and $d(x, y) < 1/n(n+1)$ the triangle inequality shows that $d(y, F) < 1/n$, hence $y \notin A_n$. Therefore

$$\mu_*(A_{2k+1}) \geq \mu_*(B_{2k} \cup A_{2k-1}) = \mu_*(B_{2k}) + \mu_*(A_{2k-1}),$$

and this implies that

$$\mu_*(A_{2k+1}) \geq \sum_{j=1}^{k} \mu_*(B_{2j}).$$

A similar argument also gives

$$\mu_*(A_{2k}) \geq \sum_{j=1}^{k} \mu_*(B_{2j-1}).$$

Since $\mu_*(A)$ is finite, we find that both series $\sum \mu_*(B_{2j})$ and $\sum \mu_*(B_{2j-1})$ are convergent. Finally, we note that

$$\mu_*(A_n) \leq \mu_*(F^c \cap A) \leq \mu_*(A_n) + \sum_{j=n+1}^{\infty} \mu_*(B_j),$$

and this proves the limit (3). Letting n tend to infinity in the inequality (2) we find that $\mu_*(A) \geq \mu_*(F \cap A) + \mu_*(F^c \cap A)$, and hence F is measurable, as was to be shown.

Given a metric space X, a measure μ defined on the Borel sets of X will be referred to as a **Borel measure**. Borel measures that assign a finite measure to all balls (of finite radius) also satisfy a useful regularity property. The requirement that $\mu(B) < \infty$ for all balls B is satisfied in many (but not in all) circumstances that arise in practice.[1] When it does hold, we get the following proposition.

Proposition 1.3 *Suppose the Borel measure μ is finite on all balls in X of finite radius. Then for any Borel set E and any $\epsilon > 0$, there are an open set $\mathcal{O}$ and a closed set F such that $E \subset \mathcal{O}$ and $\mu(\mathcal{O} - E) < \epsilon$, while $F \subset E$ and $\mu(E - F) < \epsilon$.*

Proof. We need the following preliminary observation. Suppose $F^* = \bigcup_{k=1}^{\infty} F_k$, where the F_k are closed sets. Then for any $\epsilon > 0$, we can find a closed set $F \subset F^*$ such that $\mu(F^* - F) < \epsilon$. To prove this we can assume that the sets $\{F_k\}$ are increasing. Fix a point $x_0 \in X$, and let B_n denote the ball $\{x : d(x, x_0) < n\}$, with $B_0 = \{\emptyset\}$. Since $\bigcup_{n=1}^{\infty} B_n = X$, we have that

$$F^* = \bigcup F^* \cap (\overline{B}_n - B_{n-1}).$$

Now for each n, $F^* \cap (\overline{B}_n - B_{n-1})$ is the limit as $k \to \infty$ of the increasing sequence of closed sets $F_k \cap (\overline{B}_n - B_{n-1})$, so (recalling that $\overline{B}_n$ has finite measure) we can find an $N = N(n)$ so that $(F^* - F_{N(n)}) \cap (\overline{B}_n - B_{n-1})$ has measure less than $\epsilon/2^n$. If we now let

$$F = \bigcup_{n=1}^{\infty} \left(F_{N(n)} \cap (\overline{B}_n - B_{n-1}) \right),$$

it follows that the measure of $F^* - F$ is less that $\sum_{n=1}^{\infty} \epsilon/2^n = \epsilon$. We also see that $F \cap \overline{B}_k$ is closed since it is the finite union of closed sets. Thus F itself is closed because, as is easily seen, any set F is closed whenever the sets $F \cap \overline{B}_k$ are closed for all k.

Having established the observation, we call $\mathcal{C}$ the collection of all sets that satisfy the conclusions of the proposition. Notice first that if E belongs to $\mathcal{C}$ then automatically so does its complement.

---

[1] This restriction is not always valid for the Hausdorff measures that are considered in the next chapter.

Suppose now that $E = \bigcup_{k=1}^{\infty} E_k$, with each $E_k \in \mathcal{C}$. Then there are open sets $\mathcal{O}_k$, $\mathcal{O}_k \supset E_k$, with $\mu(\mathcal{O}_k - E_k) < \epsilon/2^k$. However, if $\mathcal{O} = \bigcup_{k=1}^{\infty} \mathcal{O}_k$, then $\mathcal{O} - E \subset \bigcup_{k=1}^{\infty}(\mathcal{O}_k - E_k)$, and so $\mu(\mathcal{O} - E) \leq \sum_{k=1}^{\infty} \epsilon/2^k = \epsilon$.

Next, there are closed sets $F_k \subset E_k$ with $\mu(E_k - F_k) < \epsilon/2^k$. Thus if $F^* = \bigcup_{k=1}^{\infty} F_k$, we see as before that $\mu(E - F^*) < \epsilon$. However, F^* is not necessarily closed, so we can use our preliminary observation to find a closed set $F \subset F^*$ with $\mu(F^* - F) < \epsilon$. Thus $\mu(E - F) < 2\epsilon$. Since ϵ is arbitrary, this proves that $\bigcup_{k=1}^{\infty} E_k$ belongs to $\mathcal{C}$.

Let us finally note that any open set $\mathcal{O}$ is in $\mathcal{C}$. The property regarding containment by open sets is immediate. To find a closed $F \subset \mathcal{O}$, so that $\mu(\mathcal{O} - F) < \epsilon$, let $F_k = \{x \in \overline{B}_k : d(x, \mathcal{O}^c) \geq 1/k\}$. Then it is clear that each F_k is closed and $\mathcal{O} = \bigcup_{k=1}^{\infty} F_k$. We then need only apply the observation again to find the required set F. Thus we have shown that $\mathcal{C}$ is a σ-algebra that contains the open sets, and hence all Borel sets. The proposition is therefore proved.

1.3 The extension theorem

As we have seen, a class of measurable sets on X can be constructed once we start with a given exterior measure. However, the definition of an exterior measure usually depends on a more primitive idea of measure defined on a simpler class of sets. This is the role of a premeasure defined below. As we will show, any premeasure can be extended to a measure on X. We begin with several definitions.

Let X be a set. An **algebra** in X is a non-empty collection of subsets of X that is closed under complements, *finite* unions, and *finite* intersections. Let $\mathcal{A}$ be an algebra in X. A **premeasure** on an algebra $\mathcal{A}$ is a function $\mu_0 : \mathcal{A} \to [0, \infty]$ that satisfies:

(i) $\mu_0(\emptyset) = 0$.

(ii) If $E_1, E_2, \ldots$ is a countable collection of disjoint sets in $\mathcal{A}$ with $\bigcup_{k=1}^{\infty} E_k \in \mathcal{A}$, then

$$\mu_0\left(\bigcup_{k=1}^{\infty} E_k\right) = \sum_{k=1}^{\infty} \mu_0(E_k).$$

In particular, μ_0 is finitely additive on $\mathcal{A}$.

Premeasures give rise to exterior measures in a natural way.

Lemma 1.4 *If μ_0 is a premeasure on an algebra $\mathcal{A}$, define μ_* on any subset E of X by*

$$\mu_*(E) = \inf \left\{ \sum_{j=1}^{\infty} \mu_0(E_j) : \ E \subset \bigcup_{j=1}^{\infty} E_j, \ where \ E_j \in \mathcal{A} \ for \ all \ j \right\}.$$

Then, μ_* is an exterior measure on X that satisfies:

(i) $\mu_*(E) = \mu_0(E)$ *for all $E \in \mathcal{A}$.*

(ii) *All sets in $\mathcal{A}$ are measurable in the sense of* (1).

Proof. Proving that μ_* is an exterior measure presents no difficulty. To see why the restriction of μ_* to $\mathcal{A}$ coincides with μ_0, suppose that $E \in \mathcal{A}$. Clearly, one always has $\mu_*(E) \leq \mu_0(E)$ since E covers itself. To prove the reverse inequality let $E \subset \bigcup_{j=1}^{\infty} E_j$, where $E_j \in \mathcal{A}$ for all j. Then, if we set

$$E_k' = E \cap \left(E_k - \bigcup_{j=1}^{k-1} E_j \right),$$

the sets E_k' are disjoint elements of $\mathcal{A}$, $E_k' \subset E_k$ and $E = \bigcup_{k=1}^{\infty} E_k'$. By (ii) in the definition of a premeasure, we have

$$\mu_0(E) = \sum_{k=1}^{\infty} \mu_0(E_k') \leq \sum_{k=1}^{\infty} \mu_0(E_k).$$

Therefore, we find that $\mu_0(E) \leq \mu_*(E)$, as desired.

Finally, we must prove that sets in $\mathcal{A}$ are measurable for μ_*. Let A be any subset of X, $E \in \mathcal{A}$, and $\epsilon > 0$. By definition, there exists a countable collection $E_1, E_2, \ldots$ of sets in $\mathcal{A}$ such that $A \subset \bigcup_{j=1}^{\infty} E_j$ and

$$\sum_{j=1}^{\infty} \mu_0(E_j) \leq \mu_*(A) + \epsilon.$$

Since μ_0 is a premeasure, it is finitely additive on $\mathcal{A}$ and therefore

$$\sum_{j=1}^{\infty} \mu_0(E_j) = \sum_{j=1}^{\infty} \mu_0(E \cap E_j) + \sum_{j=1}^{\infty} \mu_0(E^c \cap E_j)$$
$$\geq \mu_*(E \cap A) + \mu_*(E^c \cap A).$$

Since ϵ is arbitrary, we conclude that $\mu_*(A) \geq \mu_*(E \cap A) + \mu_*(E^c \cap A)$, as desired.

The σ-algebra generated by an algebra $\mathcal{A}$ is by definition the smallest σ-algebra that contains $\mathcal{A}$. The above lemma then provides the necessary step for extending μ_0 on $\mathcal{A}$ to a measure on the σ-algebra generated by $\mathcal{A}$.

Theorem 1.5 *Suppose that $\mathcal{A}$ is an algebra of sets in X, μ_0 a premeasure on $\mathcal{A}$, and $\mathcal{M}$ the σ-algebra generated by $\mathcal{A}$. Then there exists a measure μ on $\mathcal{M}$ that extends μ_0.*

One notes below that μ is the only such extension of μ_0 under the assumption that μ is σ-finite.

Proof. The exterior measure μ_* induced by μ_0 defines a measure μ on the σ-algebra of Carathéodory measurable sets. Therefore, by the result in the previous lemma, μ is also a measure on $\mathcal{M}$ that extends μ_0. (We should observe that in general the class $\mathcal{M}$ is not as large as the class of all sets that are measurable in the sense of (1).)

To prove that this extension is unique whenever μ is σ-finite, we argue as follows. Suppose that ν is another measure on $\mathcal{M}$ that coincides with μ_0 on $\mathcal{A}$, and suppose that $F \in \mathcal{M}$ has finite measure. We claim that $\mu(F) = \nu(F)$. If $F \subset \bigcup E_j$, where $E_j \in \mathcal{A}$, then

$$\nu(F) \leq \sum_{j=1}^{\infty} \nu(E_j) = \sum_{j=1}^{\infty} \mu_0(E_j),$$

so that $\nu(F) \leq \mu(F)$. To prove the reverse inequality, note that if $E = \bigcup E_j$, then the fact that ν and μ are two measures that agree on $\mathcal{A}$ gives

$$\nu(E) = \lim_{n \to \infty} \nu\left(\bigcup_{j=1}^{n} E_j\right) = \lim_{n \to \infty} \mu\left(\bigcup_{j=1}^{n} E_j\right) = \mu(E).$$

If the sets E_j are chosen so that $\mu(E) \leq \mu(F) + \epsilon$, then the fact that $\mu(F) < \infty$ implies $\mu(E - F) \leq \epsilon$, and therefore

$$\mu(F) \leq \mu(E) = \nu(E) = \nu(F) + \nu(E - F) \leq \nu(F) + \mu(E - F)$$
$$\leq \mu(F) + \epsilon.$$

Since ϵ is arbitrary, we find that $\mu(F) \leq \nu(F)$, as desired.

Finally, we use this last result to prove that if μ is σ-finite, then $\mu = \nu$. Indeed, we may write $X = \bigcup E_j$, where $E_1, E_2, \ldots$ is a countable collection of disjoint sets in $\mathcal{A}$ with $\mu(E_j) < \infty$. Then for any $F \in \mathcal{M}$ we have

$$\mu(F) = \sum \mu(F \cap E_j) = \sum \nu(F \cap E_j) = \nu(F),$$

and the uniqueness is proved.

For later use we record the following observation about the premeasure μ_0 on the algebra $\mathcal{A}$ and the resulting measure μ_* that is implicit in the argument given above. The details of the proof may be left to the reader.

We define $\mathcal{A}_\sigma$ as the collection of sets that are countable unions of sets in $\mathcal{A}$, and $\mathcal{A}_{\sigma\delta}$ as the sets that arise as countable intersections of sets in $\mathcal{A}_\sigma$.

Proposition 1.6 *For any set E and any $\epsilon > 0$, there are sets $E_1 \in \mathcal{A}_\sigma$ and $E_2 \in \mathcal{A}_{\sigma\delta}$, such that $E \subset E_1$, $E \subset E_2$, and $\mu_*(E_1) \le \mu_*(E) + \epsilon$, while $\mu_*(E_2) = \mu_*(E)$.*

2 Integration on a measure space

Once we have established the basic properties of a measure space X, the fundamental facts about measurable functions and integration of such functions on X can be deduced as in the case of the Lebesgue measure on $\mathbb{R}^d$. Indeed, the results in Section 4 of Chapter 1 and all of Chapter 2 go over to the general case, with proofs remaining almost word-for-word the same. For this reason we shall not repeat these arguments but limit ourselves to the bare statement of the main points. The reader should have no difficulty in filling in the missing details.

To avoid unnecessary complications we will assume throughout that the measure space $(X, \mathcal{M}, \mu)$ under consideration is σ-finite.

Measurable functions

A function f on X with values in the extended real numbers is **measurable** if

$$f^{-1}([-\infty, a)) = \{x \in X : f(x) < a\} \in \mathcal{M} \quad \text{for all } a \in \mathbb{R}.$$

With this definition, the basic properties of measurable functions obtained in the case of $\mathbb{R}^d$ with the Lebesgue measure continue to hold. (See Properties 3 through 6 for measurable functions in Chapter 1.) For instance, the collection of measurable functions is closed under the basic algebraic manipulations. Also, the pointwise limits of measurable functions are measurable.

The notion of "almost everywhere" that we use now is with respect to the measure μ. For instance, if f and g are measurable functions on X, we write $f = g$ a.e. to say that

$$\mu\left(\{x \in X : f(x) \neq g(x)\}\right) = 0.$$

A **simple function** on X takes the form

$$\sum_{k=1}^{N} a_k \chi_{E_k},$$

where E_k are measurable sets of finite measure and a_k are real numbers. Approximations by simple functions played an important role in the definition of the Lebesgue integral. Fortunately, this result continues to hold in our abstract setting.

- *Suppose f is a non-negative measurable function on a measure space $(X, \mathcal{M}, \mu)$. Then there exists a sequence of simple functions $\{\varphi_k\}_{k=1}^{\infty}$ that satisfies*

$$\varphi_k(x) \leq \varphi_{k+1}(x) \quad and \quad \lim_{k \to \infty} \varphi_k(x) = f(x) \text{ for all } x.$$

  In general, if f is only measurable, there exists a sequence of simple functions $\{\varphi_k\}_{k=1}^{\infty}$ that satisfies

$$|\varphi_k(x)| \leq |\varphi_{k+1}(x)| \quad and \quad \lim_{k \to \infty} \varphi_k(x) = f(x) \text{ for all } x.$$

The proof of this result can be obtained with some obvious minor modifications of the proofs of Theorems 4.1 and 4.2 in Chapter 1. Here, one makes use of the technical condition imposed on X, that of being σ-finite. Indeed, if we write $X = \bigcup F_k$, where $F_k \in \mathcal{M}$ are of finite measure, then the sets F_k play the role of the cubes Q_k in the proof of Theorem 4.1, Chapter 1.

Another important result that generalizes immediately is Egorov's theorem.

- *Suppose $\{f_k\}_{k=1}^{\infty}$ is a sequence of measurable functions defined on a measurable set $E \subset X$ with $\mu(E) < \infty$, and $f_k \to f$ a.e. Then for each $\epsilon > 0$ there is a set A_ϵ with $A_\epsilon \subset E$, $\mu(E - A_\epsilon) \leq \epsilon$, and such that $f_k \to f$ uniformly on A_ϵ.*

Definition and main properties of the integral

The four-step approach to the construction of the Lebesgue integral that begins with its definition on simple functions given in Chapter 2 carries over to the situation of a σ-finite measure space $(X, \mathcal{M}, \mu)$. This leads to the notion of the integral, with respect to the measure μ, of a non-negative measurable function f on X. This integral is denoted by

$$\int_X f(x) \, d\mu(x),$$

which we sometimes simplify as $\int_X f \, d\mu$, $\int f \, d\mu$ or $\int f$, when no confusion is possible. Finally, we say that a measurable function f is **integrable** if

$$\int_X |f(x)| \, d\mu(x) < \infty.$$

The elementary properties of the integral, such as linearity and monotonicity, continue to hold in this general setting, as well as the following basic limit theorems.

(i) Fatou's lemma. *If $\{f_n\}$ is a sequence of non-negative measurable functions on X, then*

$$\int \liminf_{n \to \infty} f_n \, d\mu \le \liminf_{n \to \infty} \int f_n \, d\mu.$$

(ii) Monotone convergence. *If $\{f_n\}$ is a sequence of non-negative measurable functions with $f_n \nearrow f$, then*

$$\lim_{n \to \infty} \int f_n = \int f.$$

(iii) Dominated convergence. *If $\{f_n\}$ is a sequence of measurable functions with $f_n \to f$ a.e., and such that $|f_n| \le g$ for some integrable g, then*

$$\int |f_n - f| \, d\mu \to 0 \qquad \text{as } n \to \infty,$$

and consequently

$$\int f_n \, d\mu \to \int f \, d\mu \qquad \text{as } n \to \infty.$$

The spaces $L^1(X, \mu)$ and $L^2(X, \mu)$

The equivalence classes (modulo functions that vanish almost everywhere) of integrable functions on $(X, \mathcal{M}, \mu)$ form a vector space equipped with a norm. This space is denoted by $L^1(X, \mu)$ and its norm is

$$(4) \qquad \|f\|_{L^1(X,\mu)} = \int_X |f(x)| \, d\mu(x).$$

Similarly we can define $L^2(X, \mu)$ to be the equivalence class of measurable functions for which $\int_X |f(x)|^2 \, d\mu(x) < \infty$. The norm is then

$$(5) \qquad \|f\|_{L^2(X,\mu)} = \left(\int_X |f(x)|^2 \, d\mu(x) \right)^{1/2}.$$

There is also an inner product on this space given by

$$(f, g) = \int_X f(x)\overline{g(x)} \, d\mu(x).$$

The proofs of Proposition 2.1 and Theorem 2.2 in Chapter 2, as well as the results in Section 1 of Chapter 4, extend to this general case and give:

- *The space $L^1(X, \mu)$ is a complete normed vector space.*

- *The space $L^2(X, \mu)$ is a (possibly non-separable) Hilbert space.*

3 Examples

We now discuss some useful examples of the general theory.

3.1 Product measures and a general Fubini theorem

Our first example concerns the construction of product measures, and leads to a general form of the theorem that expresses a multiple integral as a repeated integral, extending the case of Euclidean space considered in Section 3 of Chapter 2.

Suppose $(X_1, \mathcal{M}_1, \mu_1)$ and $(X_2, \mathcal{M}_2, \mu_2)$ are a pair of measure spaces. We want to describe the **product measure** $\mu_1 \times \mu_2$ on the space $X = X_1 \times X_2 = \{(x_1, x_2) : x_1 \in X_1, x_2 \in X_2\}$.

We will assume here that the two measure spaces are each complete and σ-finite.

We begin by considering **measurable rectangles**: these are subsets of X of the form $A \times B$, with A and B measurable sets, that is, $A \in \mathcal{M}_1$

and $B \in \mathcal{M}_2$. We then let $\mathcal{A}$ denote the collection of all sets in X that are finite unions of disjoint measurable rectangles. It is easy to check that $\mathcal{A}$ is an algebra of subsets of X. (Indeed, the complement of a measurable rectangle is the union of three disjoint such rectangles, while the union of two measurable rectangles is the disjoint union of at most six such rectangles.) From now on we abbreviate our terminology by referring to measurable rectangles simply as "rectangles."

On the rectangles we define the function μ_0 by $\mu_0(A \times B) = \mu_1(A)\mu_2(B)$. Now the fact that μ_0 has a unique extension to the algebra $\mathcal{A}$ for which μ_0 becomes a premeasure is a consequence of the following fact: whenever a rectangle $A \times B$ is the disjoint union of a countable collection of rectangles $\{A_j \times B_j\}$, $A \times B = \bigcup_{j=1}^{\infty} A_j \times B_j$, then

$$(6) \qquad \mu_0(A \times B) = \sum_{j=1}^{\infty} \mu_0(A_j \times B_j).$$

To prove this, observe that if $x_1 \in A$, then for each $x_2 \in B$ the point (x_1, x_2) belongs to exactly one $A_j \times B_j$. Therefore we see that B is the disjoint union of the B_j for which $x_1 \in A_j$. By the countable additivity property of the measure μ_2 this has as an immediate consequence the fact that

$$\chi_A(x_1)\mu_2(B) = \sum_{j=1}^{\infty} \chi_{A_j}(x_1)\mu_2(B_j).$$

Hence integrating in x_1 and using the monotone convergence theorem we get $\mu_1(A)\mu_2(B) = \sum_{j=1}^{\infty} \mu_1(A_j)\mu_2(B_j)$, which is (6).

Now that we know that μ_0 is a premeasure on $\mathcal{A}$, we obtain from Theorem 1.5 a measure (which we denote by $\mu = \mu_1 \times \mu_2$) on the σ-algebra $\mathcal{M}$ of sets generated by the algebra $\mathcal{A}$ of measurable rectangles. In this way, we have defined the product measure space $(X_1 \times X_2, \mathcal{M}, \mu_1 \times \mu_2)$.

Given a set E in $\mathcal{M}$ we shall now consider slices

$$E_{x_1} = \{x_2 \in X_2 : (x_1, x_2) \in E\} \quad \text{and} \quad E^{x_2} = \{x_1 \in X_1 : (x_1, x_2) \in E\}.$$

We recall the definitions according to which $\mathcal{A}_\sigma$ denotes the collection of sets that are countable unions of elements of $\mathcal{A}$, and $\mathcal{A}_{\sigma\delta}$ the sets that arise as countable intersections of sets from $\mathcal{A}_\sigma$. We then have the following key fact.

Proposition 3.1 *If E belongs to $\mathcal{A}_{\sigma\delta}$, then E^{x_2} is μ_1-measurable for every x_2; moreover, $\mu_1(E^{x_2})$ is a μ_2-measurable function. In addition*

$$(7) \qquad \int_{X_2} \mu_1(E^{x_2})\, d\mu_2 = (\mu_1 \times \mu_2)(E).$$

Proof. One notes first that all the assertions hold immediately when E is a (measurable) rectangle. Next suppose E is a set in $\mathcal{A}_\sigma$. Then we can decompose it as a countable union of *disjoint* rectangles E_j. (If the E_j are not already disjoint we only need to replace the E_j by $\bigcup_{k \leq j} E_k - \bigcup_{k \leq j-1} E_k$.) Then for each x_2 we have $E^{x_2} = \bigcup_{j=1}^\infty E_j^{x_2}$, and we observe that $\{E_j^{x_2}\}$ are disjoint sets. Thus by (7) applied to each rectangle E_j and the monotone convergence theorem we get our conclusion for each set $E \in \mathcal{A}_\sigma$.

Next assume $E \in \mathcal{A}_{\sigma\delta}$ and that $(\mu_1 \times \mu_2)(E) < \infty$. Then there is a sequence $\{E_j\}$ of sets with $E_j \in \mathcal{A}_\sigma$, $E_{j+1} \subset E_j$, and $E = \bigcap_{j=1}^\infty E_j$. We let $f_j(x_2) = \mu_1(E_j^{x_2})$ and $f(x_2) = \mu_1(E^{x_2})$. To see that E^{x_2} is μ_1-measurable and $f(x_2)$ is well-defined, note that E^{x_2} is the decreasing limit of the sets $E_j^{x_2}$, which we have seen by the above are measurable. Moreover, since $E_1 \in \mathcal{A}_\sigma$ and $(\mu_1 \times \mu_2)(E_1) < \infty$, we see that $f_j(x_2) \to f(x_2)$, as $j \to \infty$ for each x_2. Thus $f(x_2)$ is measurable. However, $\{f_j(x_2)\}$ is a decreasing sequence of non-negative functions, hence

$$\int_{X_2} f(x_2)\, d\mu_2(x) = \lim_{j \to \infty} \int_{X_2} f_j(x_2)\, d\mu_2(x),$$

and therefore (7) is proved in the case when $(\mu_1 \times \mu_2)(E) < \infty$. Now since we assumed both μ_1 and μ_2 are σ-finite, we can find sequences $F_1 \subset F_2 \subset \cdots \subset F_j \subset \cdots \subset X_1$ and $G_1 \subset G_2 \subset \cdots \subset G_j \subset \cdots \subset X_2$, with $\bigcup_{j=1}^\infty F_j = X_1$, $\bigcup_{j=1}^\infty G_j = X_2$, $\mu_1(F_j) < \infty$, and $\mu_2(G_j) < \infty$ for all j. Then we merely need to replace E by $E_j = E \cap (F_j \times G_j)$, and let $j \to \infty$ to obtain the general result.

We now extend the result in the above proposition to an arbitrary measurable set E in $X_1 \times X_2$, that is, $E \in \mathcal{M}$, the σ-algebra generated by the measurable rectangles.

Proposition 3.2 *If E is an arbitrary measurable set in X, then the conclusion of Proposition 3.1 are still valid except that we only assert that E^{x_2} is μ_1-measurable and $\mu_1(E^{x_2})$ is defined for almost every $x_2 \in X_2$.*

Proof. Consider first the case when E is a set of measure zero. Then we know by Proposition 1.6 that there is a set $F \in \mathcal{A}_{\sigma\delta}$ such that

$E \subset F$ and $(\mu_1 \times \mu_2)(F) = 0$. Since $E^{x_2} \subset F^{x_2}$ for every x_2 and F^{x_2} has μ_1-measure zero for almost every x_2 by (7) applied to F, the assumed completeness of the measure μ_2 shows that E^{x_2} is measurable and has measure zero for those x_2. Thus the desired conclusion holds when E has measure zero.

If we drop this assumption on E, we can invoke Proposition 1.6 again to find an $F \in \mathcal{A}_{\sigma\delta}$, $F \supset E$, such that $F - E = Z$ has measure zero. Since $F^{x_2} - E^{x_2} = Z^{x_2}$ we can apply the case we have just proved, and find that for almost all x_2 the set E^{x_2} is measurable and $\mu_1(E^{x_2}) = \mu_1(F^{x_2}) - \mu_1(Z^{x_2})$. From this the proposition follows.

We now obtain the main result, generalizing Fubini's theorem in Chapter 2.

Theorem 3.3 *In the setting above, suppose $f(x_1, x_2)$ is an integrable function on $(X_1 \times X_2, \mu_1 \times \mu_2)$.*

(i) *For almost every $x_2 \in X_2$, the slice $f^{x_2}(x_1) = f(x_1, x_2)$ is integrable on (X_1, μ_1).*

(ii) *$\int_{X_1} f(x_1, x_2) \, d\mu_1$ is an integrable function on X_2.*

(iii) *$\int_{X_2} \left(\int_{X_1} f(x_1, x_2) \, d\mu_1 \right) d\mu_2 = \int_{X_1 \times X_2} f \, d\mu_1 \times \mu_2$.*

Proof. Note that if the desired conclusions hold for finitely many functions, they also hold for their linear combinations. In particular it suffices to assume that f is non-negative. When $f = \chi_E$, where E is a set of finite measure, what we wish to prove is contained in Proposition 3.2. Hence the desired result also holds for simple functions. Therefore by the monotone convergence theorem it is established for all non-negative functions, and the theorem is proved.

We remark that in general the product space $(X, \mathcal{M}, \mu)$ constructed above is not complete. However, if we define the completed space $(\overline{X}, \overline{\mathcal{M}}, \mu)$ as in Exercise 2, the theorem continues to hold in this completed space. The proof requires only a simple modification of the argument in Proposition 3.2.

3.2 Integration formula for polar coordinates

The polar coordinates of a point $x \in \mathbb{R}^d - \{0\}$ are the pair (r, γ), where $0 < r < \infty$ and γ belongs to the unit sphere $S^{d-1} = \{x \in \mathbb{R}^d, |x| = 1\}$. These are determined by

(8) $$r = |x|, \quad \gamma = \frac{x}{|x|}, \quad \text{and reciprocally by } x = r\gamma.$$

Our intention here is to deal with the formula that, with appropriate definitions and under suitable hypotheses, states:

$$(9) \qquad \int_{\mathbb{R}^d} f(x)\, dx = \int_{S^{d-1}} \left(\int_0^\infty f(r\gamma) r^{d-1}\, dr \right) d\sigma(\gamma).$$

For this we consider the following pair of measure spaces. First, $(X_1, \mathcal{M}_1, \mu_1)$, where $X_1 = (0, \infty)$, $\mathcal{M}_1$ is the collection of Lebesgue measurable sets in $(0, \infty)$, and $d\mu_1(r) = r^{d-1} dr$ in the sense that $\mu_1(E) = \int_E r^{d-1}\, dr$. Next, X_2 is the unit sphere S^{d-1}, and the measure μ_2 is the one in effect determined by (9) with $\mu_2 = \sigma$. Indeed given any set $E \subset S^{d-1}$ we let $\tilde{E} = \{x \in \mathbb{R}^d : x/|x| \in E,\ 0 < |x| < 1\}$ be the "sector" in the unit ball whose "end-points" are in E. We shall say $E \in \mathcal{M}_2$ exactly when $\tilde{E}$ is a Lebesgue measurable subset of $\mathbb{R}^d$, and define $\mu_2(E) = \sigma(E) = d \cdot m(\tilde{E})$, where m is Lebesgue measure in $\mathbb{R}^d$.

With this it is clear that both $(X_1, \mathcal{M}_1, \mu_1)$ and $(X_2, \mathcal{M}_2, \mu_2)$ satisfy all the properties of complete and σ-finite measure spaces. We note also that the sphere S^{d-1} has a metric on it given by $d(\gamma, \gamma') = |\gamma - \gamma'|$, for $\gamma, \gamma' \in S^{d-1}$. If E is an open set (with respect to this metric) in S^{d-1}, then $\tilde{E}$ is open in $\mathbb{R}^d$, and hence E is a measurable set in S^{d-1}.

Theorem 3.4 *Suppose f is an integrable function on $\mathbb{R}^d$. Then for almost every $\gamma \in S^{d-1}$ the slice f^γ defined by $f^\gamma(r) = f(r\gamma)$ is an integrable function with respect to the measure $r^{d-1} dr$. Moreover, $\int_0^\infty f^\gamma(r) r^{d-1}\, dr$ is integrable on S^{d-1} and the identity (9) holds.*

There is a corresponding result with the order of integration of r and γ reversed.

Proof. We consider the product measure $\mu = \mu_1 \times \mu_2$ on $X_1 \times X_2$ given by Theorem 3.3. Since the space $X_1 \times X_2 = \{(r, \gamma) : 0 < r < \infty \text{ and } \gamma \in S^{d-1}\}$ can be identified with $\mathbb{R}^d - \{0\}$, we can think of μ as a measure of the latter space, and our main task is to identify it with the (restriction of) Lebesgue measure on that space. We claim first that

$$(10) \qquad\qquad m(E) = \mu(E)$$

whenever E is a measurable rectangle $E = E_1 \times E_2$, and in this case $\mu(E) = \mu_1(E_1)\mu_2(E_2)$. In fact this holds for E_2 an arbitrary measurable subset of S^{d-1} and $E_1 = (0, 1)$, because then $E = E_1 \times E_2$ is the sector $\tilde{E}_2$, while $\mu_1(E_1) = 1/d$.

Because of the relative dilation-invariance of Lebesgue measure, (10) also holds when $E = (0, b) \times E_2$, $b > 0$. A simple limiting argument then proves the result for sets $E_1 = (0, a]$, and by subtraction to all open intervals $E_1 = (a, b)$, and thus for all open sets. Thus we have $m(E_1 \times E_2) = \mu_1(E_1)\mu_2(E_2)$ for all open sets E_1, and hence for all closed sets, and therefore for all Lebesgue measurable sets. (In fact, we can find sets $F_1 \subset E_1 \subset \mathcal{O}_1$ with F_1 closed and $\mathcal{O}_1$ open, such that $m_1(\mathcal{O}_1) - \epsilon \leq m_1(E_1) \leq m_1(F_1) + \epsilon$, and apply the above to $F_1 \times E_2$ and $\mathcal{O}_1 \times E_2$.) So we have established the identity (10) for all measurable rectangles and as a result for all finite unions of measurable rectangles. This is the algebra $\mathcal{A}$ that occurs in the proof of Theorem 3.3, and hence by the uniqueness in Theorem 1.5, the identity extends to the σ-algebra generated by $\mathcal{A}$, which is the σ-algebra $\mathcal{M}$ on which the measure μ is defined. To summarize, whenever $E \in \mathcal{M}$, the assertion (9) holds for $f = \chi_E$.

To go further we note that any open set in $\mathbb{R}^d - \{0\}$ can be written as a countable union of rectangles $\bigcup_{j=1}^{\infty} A_j \times B_j$, where A_j and B_j are open in $(0, \infty)$ and S^{d-1}, respectively. (This small technical point is taken up in Exercise 12.) It follows that any open set is in $\mathcal{M}$, and therefore so is any Borel set. Thus (9) is valid for χ_E whenever E is any Borel set in $\mathbb{R}^d - \{0\}$. The result then goes over to any Lebesgue set $E' \subset \mathbb{R}^d - \{0\}$, since such a set can be written as a disjoint union $E' = E \cup Z$, where E is a Borel set and $Z \subset F$, with F a Borel set of measure zero. To finish the proof we follow the familiar steps of deducing (9) for simple functions, and then by monotonic convergence for non-negative integrable functions, and from that for the general case.

3.3 Borel measures on $\mathbb{R}$ and the Lebesgue-Stieltjes integral

The Stieltjes integral was introduced to provide a generalization of the Riemann integral $\int_a^b f(x)\, dx$, where the increments dx were replaced by the increments $dF(x)$ for a given increasing function F on $[a, b]$. We wish to pursue this idea from the general point of view taken in this chapter. The question that is then raised is that of characterizing the measures on $\mathbb{R}$ that arise in this way, and in particular measures defined on the Borel sets on the real line.

To have a unique correspondence between measures and increasing functions as we shall have below, we need first to normalize these functions appropriately. Recall that an increasing function F can have at most a countable number of discontinuities. If x_0 is such a discontinuity, then

$$\lim_{\substack{x < x_0 \\ x \to x_0}} F(x) = F(x_0^-) \quad \text{and} \quad \lim_{\substack{x > x_0 \\ x \to x_0}} F(x) = F(x_0^+)$$

both exist, while $F(x_0^-) < F(x_0^+)$ and $F(x_0)$ is some value between $F(x_0^-)$ and $F(x_0^+)$. We shall now modify F at x_0, if necessary, by setting $F(x_0) = F(x_0^+)$, and we do this for every point of discontinuity. The function F so obtained is now still increasing, yet right-continuous at every point, and we say such functions are **normalized**. The main result is then as follows.

Theorem 3.5 *Let F be an increasing function on $\mathbb{R}$ that is normalized. Then there is a unique measure μ (also denoted by dF) on the Borel sets $\mathcal{B}$ on $\mathbb{R}$ such that $\mu((a,b]) = F(b) - F(a)$ if $a < b$. Conversely, if μ is a measure on $\mathcal{B}$ that is finite on bounded intervals, then F defined by $F(x) = \mu((0,x]), x > 0, F(0) = 0$ and, $F(x) = -\mu((-x,0]), x < 0$, is increasing and normalized.*

Before we come to the proof, we remark that the condition that μ be finite on bounded intervals is crucial. In fact, the Hausdorff measures that will be considered in the next chapter provide examples of Borel measures on $\mathbb{R}$ of a very different character from those treated in the theorem.

Proof. We define a function μ_* on all subsets of $\mathbb{R}$ by

$$\mu_*(E) = \inf \sum_{j=1}^{\infty} (F(b_j) - F(a_j)),$$

where the infimum is taken over all coverings of E of the form $\bigcup_{j=1}^{\infty}(a_j, b_j]$.

It is easy to verify that μ_* is an exterior measure on $\mathbb{R}$. We observe next that $\mu_*((a,b]) = (F(b) - F(a))$, if $a < b$. Clearly $\mu_*((a,b]) \leq F(b) - F(a)$, since $(a,b]$, then covers itself. Next, suppose that $\bigcup_{j=1}^{\infty}(a_j, b_j]$ covers $(a,b]$; then it covers $[a',b]$ for any $a < a' < b$. However, by the right-continuity of F, if $\epsilon > 0$ is given, we can always choose $b_j' > b_j$ such that $F(b_j') \leq F(b_j) + \epsilon/2^j$. Now the union of open intervals $\bigcup_{j=1}^{\infty}(a_j, b_j')$ covers $[a',b]$. By the compactness of this interval, $\bigcup_{j=1}^{N}(a_j, b_j')$ covers $[a',b]$ for some N. Thus since F is increasing we have

$$F(b) - F(a') \leq \sum_{j=1}^{N} F(b_j') - F(a_j) \leq \sum_{j=1}^{N}(F(b_j) - F(a_j) + \epsilon/2^j)$$

$$\leq \mu_*((a,b]) + \epsilon.$$

Thus letting $a' \to a$, and using the right-continuity of F again, we see that $F(b) - F(a) \leq \mu_*((a,b]) + \epsilon$. Since ϵ was arbitrary this then proves $F(b) - F(a) = \mu_*((a,b])$.

Next we show that μ_* is a metric exterior measure (for the usual metric $d(x, x') = |x - x'|$ on the real line). Since μ_* is an exterior measure we have $\mu_*(E_1 \cup E_2) \leq \mu_*(E_1) + \mu_*(E_2)$; thus it suffices to see that the reverse inequality holds whenever $d(E_1, E_2) \geq \delta$, for some $\delta > 0$.

Suppose that we are given a positive ϵ, and that $\bigcup_{j=1}^{\infty}(a_j, b_j]$ is a covering of $E_1 \cup E_2$ such that

$$\sum_{j=1}^{\infty} F(b_j) - F(a_j) \leq \mu_*(E_1 \cup E_2) + \epsilon.$$

We may assume, after subdividing the intervals $(a_j, b_j]$ into smaller half-open intervals, that each interval in the covering has length less than δ. When this is so each interval can intersect at most one of the two sets E_1 or E_2. If we denote by J_1 and J_2 the sets of those indices for which $(a_j, b_j]$ intersects E_1 and E_2, respectively, then $J_1 \cap J_2$ is empty; moreover, we have $E_1 \subset \bigcup_{j \in J_1}(a_j, b_j]$ as well as $E_2 \subset \bigcup_{j \in J_2}(a_j, b_j]$. Therefore

$$\mu_*(E_1) + \mu_*(E_2) \leq \sum_{j \in J_1} F(b_j) - F(a_j) + \sum_{j \in J_2} F(b_j) - F(a_j)$$

$$\leq \sum_{j=1}^{\infty} F(b_j) - F(a_j) \leq \mu_*(E_1 \cup E_2) + \epsilon.$$

Since ϵ was arbitrary, we see that $\mu_*(E_1) + \mu_*(E_2) \leq \mu_*(E_1 \cup E_2)$, as we intended to show.

We can now invoke Theorem 1.5. This guarantees the existence of a measure μ for which the Borel sets are measurable; moreover, we have $\mu((a, b]) = F(b) - F(a)$, since clearly $(a, b]$ is a Borel set and we have previously seen that $\mu_*((a, b]) = F(b) - F(a)$.

To prove that μ is the unique Borel measure on $\mathbb{R}$ for which $\mu((a, b]) = F(b) - F(a)$, let us suppose that ν is another Borel measure with this property. It now suffices to show that $\nu = \mu$ on all Borel sets. We can write any open interval as a disjoint union $(a, b) = \bigcup_{j=1}^{\infty}(a_j, b_j]$, by choosing $\{b_j\}_{j=1}^{\infty}$ to be a strictly increasing sequence with $a < b_j < b$, $b_j \to b$ as $j \to \infty$, and taking $a_1 = a$, $a_{j+1} = b_j$. Since ν and μ agree on each $(a_j, b_j]$, it follows that ν and μ agree on (a, b), and hence on all open intervals, and therefore on all open sets. Moreover, clearly ν and μ are finite on all bounded intervals; thus the regularity in Proposition 1.3 allows one to conclude that $\mu = \nu$ on all Borel sets.

Conversely, if we start with a Borel measure μ on $\mathbb{R}$ that is finite on bounded intervals, we can define the function F as in the statement of the theorem. Then clearly F is increasing. To see that it is right-continuous, note that if, for instance, $x_0 > 0$, the sets $E_n = (0, x_0 + 1/n]$ decrease to $E = (0, x_0]$ as $n \to \infty$, hence $\mu(E_n) \to \mu(E)$, since $\mu(E_1) < \infty$. This means that $F(x_0 + 1/n) \to F(x_0)$. Since F is increasing, this implies that F is right-continuous at x_0. The argument for any $x_0 \leq 0$ is similar, and thus the theorem is proved.

Remarks. Several comments about the theorem are in order.

(i) Two increasing functions F and G give the same measure if $F - G$ is constant. The converse if also true because $F(b) - F(a) = G(b) - G(a)$ for all $a < b$ exactly when $F - G$ is constant.

(ii) The measure μ constructed in the proof of the theorem is defined on a larger σ-algebra than the Borel sets, and is actually complete. However, in applications, its restriction to the Borel sets often suffices.

(iii) If F is an increasing normalized function given on a closed interval $[a, b]$, we can extend it to $\mathbb{R}$ by setting $F(x) = F(a)$ for $x < a$, and $F(x) = F(b)$ for $x > b$. For the resulting measure μ, the intervals $(-\infty, a]$ and (b, ∞) have measure zero. One then often writes

$$\int_{\mathbb{R}} f(x) \, d\mu(x) = \int_a^b f(x) \, dF(x),$$

for every f that is integrable with respect to μ. If F arises from an increasing function F_0 defined on $\mathbb{R}$, one may wish to account for the possible jump of F_0 at a. In this case it is sometimes useful to define

$$\int_{a^-}^b f(x) \, dF(x) \quad \text{as} \quad \int_a^b f(x) \, d\mu_0(x),$$

where μ_0 is the measure on $\mathbb{R}$ corresponding to F_0.

(iv) Note that the above definition of the Lebesgue-Stieltjes integral extends to the case when F is of bounded variation. Indeed suppose F is a complex-valued function on $[a, b]$ such that $F = \sum_{j=1}^4 \epsilon_j F_j$, where each F_j is increasing and normalized, and ϵ_j are ± 1 or $\pm i$. Then we can define $\int_a^b f(x) \, dF(x)$ as $\sum_{j=1}^4 \epsilon_j \int_a^b f(x) \, dF_j(x)$; here we require that f be integrable with respect to the Borel measure $\mu = \sum_{j=1}^4 \mu_j$, where μ_j is the measure corresponding to F_j.

(v) The value of these integrals can be calculated more directly in the following cases.

(a) If F is an absolutely continuous function on $[a, b]$, then

$$\int_a^b f(x)\, dF(x) = \int_a^b f(x) F'(x)\, dx$$

for every Borel measurable function f that is integrable with respect to $\mu = dF$.

(b) Suppose F is a pure jump function as in Section 3.3, Chapter 3, with jumps $\{\alpha_n\}_{n=1}^\infty$ at the points $\{x_n\}_{n=1}^\infty$. Then whenever f is, say, continuous and vanishes outside some finite interval we have

$$\int_a^b f(x)\, dF(x) = \sum_{n=1}^\infty f(x_n)\alpha_n.$$

In particular, for the measure μ we have $\mu(\{x_n\}) = \alpha_n$ and $\mu(E) = 0$ for all sets that do not contain any of the x_n.

(c) A special instance arises when $F = H$, the Heaviside function defined by $H(x) = 1$ for $x \geq 0$, and $H(x) = 0$ for $x < 0$. Then

$$\int_{-\infty}^\infty f(x)\, dH(x) = f(0),$$

which is another expression for the Dirac delta function arising in Section 2 of Chapter 3.

Further details about (v) can be found in Exercise 11.

4 Absolute continuity of measures

The generalization of the notion of absolute continuity considered in Chapter 3 requires that we extend the ideas of a measure to encompass set functions that may be positive or negative. We describe this notion first.

4.1 Signed measures

Loosely speaking, a signed measure possesses all the properties of a measure, except that it may take positive or negative values. More precisely, a **signed measure** ν on a σ-algebra $\mathcal{M}$ is a mapping that satisfies:

(i) The set function ν is extended-valued in the sense that $-\infty < \nu(E) \leq \infty$ for all $E \in \mathcal{M}$.

(ii) If $\{E_j\}_{j=1}^{\infty}$ are disjoint subsets of $\mathcal{M}$, then

$$\nu\left(\bigcup_{j=1}^{\infty} E_j\right) = \sum_{j=1}^{\infty} \nu(E_j).$$

Note that for this to hold the sum $\sum \nu(E_j)$ must be independent of the rearrangements of terms, so that if $\nu(\bigcup_{j=1}^{\infty} E_j)$ is finite, it implies that the sum converges absolutely.

Examples of signed measures arise naturally if we drop the assumption that f be non-negative in the expression

$$\nu(E) = \int_E f \, d\mu,$$

where $(X, \mathcal{M}, \mu)$ is a measure space and f is μ-measurable. In fact, to ensure that ν satisfies (i) and (ii) the function f is required to be "integrable" with respect to μ in the extended sense that $\int f^- \, d\mu$ must be finite, while $\int f^+ \, d\mu$ may be infinite.

Given a signed measure ν on $(X, \mathcal{M})$ it is always possible to find a (positive) measure μ that dominates ν, in the sense that

$$\nu(E) \le \mu(E) \quad \text{for all } E,$$

and that in addition is the "smallest" μ that has this property.

The construction is in effect an abstract version of the decomposition of a function of bounded variation as the difference of two increasing functions, as carried out in Chapter 3. We proceed as follows. We define a function $|\nu|$ on $\mathcal{M}$, called the **total variation** of ν, by

$$|\nu|(E) = \sup \sum_{j=1}^{\infty} |\nu(E_j)|,$$

where the supremum is taken over all partitions of E, that is, over all countable unions $E = \bigcup_{j=1}^{\infty} E_j$, where the sets E_j are disjoint and belong to $\mathcal{M}$.

The fact that $|\nu|$ is actually additive is not obvious, and is given in the proof below.

Proposition 4.1 *The total variation $|\nu|$ of a signed measure ν is itself a (positive) measure that satisfies $\nu \le |\nu|$.*

Proof. Suppose $\{E_j\}_{j=1}^{\infty}$ is a countable collection of disjoints sets in $\mathcal{M}$, and let $E = \bigcup E_j$. It suffices to prove:

$$\text{(11)} \qquad \sum |\nu|(E_j) \le |\nu|(E) \quad \text{and} \quad |\nu|(E) \le \sum |\nu|(E_j).$$

Let α_j be a real number that satisfies $\alpha_j < |\nu|(E_j)$. By definition, each E_j can be written as $E_j = \bigcup_i F_{i,j}$, where the $F_{i,j}$ are disjoint, belong to $\mathcal{M}$, and

$$\alpha_j \le \sum_{i=1}^{\infty} |\nu(F_{i,j})|.$$

Since $E = \bigcup_{i,j} F_{i,j}$, we have

$$\sum \alpha_j \le \sum_{j,i} |\nu(F_{i,j})| \le |\nu|(E).$$

Consequently, taking the supremum over the numbers α_j gives the first inequality in (11).

For the reverse inequality, let F_k be any other partition of E. For a fixed k, $\{F_k \cap E_j\}_j$ is a partition of F_k, so

$$\sum_k |\nu(F_k)| = \sum_k \left| \sum_j \nu(F_k \cap E_j) \right|,$$

since ν is a signed measure. An application of the triangle inequality and the fact that $\{F_k \cap E_j\}_k$ is a partition of E_j gives

$$\sum_k |\nu(F_k)| \le \sum_k \sum_j |\nu(F_k \cap E_j)|$$

$$= \sum_j \sum_k |\nu(F_k \cap E_j)|$$

$$\le \sum_j |\nu|(E_j).$$

Since $\{F_k\}$ was an arbitrary partition of E, we obtain the second inequality in (11) and the proof is complete.

It is now possible to write ν as the difference of two (positive) measures. To see this, we define the **positive variation** and **negative variation** of ν by

$$\nu^+ = \frac{1}{2}(|\nu| + \nu) \quad \text{and} \quad \nu^- = \frac{1}{2}(|\nu| - \nu).$$

By the proposition we see that ν^+ and ν^- are measures, and they clearly satisfy

$$\nu = \nu^+ - \nu^- \quad \text{and} \quad |\nu| = \nu^+ + \nu^-.$$

In the above if $\nu(E) = \infty$ for a set E, then $|\nu|(E) = \infty$, and $\nu^-(E)$ is defined to be zero.

We also make the following definition: we say that the signed measure ν is σ-**finite** if the measure $|\nu|$ is σ-finite. Since $\nu \le |\nu|$ and $|-\nu| = |\nu|$, we find that

$$-|\nu| \le \nu \le |\nu|.$$

As a result, if ν is σ-finite, then so are ν^+ and ν^-.

4.2 Absolute continuity

Given two measures defined on a common σ-algebra we describe here the relationships that can exist between them. More concretely, consider two measures ν and μ defined on the σ-algebra $\mathcal{M}$; two extreme scenarios are

(a) ν and μ are "supported" on separate parts of $\mathcal{M}$.

(b) The support of ν is an essential part of the support of μ.

Here we adopt the terminology that the measure ν is **supported** on a set A, if $\nu(E) = \nu(E \cap A)$ for all $E \in \mathcal{M}$.

The Lebesgue-Radon-Nikodym theorem below states that in a precise sense the relationship between any two measures ν and μ is a combination of the above two possibilities.

Mutually singular and absolutely continuous measures

Two signed measures ν and μ on $(X, \mathcal{M})$ are **mutually singular** if there are disjoint subsets A and B in $\mathcal{M}$ so that

$$\nu(E) = \nu(A \cap E) \quad \text{and} \quad \mu(E) = \mu(B \cap E) \quad \text{for all } E \in \mathcal{M}.$$

Thus ν and μ are supported on disjoint subsets. We use the symbol $\nu \perp \mu$ to denote the fact that the measures are mutually singular.

In contrast, if ν is a signed measure and μ a (positive) measure on $\mathcal{M}$, we say that ν is **absolutely continuous** with respect to μ if

(12) $\nu(E) = 0$ whenever $E \in \mathcal{M}$ and $\mu(E) = 0$.

Thus if ν is supported in a set A, then A must be an essential part of the support of μ in the sense that $\mu(A) > 0$. We use the symbol $\nu \ll \mu$ to indicate that ν is absolutely continuous with respect to μ. Note that if ν and μ are mutually singular, and ν is also absolutely continuous with respect to μ, then ν vanishes identically.

An important example is given by integration with respect to μ. Indeed, if $f \in L^1(X, \mu)$, or if f is merely integrable in the extended sense (where $\int f^- < \infty$, but possibly $\int f^+ = \infty$), then the signed measure ν defined by

$$(13) \qquad\qquad \nu(E) = \int_E f \, d\mu$$

is absolutely continuous with respect to μ. We shall use the shorthand $d\nu = f \, d\mu$ to indicate that ν is defined by (13).

This is a variant of the notion of absolute continuity that arose in Chapter 3 in the special case of $\mathbb{R}$ (with $\mathcal{M}$ the Lebesgue measurable sets and $d\mu = dx$ the Lebesgue measure). In fact, with ν defined by (13) and f an integrable function, we saw that in place of (12) we had the following stronger assertion:

(14)
For each $\epsilon > 0$, there is a $\delta > 0$ such that $\mu(E) < \delta$ implies $|\nu(E)| < \epsilon$.

In the general situation the relation between the two conditions (12) and (14) is clarified by the following observation.

Proposition 4.2 *The assertion (14) implies (12). Conversely, if $|\nu|$ is a finite measure, then (12) implies (14).*

That (12) is a consequence of (14) is obvious because $\mu(E) = 0$ gives $|\nu(E)| < \epsilon$ for every $\epsilon > 0$. To prove the converse, it suffices to consider the case when ν is positive, upon replacing ν by $|\nu|$. We then assume that (14) does not hold. This means that it fails for some fixed $\epsilon > 0$. Hence for each n, there is a measurable set E_n with $\mu(E_n) < 2^{-n}$ while $\nu(E_n) \geq \epsilon$. Now let $E^* = \limsup_{n \to \infty} E_n = \bigcap_{n=1}^{\infty} E_n^*$, where $E_n^* = \bigcup_{k \geq n} E_k$. Then since $\mu(E_n^*) \leq \sum_{k \geq n} 1/2^k = 1/2^{n-1}$, and the decreasing sets $\{E_k^*\}$ are contained in a set of finite measure (E_1^*), we get $\mu(E^*) = 0$. However $\nu(E_n^*) \geq \nu(E_n) \geq \epsilon$, and the ν measure is assumed finite. So $\nu(E^*) = \lim_{n \to \infty} \nu(E_n^*) \geq \epsilon$, which gives a contradiction.

After these preliminaries we can come to the main result. It guarantees among other things a converse to the representation (13); it was proved in the case of $\mathbb{R}$ by Lebesgue, and in the general case by Radon and Nikodym.

Theorem 4.3 *Suppose μ is a σ-finite positive measure on the measure space $(X, \mathcal{M})$ and ν a σ-finite signed measure on $\mathcal{M}$. Then there exist unique signed measures ν_a and ν_s on $\mathcal{M}$ such that $\nu_a \ll \mu$, $\nu_s \perp \mu$ and $\nu = \nu_a + \nu_s$. In addition, the measure ν_a takes the form $d\nu_a = f d\mu$; that is,*

$$\nu_a(E) = \int_E f(x) \, d\mu(x)$$

for some extended μ-integrable function f.

Note the following consequence. If ν is absolutely continuous with respect to μ, then $d\nu = f d\mu$, and this assertion can be viewed as a generalization of Theorem 3.11 in Chapter 3.

There are several known proofs of the above theorem. The argument given below, due to von Neumann, has the virtue that it exploits elegantly the application of a simple Hilbert space idea.

We start with the case when both ν and μ are positive and finite. Let $\rho = \nu + \mu$, and consider the transformation on $L^2(X, \rho)$ defined by

$$\ell(\psi) = \int_X \psi(x) \, d\nu(x).$$

The mapping ℓ defines a bounded linear functional on $L^2(X, \rho)$ since

$$|\ell(\psi)| \leq \int_X |\psi(x)| \, d\nu(x) \leq \int_X |\psi(x)| \, d\rho(x)$$

$$\leq (\rho(X))^{1/2} \left(\int_X |\psi(x)|^2 \, d\rho(x) \right)^{1/2},$$

where the last inequality follows by the Cauchy-Schwarz inequality. But $L^2(X, \rho)$ is a Hilbert space, so the Riesz representation theorem (in Chapter 4) guarantees the existence of $g \in L^2(X, \rho)$ such that

$$(15) \quad \int_X \psi(x) \, d\nu(x) = \int_X \psi(x) g(x) \, d\rho(x) \quad \text{for all } \psi \in L^2(X, \rho).$$

If $E \in \mathcal{M}$ with $\rho(E) > 0$, when we set $\psi = \chi_E$ in (15) and recall that $\nu \leq \rho$, we find

$$0 \leq \frac{1}{\rho(E)} \int_E g(x) \, d\rho(x) \leq 1,$$

from which we conclude that $0 \leq g(x) \leq 1$ for a.e. x (with respect to the measure ρ). In fact, $0 \leq \int_E g(x) \, d\rho(x)$ for all sets $E \in \mathcal{M}$ implies that

$g(x) \geq 0$ almost everywhere. In the same way, $0 \leq \int_E (1 - g(x)) \, d\rho(x)$ for all $E \in \mathcal{M}$ guarantees that $g(x) \leq 1$ almost everywhere. Therefore we may clearly assume $0 \leq g(x) \leq 1$ for all x without disturbing the identity (15), which we rewrite as

$$(16) \qquad \int \psi(1 - g) \, d\nu = \int \psi g \, d\mu.$$

Consider now the two sets

$$A = \{x \in X : 0 \leq g(x) < 1\} \quad \text{and} \quad B = \{x \in X : g(x) = 1\},$$

and define two measures ν_a and ν_s on $\mathcal{M}$ by

$$\nu_a(E) = \nu(A \cap E) \quad \text{and} \quad \nu_s(E) = \nu(B \cap E).$$

To see why $\nu_s \perp \mu$, it suffices to note that setting $\psi = \chi_B$ in (16) gives

$$0 = \int \chi_B \, d\mu = \mu(B).$$

Finally, we set $\psi = \chi_E(1 + g + \cdots + g^n)$ in (16) :

$$(17) \qquad \int_E (1 - g^{n+1}) \, d\nu = \int_E g(1 + \cdots + g^n) \, d\mu.$$

Since $(1 - g^{n+1})(x) = 0$ if $x \in B$, and $(1 - g^{n+1})(x) \to 1$ if $x \in A$, the dominated convergence theorem implies that the left-hand side of (17) converges to $\nu(A \cap E) = \nu_a(E)$. Also, $1 + g + \cdots + g^n$ converges to $\frac{1}{1-g}$, so we find in the limit that

$$\nu_a(E) = \int_E f \, d\mu, \quad \text{where } f = \frac{g}{1-g}.$$

Note that $f \in L^1(X, \mu)$, since $\nu_a(X) \leq \nu(X) < \infty$. If μ and ν are σ-finite and positive we may clearly find sets $E_j \in \mathcal{M}$ such that $X = \bigcup E_j$ and

$$\mu(E_j) < \infty, \quad \nu(E_j) < \infty \quad \text{for all } j.$$

We may define positive and finite measures on $\mathcal{M}$ by

$$\mu_j(E) = \mu(E \cap E_j) \quad \text{and} \quad \nu_j(E) = \nu(E \cap E_j),$$

and then we can write for each j, $\nu_j = \nu_{j,a} + \nu_{j,s}$ where $\nu_{j,s} \perp \mu_j$ and $\nu_{j,a} = f_j \, d\mu_j$. Then it suffices to set

$$f = \sum f_j, \quad \nu_s = \sum \nu_{j,s}, \quad \text{and} \quad \nu_a = \sum \nu_{j,a}.$$

Finally, if ν is signed, then we apply the argument separately to the positive and negative variations of ν.

To prove the uniqueness of the decomposition, suppose we also have $\nu = \nu'_a + \nu'_s$, where $\nu'_a \ll \mu$ and $\nu'_s \perp \mu$. Then

$$\nu_a - \nu'_a = \nu'_s - \nu_s.$$

The left-hand side is absolutely continuous with respect to μ, and the right-hand side is singular with respect to μ. Thus both sides are zero and the theorem is proved.

5* Ergodic theorems

Ergodic theory had its beginnings in certain problems in statistical mechanics studied in the late nineteenth century. Since then it has grown rapidly and has gained wide influence in a number of mathematical disciplines, in particular those related to dynamical systems and probability theory. It is not our purpose to try to give an account of this broad and fascinating theory. Rather, we restrict our presentation to some of the basic limit theorems that lie at its foundation. These theorems are most naturally formulated in the general context of abstract measure spaces, and thus for us they serve as excellent illustrations of the general framework developed in this chapter.

The setting for the theory is a σ-finite measure space $(X, \mathcal{M}, \mu)$ endowed with a mapping $\tau : X \to X$ such that whenever E is a measurable subset of X, then so is $\tau^{-1}(E)$, and $\mu(\tau^{-1}(E)) = \mu(E)$. Here $\tau^{-1}(E)$ is the pre-image of E under τ; that is, $\tau^{-1}(E) = \{x \in X : \tau(x) \in E\}$. A mapping τ with these properties is called a **measure-preserving transformation**. If in addition for such a τ we have the feature that it is a bijection and τ^{-1} is also a measure-preserving transformation, then τ is referred to as a **measure-preserving isomorphism**.

Let us note that if τ is a measure-preserving transformation, then $f(\tau(x))$ is measurable if $f(x)$ is measurable, and is integrable if f is integrable; moreover, then

$$(18) \qquad \int_X f(\tau(x)) \, d\mu(x) = \int_X f(x) \, d\mu(x).$$

Indeed, if χ_E is the characteristic function of the set E, we note that $\chi_E(\tau(x)) = \chi_{\tau^{-1}(E)}(x)$, and so the assertion holds for characteristic functions of measurable sets and thus for simple functions, and hence by the usual limiting arguments for all non-negative measurable functions, and then integrable functions. For later purposes we record here an equivalent statement: whenever f is a real-valued measurable function and α is any real number, then

$$\mu(\{x : f(x) > \alpha\}) = \mu(\{x : f(\tau(x)) > \alpha\}).$$

Before we proceed further, we describe several examples of measure-preserving transformations:

(i) Here $X = \mathbb{Z}$, the integers, with μ its counting measure; that is, $\mu(E) = \#(E) = $ the number of integers in E, for any $E \subset \mathbb{Z}$. We define τ to be the unit translation, $\tau : n \mapsto n + 1$. Note that τ gives a measure-preserving isomorphism of $\mathbb{Z}$.

(ii) Another easy example is $X = \mathbb{R}^d$ with Lebesgue measure, and τ a translation, $\tau : x \mapsto x + h$ for some fixed $h \in \mathbb{R}^d$. This is of course a measure-preserving isomorphism. (See the section on invariance properties of the Lebesgue measure in Chapter 1.)

(iii) Here X is the unit circle, given as $\mathbb{R}/\mathbb{Z}$, with the measure induced from Lebesgue measure on $\mathbb{R}$. That is, we may realize X as the unit interval $(0, 1]$, and take μ to be the Lebesgue measure restricted to this interval. For any real number α, the translation $x \mapsto x + \alpha$, taken modulo $\mathbb{Z}$, is well defined on $X = \mathbb{R}/\mathbb{Z}$, and is measure-preserving. (See the related Exercise 3 in Chapter 2.) It can be interpreted as a rotation of the circle by angle $2\pi\alpha$.

(iv) In this example X is again $(0, 1]$ with Lebesgue measure μ, but τ is the doubling map $\tau(x) = 2x$ mod 1. It is easy to verify that τ is a measure-preserving transformation. Indeed, any set $E \subset (0, 1]$ has two pre-images E_1 and E_2, the first in $(0, 1/2]$ and the second in $(1/2, 1]$, both of measure $\mu(E)/2$, if E is measurable. (See Figure 1.) However, τ is not an isomorphism, since τ is not injective.

(v) A trickier example is given by the transformation that is key in the theory of continued fractions. Here $X = [0, 1)$ and τ is defined by $\tau(x) = \langle 1/x \rangle$, the fractional part of $1/x$; when $x = 0$ we set $\tau(0) = 0$. Gauss observed, in effect, that the measure $d\mu = \frac{1}{1+x} dx$ is preserved by the transformation τ. Note that each $x \in (0, 1)$ has infinitely may pre-images under τ; that is, the sequence $\{1/(x + k)\}_{k=1}^{\infty}$. More about this example can be found in Problems 8 through 10 below.

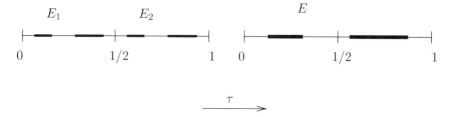

Figure 1. Pre-images E_1 and E_2 under the doubling map

Having pointed out these examples, we can now return to the general theory. The notions described above are of interest, in part, because they abstract the idea of a dynamical system, one whose totality of states is represented by the space X, with each point $x \in X$ giving a particular state of the system. The mapping $\tau : X \to X$ then describes the transformation of the system after a unit of time has elapsed. For such a system there is often associated a notion of "volume" or "mass" that is unchanged by the evolution, and this is the role of the invariant measure μ. The iterates, $\tau^n = \tau \circ \tau \circ \cdots \circ \tau$ (n times) describe the evolution of the system after n units of time, and a principal concern is the average behaviour, as $n \to \infty$, of various quantities associated with the system. Thus one is led to study averages

$$(19) \qquad A_n(f)(x) = \frac{1}{n} \sum_{k=0}^{n-1} f(\tau^k(x)),$$

and their limits as $n \to \infty$. To this we now turn.

5.1 Mean ergodic theorem

The first theorem dealing with the averages (19) that we consider is purely Hilbert-space in character. Historically it preceded both Theorems 5.3 and 5.4 which will be proved below.

For the specific application of the theorem below, one takes the Hilbert space $\mathcal{H}$ to be $L^2(X, \mathcal{M}, \mu)$. Given the measure-preserving transformation τ on X, we define the linear operator T on $\mathcal{H}$ by

$$(20) \qquad T(f)(x) = f(\tau(x)).$$

Then T is an isometry; that is,

$$(21) \qquad \|Tf\| = \|f\|,$$

where $\| \cdot \|$ denotes the Hilbert space (that is, the L^2) norm. This is clear from (18) with f replaced by $|f|^2$. Observe that if τ were also supposed to be a measure-preserving isomorphism, then T would be invertible and hence unitary; but we do not assume this.

Now with T as above, consider the subspace S of **invariant vectors**, $S = \{f \in \mathcal{H} : T(f) = f\}$. Clearly, because of (21), the subspace S is closed. Let P denote the orthogonal projection on this subspace. The theorem that follows deals with the "mean" convergence, meaning convergence in the norm.

Theorem 5.1 *Suppose T is an isometry of the Hilbert space $\mathcal{H}$, and let P be the orthogonal projection on the subspace of the invariant vectors of T. Let $A_n = \frac{1}{n}(I + T + T^2 + \cdots + T^{n-1})$. Then for each $f \in \mathcal{H}$, $A_n(f)$ converges to $P(f)$ in norm, as $n \to \infty$.*

Together with the subspace S defined above we consider the subspaces $S_* = \{f \in \mathcal{H} : T^*(f) = f\}$ and $S_1 = \{f \in \mathcal{H} : f = g - Tg, \ g \in \mathcal{H}\}$; here T^* denotes the adjoint of T. Then S_*, like S, is closed, but S_1 is not necessarily closed. We denote its closure by $\overline{S_1}$. The proof of the theorem is based on the following lemma.

Lemma 5.2 *The following relations hold among the subspaces S, S_*, and $\overline{S_1}$.*

(i) $S = S_*$.

(ii) *The orthogonal complement of $\overline{S_1}$ is S.*

Proof. First, since T is an isometry, we have that $(Tf, Tg) = (f, g)$ for all $f, g \in \mathcal{H}$, and thus $T^*T = I$. (See Exercise 22 in Chapter 4.) So if $Tf = f$ then $T^*Tf = T^*f$, which means that $f = T^*f$. To prove the converse inclusion, assume $T^*f = f$. As a consequence $(f, T^*f - f) = 0$, and thus $(f, T^*f) - (f, f) = 0$; that is, $(Tf, f) = \|f\|^2$. However, $\|Tf\| = \|f\|$, so we have in the above an instance of equality for the Cauchy-Schwarz inequality. As a result of Exercise 2 in Chapter 4 we get $Tf = cf$, which by the above gives $Tf = f$. Thus part (i) is proved.

Next we observe that f is in the orthogonal complement of $\overline{S_1}$ exactly when $(f, g - Tg) = 0$, for all $g \in \mathcal{H}$. However, this means that $(f - T^*f, g) = 0$ for all g, and hence $f = T^*f$, which by part (i) means $f \in S$.

Having established the lemma we can finish the proof of the theorem. Given any $f \in \mathcal{H}$, we write $f = f_0 + f_1$, where $f_0 \in S$ and $f_1 \in \overline{S_1}$ (since S and $\overline{S_1}$ are orthogonal complements). We also fix $\epsilon > 0$ and pick $f_1' \in$

S_1 such that $\|f_1 - f_1'\| < \epsilon$. We then write

$$(22) \qquad A_n(f) = A_n(f_0) + A_n(f_1') + A_n(f_1 - f_1'),$$

and consider each term separately.

For the first term, we recall that P is the orthogonal projection on S, so $P(f) = f_0$, and since $Tf_0 = f_0$ we deduce

$$A_n(f_0) = \frac{1}{n} \sum_{k=0}^{n-1} T^k(f_0) = f_0 = P(f) \qquad \text{for every } n \geq 1.$$

For the second term, we recall the definition of S_1 and pick a $g \in \mathcal{H}$ with $f_1' = g - Tg$. Thus

$$A_n(f_1') = \frac{1}{n} \sum_{k=0}^{n-1} T^k(1-T)(g) = \frac{1}{n} \sum_{k=0}^{n-1} T^k(g) - T^{k+1}(g)$$
$$= \frac{1}{n}(g - T^n(g)).$$

Since T is an isometry, the above identity shows that $A_n(f_1')$ converges to 0 in the norm as $n \to \infty$.

For the last term, we use once again the fact that each T^k is an isometry to obtain

$$\|A_n(f_1 - f_1')\| \leq \frac{1}{n} \sum_{k=0}^{n-1} \|T^k(f_1 - f_1')\| \leq \|f_1 - f_1'\| < \epsilon.$$

Finally, from (22) and the above three observations, we deduce that $\limsup_{n \to \infty} \|A_n(f) - P(f)\| \leq \epsilon$, and this concludes the proof of the theorem.

5.2 Maximal ergodic theorem

We now turn to the question of almost everywhere convergence of the averages (19). As in the case of the averages that occur in the differentiation theorems of Chapter 3, the key to dealing with such pointwise limits lies in estimates for their corresponding maximal functions. In the present case this function is defined by

$$(23) \qquad f^*(x) = \sup_{1 \leq m < \infty} \frac{1}{m} \sum_{k=0}^{m-1} |f(\tau^k(x))|.$$

Theorem 5.3 *Whenever $f \in L^1(X, \mu)$, the maximal function $f^*(x)$ is finite for almost every x. Moreover, there is a universal constant A so that*

$$(24) \qquad \mu(\{x : f^*(x) > \alpha\}) \le \frac{A}{\alpha} \|f\|_{L^1(X,\mu)} \qquad \text{for all } \alpha > 0.$$

There are several proofs of this theorem. The one we choose emphasizes the close connection to the maximal function given in Section 1.1 of Chapter 3, and we shall in fact deduce the present theorem from the one-dimensional case of that chapter. This argument gives the value $A = 6$ for the constant in (24). By a different argument one can obtain $A = 1$, but this improvement is not relevant in what follows.

Before beginning the proof, we make some preliminary remarks. Note that in the present case the function f^* is automatically measurable, since it is the supremum of a countable number of measurable functions. Also, we may assume that our function f is non-negative, since otherwise we may replace it by $|f|$.

Step 1. The case when $X = \mathbb{Z}$ and $\tau : n \mapsto n + 1$.

For each function f on $\mathbb{Z}$, we consider its extension $\tilde{f}$ to $\mathbb{R}$ defined by $\tilde{f}(x) = f(n)$ for $n \le x < n + 1$, $n \in \mathbb{Z}$. (See Figure 2.)

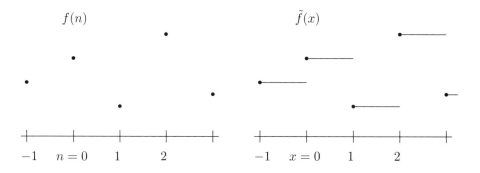

Figure 2. Extension of f to $\mathbb{R}$

Similarly, if $E \subset \mathbb{Z}$, denote by $\tilde{E}$ the set in $\mathbb{R}$ given by $\tilde{E} = \bigcup_{n \in E}[n, n + 1)$. Note that as a result of these definitions we have $m(\tilde{E}) = \#(E)$ and $\int_{\mathbb{R}} \tilde{f}(x)\, dx = \sum_{n \in \mathbb{Z}} f(n)$, and thus $\|\tilde{f}\|_{L^1(\mathbb{R})} = \|f\|_{L^1(\mathbb{Z})}$. Here m is the Lebesgue measure on $\mathbb{R}$, and $\#$ is the counting measure on $\mathbb{Z}$. Note also that

$$\sum_{k=0}^{m-1} f(n+k) = \int_0^m \tilde{f}(n+t) \, dt.$$

However, because $\int_0^m \tilde{f}(n+t) \, dt \le \int_{-1}^m \tilde{f}(x+t) \, dt$ whenever $x \in [n, n+1)$, we see that

$$\frac{1}{m} \sum_{k=0}^{m-1} f(n+k) \le \left(\frac{m+1}{m} \right) \frac{1}{m+1} \int_{-1}^m \tilde{f}(x+t) \, dt \quad \text{if } x \in [n, n+1).$$

Taking the supremum over all $m \ge 1$ in the above and noting that $(m+1)/m \le 2$, we obtain

(25) $$f^*(n) \le 2(\tilde{f})^*(x) \quad \text{whenever } x \in [n, n+1).$$

To be clear about the notation here: $f^*(n)$ denotes the maximal function of f on $\mathbb{Z}$ defined by (23), with $f(\tau^k(n)) = f(n+k)$, while $(\tilde{f})^*$ is the maximal function as defined in Chapter 3, of the extended function $\tilde{f}$ on $\mathbb{R}$.

By (25)

$$\#(\{n : f^*(n) > \alpha\}) \le m(\{x \in \mathbb{R} : (\tilde{f})^*(x) > \alpha/2\}),$$

and thus the latter is majorized by $A'/(\alpha/2) \int \tilde{f}(x) \, dx = 2A'/\alpha \|\tilde{f}\|_{L^1(\mathbb{R})}$, according to the maximal theorem for $\mathbb{R}$. The constant A' that occurs in that theorem (there denoted by A) can be taken to be 3. Hence we have

(26) $$\#(\{n : f^*(n) > \alpha\}) \le \frac{6}{\alpha} \|f\|_{L^1(\mathbb{Z})},$$

since $\|\tilde{f}\|_{L^1(\mathbb{R})} = \|f\|_{L^1(\mathbb{Z})}$. This disposes of the special case when $X = \mathbb{Z}$.

Step 2. The general case.

By a sleight-of-hand we shall "transfer" the result for $\mathbb{Z}$ just proved to the general case. We proceed as follows.

For every positive integer N, we consider the truncated maximal function f_N^* defined as

$$f_N^*(x) = \sup_{1 \le m \le N} \frac{1}{m} \sum_{k=0}^{m-1} f(\tau^k(x)).$$

Since $\{f_N^*(x)\}$ forms an increasing sequence with N, and $\lim_{N\to\infty} f_N^*(x) = f^*(x)$ for every x, it suffices to show that

(27) $$\mu\{x:\ f_N^*(x) > \alpha\} \le \frac{A}{\alpha}\, \|f\|_{L^1(X,\mu)},$$

with constant A independent of N. Letting $N \to \infty$ will then give the desired result.

So in place of f^* we estimate f_N^*, and to simplify our notation we write the latter as f^*, dropping the N subscript. Our argument will compare the maximal function f^* with the special case arising for $\mathbb{Z}$. To clarify the formula below we temporarily adopt the expedient of denoting the second maximal function by $\mathcal{M}(f)$. Thus for a positive function f on $\mathbb{Z}$ we set

$$\mathcal{M}(f)(n) = \sup_{1\le m} \frac{1}{m} \sum_{k=0}^{m-1} f(n+k).$$

Now starting with a function f on X that is integrable, we define the function F on $X \times \mathbb{Z}$ by

$$F(x,n) = \begin{cases} f(\tau^n(x)) & \text{if } n \ge 0, \\ 0 & \text{if } n < 0. \end{cases}$$

Then

$$A_m(f)(x) = \frac{1}{m} \sum_{k=0}^{m-1} f(\tau^k(x)) = \frac{1}{m} \sum_{k=0}^{m-1} F(x,k).$$

In the above we replace x by $\tau^n(x)$; then since $\tau^k(\tau^n(x)) = \tau^{n+k}(x)$, we have

$$A_m(f)(\tau^n(x)) = \frac{1}{m} \sum_{k=0}^{m-1} F(x, n+k).$$

Now we fix a large positive a and set $b = a + N$. We also write F_b for the truncated function on $X \times \mathbb{Z}$ defined by $F_b(x,n) = F(x,n)$ if $n < b$, $F_b(x,n) = 0$ otherwise. We then have

$$A_m(f)(\tau^n(x)) = \frac{1}{m} \sum_{k=0}^{m-1} F_b(x, n+k) \quad \text{if } m \le N \text{ and } n < a.$$

Thus

(28) $$f^*(\tau^n(x)) \le \mathcal{M}(F_b)(x,n) \quad \text{if } n < a.$$

(Recall that f^* is actually f_N^*!) This is the comparison of the two maximal functions we wished to obtain. Now set $E_\alpha = \{x : f^*(x) > \alpha\}$. Then
by the measure-preserving character of τ, $\mu(\{x : f^*(\tau^n(x)) > \alpha\}) =$
$\mu(E_\alpha)$. Hence on the product space $X \times \mathbb{Z}$ the product measure $\mu \times \#$
of the set $\{(x, n) \in X \times \mathbb{Z} : f^*(\tau^n(x)) > \alpha, \ 0 \le n < a\}$ equals $a\mu(E_\alpha)$.
However, because of (28) the $\mu \times \#$ measure of this set is no more than

$$\int_X \#(\{n \in \mathbb{Z} : \mathcal{M}(F_b)(x, n) > \alpha\}) \, d\mu.$$

Because of the maximal estimate (26) for $\mathbb{Z}$, we see that the integrand
above is no more than

$$\frac{A}{\alpha} \, \|F_b(x, n)\|_{L^1(\mathbb{Z})} = \frac{A}{\alpha} \sum_{n=0}^{b-1} f(\tau^n(x)),$$

with of course $A = 6$.

Hence, integrating this over X and recalling that $\int_X f(\tau^n(x)) \, d\mu =$
$\int_X f(x) \, d\mu$ gives us

$$a\mu(E_\alpha) \le \frac{A}{\alpha} b \, \|f\|_{L^1(X)} = \frac{A}{\alpha} (a + N) \, \|f\|_{L^1(X)}.$$

Thus $\mu(E_\alpha) \le \frac{A}{\alpha} \left(1 + \frac{N}{a}\right) \|f\|_{L^1(X)}$, and letting $a \to \infty$ yields estimate (27).
As we have seen, a final limit as $N \to \infty$ then completes the proof.

5.3 Pointwise ergodic theorem

The last of the series of limit theorems we will study is the **pointwise**
(or **individual**) **ergodic** theorem, which combines ideas of the first two
theorems. At this stage it will be convenient to assume that the measure
space (X, μ) is finite; we can then normalize the measure and suppose
$\mu(X) = 1$.

Theorem 5.4 *Suppose f is integrable over X. Then for almost every
$x \in X$ the averages $A_m(f) = \frac{1}{m} \sum_{k=0}^{m-1} f(\tau^k(x))$ converge to a limit as
$m \to \infty$.*

Corollary 5.5 *If we denote this limit by $P'(f)$, we have that*

$$\int_X |P'(f)(x)| \, d\mu(x) \le \int_X |f(x)| \, d\mu(x).$$

Moreover $P'(f) = P(f)$ whenever $f \in L^2(X, \mu)$.

The idea of the proof is as follows. We first show that $A_m(f)$ converges to a limit almost everywhere for a set of functions f that is dense in $L^1(X, \mu)$. We then use the maximal theorem to show that this implies the conclusion for all integrable functions.

We remark to begin with that because the total measure of X is 1, we have $L^2(X, \mu) \subset L^1(X, \mu)$ and $\|f\|_{L^1} \leq \|f\|_{L^2}$, and moreover $L^2(X, \mu)$ is dense in $L^1(X, \mu)$. In fact, if f belongs to L^1, consider the sequence $\{f_n\}$ defined by $f_n(x) = f(x)$ if $|f(x)| \leq n$, $f_n(x) = 0$ otherwise. Then each f_n is clearly in L^2, while by the dominated convergence theorem $\|f - f_n\|_{L^1} \to 0$.

Now starting with an integrable f and any $\epsilon > 0$ we shall see that we can write $f = F + H$, where $\|H\|_{L^1} < \epsilon$, and $F = F_0 + (1 - T)G$, where both F_0 and G belong to L^2, and $T(F_0) = F_0$, with $T(F_0) = F_0(\tau(x))$. To obtain this decomposition of f, we first write $f = f' + h'$, where $f' \in L^2$ and $\|h'\|_{L^1} < \epsilon/2$, which we can do in view of the density of L^2 in L^1 as seen above. Next, since the subspaces S and $\overline{S_1}$ of Lemma 5.2 are orthogonal complements in L^2, we can find $F_0 \in S$, $F_1 \in S_1$, such that $f' = F_0 + F_1 + h$ with $\|h\|_{L^2} < \epsilon/2$. Because $F_1 \in S_1$ is automatically of the form $F_1 = (1 - T)G$, we obtain $f = F + H$, with $F = F_0 + (1 - T)G$ and $H = h + h'$. Thus $\|H\|_{L^1} \leq \|h\|_{L^1} + \|h'\|_{L^1}$ and since $\|h\|_{L^1} \leq \|h\|_{L^2} < \epsilon/2$ we have achieved our desired decomposition of f.

Now $A_m(F) = A_m(F_0) + A_m((1 - T)G) = F_0 + \frac{1}{m}(1 - T^m(G))$, as we have already seen in the proof of Theorem 5.1. Note that $\frac{1}{m}T^m(G) = \frac{1}{m}G(\tau^m(x))$ converges to zero as $m \to \infty$ for almost every $x \in X$. Indeed, the series $\sum_{m=1}^{\infty} \frac{1}{m^2}(G(\tau^m(x)))^2$ converges almost everywhere by the monotone convergence theorem, since its integral over X is

$$\sum_{m=1}^{\infty} \frac{1}{m^2}\|T^m G\|_{L^2}^2 = \|G\|_{L^2}^2 \sum_{m=1}^{\infty} \frac{1}{m^2},$$

which is finite.

As a result, $A_m(F)(x)$ converges for almost every $x \in X$. Finally, to prove the corresponding convergence for $A_m(f)(x)$, we argue as in Theorem 1.3 in Chapter 3 and set

$$E_\alpha = \{x : \lim_{N \to \infty} \sup_{n,m \geq N} |A_n(f)(x) - A_m(f)(x)| > \alpha\}.$$

Then it suffices to see that $\mu(E_\alpha) = 0$ for all $\alpha > 0$. However, since $A_n(f) - A_m(f) = A_n(F) - A_m(F) + A_n(H) - A_m(H)$, and $A_m(F)(x)$ converges almost everywhere as $m \to \infty$, it follows that almost every point in the set E_α is contained in E'_α, where

$$E'_\alpha = \{x : \sup_{n,m \geq N} |A_n(H)(x) - A_m(H)(x)| > \alpha\},$$

and thus $\mu(E_\alpha) \leq \mu(E'_\alpha) \leq \mu(\{x : 2\sup_m |A_m(H)(x)| > \alpha\})$. The last quantity is majorized by $A/(\alpha/2)\|H\|_{L^1} \leq 2\epsilon A/\alpha$ by Theorem 5.3. Since ϵ was arbitrary we see that $\mu(E_\alpha) = 0$, and hence $A_m(f)(x)$ is a Cauchy sequence for almost every x, and the theorem is proved.

To establish the corollary, observe that if $f \in L^2(X)$, we know by Theorem 5.1 that $A_m(f)$ converges to $P(f)$ in the L^2-norm, and hence a subsequence converges almost everywhere to that limit, showing that $P(f) = P'(f)$ in that case. Next, for any f that is merely integrable, we have

$$\int_X |A_m(f)|\, dx \leq \frac{1}{m}\sum_{k=0}^{m-1}\int_X |f(\tau^k(x))|\, d\mu(x) = \int_X |f(x)|\, d\mu(x),$$

and thus since $A_m(f) \to P'(f)$ almost everywhere, we get by Fatou's lemma that $\int_X |P'(f)(x)|\, d\mu(x) \leq \int_X |f(x)|\, d\mu(x)$. With this the corollary is also proved.

It can be shown that the conclusions of the theorem and corollary are still valid if we drop the assumption that the space X has finite measure. The modifications of the argument needed to obtain this more general conclusion are outlined in Exercise 26.

5.4 Ergodic measure-preserving transformations

The adjective "ergodic" is commonly applied to the three limit theorems proved above. It also has a related but separate usage describing an important class of transformations of the space X.

We say that a measure-preserving transformation τ of X is **ergodic** if whenever E is a measurable set that is "invariant," that is, E and $\tau^{-1}(E)$ differ by sets of measure zero, then either E or E^c has measure zero.

There is a useful rephrasing of this condition of ergodicity. Expanding the definition used in Section 5.1 we say that a measurable function f is **invariant** if $f(x) = f(\tau(x))$ for a.e. $x \in X$. Then τ is ergodic exactly when the only invariant functions are equivalent to constants. In fact, let τ be an ergodic transformation, and assume that f is a real-valued invariant function. Then each of the sets $E_a = \{x : f(x) > a\}$ is invariant, hence $\mu(E_a) = 0$ or $\mu(E_a^c) = 0$ for each a. However, if f is not equivalent to a constant, then both $\mu(E_a)$ and $\mu(E_a^c)$ must have strictly positive measure for some a. In the converse direction we merely need to note that if all characteristic functions of measurable sets that are invariant must be constants, then τ is ergodic.

The following result subsumes the conclusion of Theorem 5.4 for ergodic transformations. We keep to the assumption of that theorem that the underlying space X has measure equal to 1.

Corollary 5.6 *Suppose τ is an ergodic measure-preserving transformation. For any integrable function f we have*

$$\frac{1}{m}\sum_{k=0}^{m-1} f(\tau^k(x)) \quad converges\ to \quad \int_X f\,d\mu \quad for\ a.e.\ x \in X\ as\ m \to \infty.$$

The result has the interpretation that the "time average" of f equals its "space average."

Proof. By Theorem 5.1 we know that the averages $A_m(f)$ converge to $P(f)$, whenever $f \in L^2$, where P is the orthogonal projection on the subspace of invariant vectors. Since in this case the invariant vectors form a one-dimensional space spanned by the constant functions, we observe that $P(f) = 1(f,1) = \int_X f\,d\mu$, where 1 designates the function identically equal to 1 on X. To verify this, note that P is the identity on constants and annihilates all functions orthogonal to constants. Next we write any $f \in L^1$ as $g + h$, where $g \in L^2$ and $\|h\|_{L^1} < \epsilon$. Then $P'(f) = P'(g) + P'(h)$. However, we also know that $P'(g) = P(g)$, and $\|P'(h)\| \le \|h\|_{L^1} < \epsilon$ by the corollary to Theorem 5.4. Thus

$$P'(f) - \int_X f\,d\mu = \int_X (g - f)\,d\mu + P'(h)$$

yields that $\|P'(f) - \int_X f\,d\mu\|_{L^1} \le \|g - f\|_{L^1} + \epsilon < 2\epsilon$. This shows that $P'(f)$ is the constant $\int_X f\,d\mu$ and the assertion is proved.

We shall now elaborate on the nature of ergodicity and illustrate its thrust in terms of several examples.

a) Rotations of the circle

Here we take up the example described in (iii) at the beginning of Section 5*. On the unit circle $\mathbb{R}/\mathbb{Z}$ with the induced Lebesgue measure, we consider the action τ given by $x \mapsto x + \alpha \mod 1$. The result is

- *The mapping τ is ergodic if and only if α is irrational.*

To begin with, if α is irrational we know by the equidistribution theorem that

$$(29) \qquad \frac{1}{n}\sum_{k=0}^{n-1} f(x+k\alpha) \to \int_0^1 f(x)\,dx \qquad \text{as } n \to \infty$$

for every x if f is continuous on $[0,1]$ and periodic $(f(0)=f(1))$. The argument used to prove this goes as follows.[2] First we verify that (29) holds whenever $f(x) = e^{2\pi i n x}$, $n \in \mathbb{Z}$, by considering the cases $n=0$ and $n \neq 0$ separately. It then follows that (29) is valid for any trigonometric polynomial (a finite linear combination of these exponentials). Finally, any continuous and periodic function can be uniformly approximated by trigonometric polynomials, so (29) goes over to the general case.

Now if P is the projection on invariant L^2-functions, then Theorem 5.1 and (29) show that P projects onto the constants, when restricted to the continuous periodic functions. Since this subspace is dense in L^2, we see that P still projects all of L^2 on constants; hence the invariant L^2-functions are constants and thus τ is ergodic.

On the other hand, suppose $\alpha = p/q$. Choose any set $E_0 \subset (0, 1/q)$, so that $0 < m(E_0) < 1/q$, and let E denote the disjoint union $\bigcup_{r=0}^{q-1}(E_0 + r/q)$. Then clearly E is invariant under $\tau : x \mapsto x + p/q$, and $0 < m(E) = qm(E_0) < 1$; thus τ is not ergodic.

The property (29) we used, which involves the existence of the limit at all points, is actually stronger than ergodicity: it implies that the measure $d\mu = dx$ is **uniquely ergodic** for this mapping τ. That means that if ν is any measure on the Borel sets of X preserved by τ and $\nu(X) = 1$, then ν must equal μ.

To see that this so in the present case, let P_ν be the orthogonal projection guaranteed by Theorem 5.1, on the space $L^2(X, \nu)$. Then (29) shows again that the range of P_ν on the continuous functions, and then on all of $L^2(X, \nu)$, is the subspace of constants, and thus $P_\nu(f) = \int_0^1 f\,d\nu$.

This means also that $\int_0^1 f(x)\,dx = \int_0^1 f\,d\nu$ whenever f is continuous and periodic. By a simple limiting argument we then get that the measure $dx = d\mu$ and ν agree on all open intervals, and thus on all open sets. As we have seen, this then proves that the two measures are then identical.

In general, uniquely ergodic measure-preserving transformations are ergodic, but the converse need not be true, as we shall see below.

b) The doubling mapping

---

[2] See also Section 2, Chapter 4 in Book I.

We now consider the mapping $x \mapsto 2x \mod 1$ for $x \in (0,1]$, with μ Lebesgue measure, that arose in example (iv) at the beginning of Section 5*. We shall prove that τ is ergodic and in fact satisfies a different and stronger property called mixing.[3] It is defined as follows.

If τ is a measure-preserving transformation on the space (X, μ), it is said to be **mixing** if whenever E and F are a pair of measurable subsets then

$$(30) \qquad \mu(\tau^{-n}(E) \cap F) \to \mu(E)\mu(F) \quad \text{as } n \to \infty.$$

The meaning of (30) can be understood as follows. In probability theory one often encounters a "universe" of possible events to which probabilities are assigned. These events are represented as measurable subsets E of some space (X, μ) with $\mu(X) = 1$. The probability of each event is then $\mu(E)$. Two events E and F are "independent" if the probability that they both occur is the product of their separate probabilities, that is, $\mu(E \cap F) = \mu(E)\mu(F)$. The assertion (30) of mixing is then that in the limit as time n tends to infinity, the sets $\tau^{-n}(E)$ and F are asymptotically independent, whatever the choices of E and F.

We shall next observe that the mixing condition is implied by the seemingly stronger condition

$$(31) \qquad (T^n f, g) \to (f, 1)(1, g) \quad \text{as } n \to \infty,$$

where $T^n(f)(x) = f(\tau^n(x))$ whenever f and g belong to $L^2(X, \mu)$. This implication follows immediately upon taking $f = \chi_E$ and $g = \chi_F$. The converse is also true, but we leave its proof as an exercise to the reader.

We now remark that the mixing condition implies the ergodicity of τ. Indeed, by (31)

$$(A_n(f), g) = \frac{1}{n} \sum_{k=0}^{n-1} (T^k f, g) \quad \text{converges to } (f, 1)(1, g).$$

This means $(P(f), g) = (f, 1)(1, g)$, and hence $P(f)$ is orthogonal to all g that are orthogonal to constants. This of course means that P is the orthogonal projection on constants, and hence τ is ergodic.

We next observe that the doubling map is mixing. Indeed, if $f(x) = e^{2\pi i m x}$, $g(x) = e^{2\pi i k x}$, then $(f, 1)(1, g) = 0$, unless both m and k are 0, in which case this product equals 1. However, in this case $(T^n f, g) = \int_0^1 e^{2\pi i m 2^n x} e^{-2\pi i k x} \, dx$, and this vanishes for sufficiently large n, unless

---

[3]This property is often referred to as a "strongly mixing" to distinguish it from still another kind of ergodicity called "weakly mixing."

both m and k are 0, in which case the integral equals 1. Thus (31) holds for all exponentials $f(x) = e^{2\pi imx}$, $g(x) = e^{2\pi ikx}$, and therefore by linearity for all trigonometric polynomials f and g. It is from there an easy step to use the completeness in Chapter 4 to pass to all f and g in $L^2((0,1])$ by approximating these functions in the L^2-norm by trigonometric polynomials.

Let us observe that the action of rotations $\tau : x \mapsto x + \alpha$ of the unit circle for irrational α, although ergodic, is not mixing. Indeed, if we take $f(x) = g(x) = e^{2\pi imx}$, $m \neq 0$, then $(T^n f, g) = e^{2\pi inm\alpha}(f, g) = e^{2\pi inm\alpha}$, while $(f, 1) = (1, g) = 0$; thus $(T^n f, g)$ does not converge to $(f, 1)(1, g)$ as $n \to \infty$.

Finally, we note that the doubling map $\tau : x \mapsto 2x \mod 1$ on $(0,1]$ is not uniquely ergodic. Besides the Lesbesgue measure, the measure ν with $\nu\{1\} = 1$ but $\nu(E) = 0$ if $1 \notin E$ is also preserved by τ.

Further examples of ergodic transformations are given below.

6* Appendix: the spectral theorem

The purpose of this appendix is to present an outline of the proof of the spectral theorem for bounded symmetric operators on a Hilbert space. Details that are not central to the proof of the theorem will be left to the interested reader to fill in. The theorem provides an interesting application of the ideas related to the Lebesgue-Stieltjes integrals that are treated in this chapter.

6.1 Statement of the theorem

A basic notion is that of a **spectral resolution** (or **spectral family**) on a Hilbert space $\mathcal{H}$. This is a function $\lambda \mapsto E(\lambda)$ from $\mathbb{R}$ to orthogonal projections on $\mathcal{H}$ that satisfies the following:

(i) $E(\lambda)$ is increasing in the sense that $\|E(\lambda)f\|$ is an increasing function of λ for every $f \in \mathcal{H}$.

(ii) There is an interval $[a, b]$ such that $E(\lambda) = 0$ if $\lambda < a$, and $E(\lambda) = I$ if $\lambda \geq b$. Here I denotes the identity operator on $\mathcal{H}$.

(iii) $E(\lambda)$ is right-continuous, that is, for every λ one has

$$\lim_{\substack{\mu \to \lambda \\ \mu > \lambda}} E(\mu)f = E(\lambda)f \quad \text{for every } f \in \mathcal{H}.$$

Observe that property (i) is equivalent with each of the following three assertions (holding for all pairs λ, μ with $\mu > \lambda$): (a) the range of $E(\mu)$ contains the range of $E(\lambda)$; (b) $E(\mu)E(\lambda) = E(\lambda)$; (c) $E(\mu) - E(\lambda)$ is an orthogonal projection.

Now given a spectral resolution $\{E(\lambda)\}$ and an element $f \in \mathcal{H}$, note that the function $\lambda \mapsto (E(\lambda)f, f) = \|E(\lambda)f\|^2$ is also increasing. As a result, the polarization identity (see Section 5 in Chapter 4) shows that for every pair $f, g \in \mathcal{H}$, the function $F(\lambda) = (E(\lambda)f, g)$ is of bounded variation, and is moreover right-continuous. With these two observations we can now state the main result.

Theorem 6.1 *Suppose T is a bounded symmetric operator on a Hilbert space $\mathcal{H}$. Then there exists a spectral resolution $\{E(\lambda)\}$ such that*

$$T = \int_{a^-}^{b} \lambda \, dE(\lambda)$$

in the sense that for every $f, g \in \mathcal{H}$

(32) $$(Tf, g) = \int_{a^-}^{b} \lambda \, d(E(\lambda)f, g) = \int_{a^-}^{b} \lambda \, dF(\lambda).$$

The integral on the right-hand side is taken in the Lebesgue-Stieltjes sense, as in (iii) and (iv) of Section 3.3.

The result encompasses the spectral theorem for compact symmetric operators T in the following sense. Let $\{\varphi_k\}$ be an orthonormal basis of eigenvectors of T with corresponding eigenvalues λ_k, as guaranteed by Theorem 6.2 in Chapter 4. In this case, we take the spectral resolution to be defined via this orthogonal expansion by

$$E(\lambda)f \sim \sum_{\lambda_k \leq \lambda} (f, \varphi_k)\varphi_k,$$

and one easily verifies that it satisfies conditions (i), (ii) and (iii) above. We also note that $\|E(\lambda)f\|^2 = \sum_{\lambda_k \leq \lambda} |(f, \varphi_k)|^2$, and thus $F(\lambda) = (E(\lambda)f, g)$ is a pure jump function as in Section 3.3 in Chapter 3.

6.2 Positive operators

The proof of the theorem depends on the concept of positivity of operators. We say that T is **positive**, written as $T \geq 0$, if T is symmetric and $(Tf, f) \geq 0$ for all $f \in \mathcal{H}$. (Note that (Tf, f) is automatically real if T is symmetric.) One then writes $T_1 \geq T_2$ to mean that $T_1 - T_2 \geq 0$. Note that for two orthogonal projections we have $E_2 \geq E_1$ if and only if $\|E_2 f\| \geq \|E_1 f\|$ for all $f \in \mathcal{H}$, and that is then equivalent with the corresponding properties (a)−(c) described above. Notice also that if S is symmetric, then $S^2 = T$ is positive. Now for T symmetric, let us write

(33) $$a = \min(Tf, f) \quad \text{and} \quad b = \max(Tf, f) \quad \text{for } \|f\| \leq 1.$$

Proposition 6.2 *Suppose T is symmetric. Then $\|T\| \leq M$ if and only if $-MI \leq T \leq MI$. As a result, $\|T\| = \max(|a|, |b|)$.*

This is a consequence of (7) in Chapter 4.

Proposition 6.3 *Suppose T is positive. Then there exists a symmetric operator S (which can be written as $T^{1/2}$) such that $S^2 = T$ and S commutes with every operator that commutes with T.*

The last assertion means that if for some operator A we have $AT = TA$, then $AS = SA$.

The existence of S is seen as follows. After multiplying by a suitable positive scalar, we may assume that $\|T\| \leq 1$. Consider the binomial expansion of $(1 - t)^{1/2}$, given by $(1 - t)^{1/2} = \sum_{k=0}^{\infty} b_k t^k$, for $|t| < 1$. The relevant fact that is needed here is that the b_k are real and $\sum_{k=0}^{\infty} |b_k| < \infty$. Indeed, by direct calculation of the power series expansion of $(1 - t)^{1/2}$ we find that $b_0 = 1$, $b_1 = -1/2$, $b_2 = -1/8$, and more generally, $b_k = -1/2 \cdot 1/2 \cdots (k - 3/2)/k!$, if $k \geq 2$, from which it follows that $b_k = O(k^{-3/2})$. Or more simply, since $b_k < 0$ when $k \geq 1$, if we let $t \to 1$ in the definition, we see that $-\sum_{k=1}^{\infty} b_k = 1$ and so $\sum_{k=0}^{\infty} |b_k| = 2$.

Now let $s_n(t)$ denote the polynomial $\sum_{k=0}^{n} b_k t^k$. Then the polynomial

$$(34) \qquad s_n^2(t) - (1 - t) = \sum_{k=0}^{2n} c_k^n t^k$$

has the property that $\sum_{k=0}^{2n} |c_k^n| \to 0$ as $n \to \infty$. In fact, $s_n(t) = (1 - t)^{1/2} - r_n(t)$, with $r_n(t) = \sum_{k=n+1}^{\infty} b_k t^k$, so $s_n^2(t) - (1 - t) = -r_n^2(t) - 2s_n(t)r_n(t)$. Now the left-hand side is clearly a polynomial of degree $\leq 2n$, and so comparing coefficients with those on the right-hand side shows that the c_k^n are majorized by $3 \sum_{j>n} |b_j| |b_{k-j}|$. From this it is immediate that $\sum_k |c_k^n| = O(\sum_{j>n} |b_j|) \to 0$ as $n \to \infty$, as asserted.

To apply this, set $T_1 = I - T$; then $0 \leq T_1 \leq I$, and thus $\|T_1\| \leq 1$, by Proposition 6.2. Let $S_n = s_n(T_1) = \sum_{k=0}^{n} b_k T_1^k$, with $T_1^0 = I$. Then in terms of operator norms, $\|S_n - S_m\| \leq \sum_{k \geq \min(n,m)} |b_k| \to 0$ as $n, m \to \infty$, because $\|T_1^k\| \leq \|T_1\|^k \leq 1$. Hence S_n converges to some operator S. Clearly S_n is symmetric for each n, and thus S is also symmetric. Moreover, by (34), $S_n^2 - T = \sum_{k=0}^{2n} c_k^n T_1^k$, therefore $\|S_n^2 - T\| \leq \sum |c_k^n| \to 0$ as $n \to \infty$, which implies that $S^2 = T$. Finally, if A commutes with T it clearly commutes with every polynomial in T, hence with S_n, and thus with S. The proof of the proposition is therefore complete.

Proposition 6.4 *If T_1 and T_2 are positive operators that commute, then $T_1 T_2$ is also positive.*

Indeed, if S is a square root of T_1 given in the previous proposition, then $T_1 T_2 = SST_2 = ST_2 S$, and hence $(T_1 T_2 f, f) = (ST_2 Sf, f) = (T_2 Sf, Sf)$, since S is symmetric, and thus the last term is positive.

Proposition 6.5 *Suppose T is symmetric and a and b are given by (33). If $p(t) = \sum_{k=0}^{n} c_k t^k$ is a real polynomial which is positive for $t \in [a, b]$, then the operator $p(T) = \sum_{k=0}^{n} c_k T^k$ is positive.*

To see this, write $p(t) = c \prod_j (t - \rho_j) \prod_k (\rho_k' - t) \prod_\ell ((t - \mu_\ell)^2 + \nu_\ell)$, where c is positive and the third factor corresponds to the non-real roots of $p(t)$ (arising in conjugate pairs), and the real roots of $p(t)$ lying in (a, b) which are necessarily of even order. The first factor contains the real roots ρ_j with $\rho_j \leq a$, and the second factor the real roots ρ_k' with $\rho_k' \geq b$. Since each of the factors $T - \rho_j I$, $\rho_k' I - T$ and $(T - \mu_\ell I)^2 + \nu_\ell^2 I$ is positive and these commute, the desired conclusion follows from the previous proposition.

Corollary 6.6 *If $p(t)$ is a real polynomial, then*

$$\|p(T)\| \leq \sup_{t \in [a,b]} |p(t)|.$$

This is an immediate consequence using Proposition 6.2, since $-M \leq p(t) \leq M$, where $M = \sup_{t \in [a,b]} |p(t)|$, and thus $-MI \leq p(T) \leq MI$.

Proposition 6.7 *Suppose $\{T_n\}$ is a sequence of positive operators that satisfy $T_n \geq T_{n+1}$ for all n. Then there is a positive operator T, such that $T_n f \to T f$ as $n \to \infty$ for every $f \in \mathcal{H}$.*

Proof. We note that for each fixed $f \in \mathcal{H}$ the sequence of positive numbers $(T_n f, f)$ is decreasing and hence convergent. Now observe that for any positive operator S with $\|S\| \leq M$ we have

(35) $$\|S(f)\|^2 \leq (Sf, f)^{1/2} M^{3/2} \|f\|.$$

In fact, the quadratic function $(S(tI + S)f, (tI + S)f) = t^2(Sf, f) + 2t(Sf, Sf) + (S^2 f, Sf)$ is positive for all real t. Hence its discriminant is negative, that is, $\|S(f)\|^4 \leq (Sf, f)(S^2 f, Sf)$, and (35) follows. We apply this to $S = T_n - T_m$ with $n \leq m$; then $\|T_n - T_m\| \leq \|T_n\| \leq \|T_1\| = M$, and since $((T_n - T_m)f, f) \to 0$ as $n, m \to \infty$ we see that $\|T_n f - T_m f\| \to 0$ as $n, m \to \infty$. Thus $\lim_{n \to \infty} T_n(f) = T(f)$ exists, and T is also clearly positive.

6.3 Proof of the theorem

Starting with a given symmetric operator T, and with a, b given by (33), we shall now exploit further the idea of associating to each suitable function Φ on $[a, b]$ a symmetric operator $\Phi(T)$. We do this in increasing order of generality. First, if Φ is a real polynomial $\sum_{k=0}^n c_k t^k$, then, as before, $\Phi(T)$ is defined as $\sum_{k=0}^n c_k T^k$. Notice that this association is a homomorphism: if $\Phi = \Phi_1 + \Phi_2$, then $\Phi(T) = \Phi_1(T) + \Phi_2(T)$; also if $\Phi = \Phi_1 \cdot \Phi_2$, then $\Phi(T) = \Phi_1(T) \cdot \Phi_2(T)$. Moreover, since Φ is real (and the c_k are real), $\Phi(T)$ is symmetric.

Next, because every real-valued continuous function Φ on $[a, b]$ can be approximated uniformly by polynomials p_n (see, for instance, Section 1.8, Chapter 5 of Book I), we see by Corollary 6.6 that the sequence $p_n(T)$ converges, in the norm of operators, to a limit which we call $\Phi(T)$, and moreover this limit does not depend on the sequence of polynomials approximating Φ. Also, $\Phi(T)$ is automatically a symmetric operator. If $\Phi(t) \geq 0$ on $[a, b]$ we can always take the approximating sequence to be positive on $[a, b]$, and as a result $\Phi(T) \geq 0$.

Finally, we define $\Phi(T)$ whenever Φ arises as a limit, $\Phi(t) = \lim_{n \to \infty} \Phi_n(t)$, where $\{\Phi_n(t)\}$ is a decreasing sequence of positive continuous functions on $[a, b]$. In fact, by Proposition 6.7 the limit $\lim_{n \to \infty} \Phi_n(T)$ exists by what we have established above for Φ_n. To show that this limit is independent of the sequence $\{\Phi_n\}$ and thus that $\Phi(t)$ is well-defined as the limit above, let $\{\Phi'_n\}$ be another sequence of decreasing continuous functions converging to Φ. Then whenever $\epsilon > 0$ is given and k is fixed, $\Phi'_n(t) \leq \Phi_k(t) + \epsilon$ for all n sufficiently large. Thus $\Phi'_n(T) \leq \Phi_k(T) + \epsilon I$ for these n, and passing to the limit first in n, then in k, and then with $\epsilon \to 0$, we get

$\lim_{n\to\infty} \Phi'_n(T) \leq \lim_{k\to\infty} \Phi_k(T)$. By symmetry, the reverse inequality holds, and the two limits are the same. Note also that for a pair of these limiting functions, if $\Phi_1(t) \leq \Phi_2(t)$ for $t \in [a,b]$, then $\Phi_1(T) \leq \Phi_2(T)$.

The basic functions Φ, $\Phi = \varphi^\lambda$, that give us the spectral resolution are defined for each real λ by

$$\varphi^\lambda(t) = 1 \quad \text{if } t \leq \lambda \quad \text{and} \quad \varphi^\lambda(t) = 0 \quad \text{if } \lambda < t.$$

We note that $\varphi^\lambda(t) = \lim \varphi_n^\lambda(t)$, where $\varphi_n^\lambda(t) = 1$ if $t \leq \lambda$, $\varphi_n^\lambda(t) = 0$ if $t \geq \lambda + 1/n$, and $\varphi_n^\lambda(t)$ is linear for $t \in [\lambda, \lambda + 1/n]$. Thus each $\varphi^\lambda(t)$ is a limit of a decreasing sequence of continuous functions. In accordance with the above we set

$$E(\lambda) = \varphi^\lambda(T).$$

Since $\lim_{n\to\infty} \varphi_n^{\lambda_1}(t)\varphi_n^{\lambda_2}(t) = \varphi_n^{\lambda_1}(t)$ whenever $\lambda_1 \leq \lambda_2$, we see that $E(\lambda_1)E(\lambda_2) = E(\lambda_1)$. Thus $E(\lambda)^2 = E(\lambda)$ for every λ, and because $E(\lambda)$ is symmetric it is therefore an orthogonal projection. Moreover, for every $f \in \mathcal{H}$

$$\|E(\lambda_1)f\| = \|E(\lambda_1)E(\lambda_2)f\| \leq \|E(\lambda_2)f\|,$$

thus $E(\lambda)$ is increasing. Clearly $E(\lambda) = 0$ if $\lambda < a$, since for those λ, $\varphi^\lambda(t) = 0$ on $[a,b]$. Similarly, $E(\lambda) = I$ for $\lambda \geq b$.

Next we note that $E(\lambda)$ is right-continuous. In fact, fix $f \in \mathcal{H}$ and $\epsilon > 0$. Then for some n, which we now keep fixed, $\|E(\lambda)f - \varphi_n^\lambda(T)f\| < \epsilon$. However, $\varphi_n^\lambda(t)$ converges to $\varphi_n^\lambda(t)$ uniformly in t as $\mu \to \lambda$. Hence $\sup_t |\varphi_n^\mu(t) - \varphi_n^\lambda(t)| < \epsilon$, if $|\mu - \lambda| < \delta$, for an appropriate δ. Thus by the corollary $\|\varphi_n^\mu(T) - \varphi_n^\lambda(T)\| < \epsilon$ and therefore $\|E(\lambda)f - \varphi_n^\lambda(T)\| < 2\epsilon$. Now with $\mu \geq \lambda$ we have that $E(\mu)E(\lambda) = E(\lambda)$ and $E(\mu)\varphi_n^\mu(T) = E(\mu)$. As a result $\|E(\lambda)f - E(\mu)f\| < 2\epsilon$, if $\lambda \leq \mu \leq \lambda + \delta$. Since ϵ was arbitrary, the right continuity is established.

Finally we verify the spectral representation (32). Let $a = \lambda_0 < \lambda_1 < \cdots < \lambda_k = b$ be any partition of $[a,b]$ for which $\sup_j(\lambda_j - \lambda_{j-1}) < \delta$. Then since

$$t = \sum_{j=1}^k t(\varphi^{\lambda_j}(t) - \varphi^{\lambda_{j-1}}(t)) + t\varphi^{\lambda_0}(t)$$

we note that

$$t \leq \sum_{j=1}^k \lambda_j(\varphi^{\lambda_j}(t) - \varphi^{\lambda_{j-1}}(t)) + \lambda_0\varphi^{\lambda_0}(t) \leq t + \delta.$$

Applying these functions to the operator T we obtain

$$T \leq \sum_{j=1}^k \lambda_j(E(\lambda_j) - E(\lambda_{j-1})) + \lambda_0 E(\lambda_0) \leq T + \delta I,$$

and thus T differs in norm from the sum above by at most δ. As a result

$$\left| (Tf, f) - \sum_{j=1}^{k} \lambda_j \int_{(\lambda_{j-1}, \lambda_j]} d(E(\lambda)f, f) - \lambda_0 (E(\lambda_0)f, f) \right| \leq \delta \|f\|^2.$$

But as we vary the partitions of $[a, b]$, letting their meshes δ tend to zero, the above sum tends to $\int_{a^-}^{b} \lambda \, d(E(\lambda)f, f)$. Therefore $(Tf, f) = \int_{a^-}^{b} \lambda \, d(E(\lambda)f, f)$, and the polarization identity gives (32).

A similar argument shows that if Φ is continuous on $[a, b]$, then the operator $\Phi(T)$ has an analogous spectral representation

$$(36) \qquad\qquad (\Phi(T)f, g) = \int_{a^-}^{b} \Phi(\lambda) \, d(E(\lambda)f, g).$$

This is because $|\Phi(t) - \sum_{j=1}^{k} \Phi(\lambda_j)(\varphi^{\lambda_j}(t) - \varphi^{\lambda_{j-1}}(t)) - \Phi(\lambda_0)\varphi^{\lambda_0}(t)| < \delta'$, where $\delta' = \sup_{|t-t'| \leq \delta} |\Phi(t) - \Phi(t')|$, which tends to zero as $\delta \to 0$.

This representation also extends to continuous Φ that are complex-valued (by considering the real and imaginary parts separately) or for Φ that are limits of decreasing pointwise continuous functions.

6.4 Spectrum

We say that a bounded operator S on $\mathcal{H}$ is **invertible** if S is a bijection of $\mathcal{H}$ and its inverse, S^{-1}, is also bounded. Note that S^{-1} satisfies $S^{-1}S = SS^{-1} = I$. The **spectrum** of S, denoted by $\sigma(S)$, is the set of complex numbers z for which $S - zI$ is *not* invertible.

Proposition 6.8 *If T is symmetric, then $\sigma(T)$ is a closed subset of the interval $[a, b]$ given by (33).*

Note that if $z \notin [a, b]$, the function $\Phi(t) = (t - z)^{-1}$ is continuous on $[a, b]$ and $\Phi(T)(T - zI) = (T - zI)\Phi(T) = I$, so $\Phi(T)$ is the inverse of $T - zI$. Now suppose $T_0 = T - \lambda_0 I$ is invertible. Then we claim that $T_0 - \epsilon I$ is invertible for all (complex) ϵ that are sufficiently small. This will prove that the complement of $\sigma(T)$ is open. Indeed, $T_0 - \epsilon I = T_0(I - \epsilon T_0^{-1})$, and we can invert the operator $(I - \epsilon T_0^{-1})$ (formally) by writing its inverse as a sum

$$\sum_{n=0}^{\infty} \epsilon^n (T_0^{-1})^{n+1}.$$

Since $\sum_{n=0}^{\infty} \|\epsilon^n (T_0^{-1})^{n+1}\| \leq \sum |\epsilon|^n \|T_0^{-1}\|^{n+1}$, the series converges when $|\epsilon| < \|T_0^{-1}\|^{-1}$, and the sum is majorized by

$$(37) \qquad\qquad \|T_0^{-1}\| \frac{1}{1 - |\epsilon| \|T_0^{-1}\|}.$$

Thus we can define the operator $(T_0 - \epsilon I)^{-1}$ as $\lim_{N \to \infty} T_0^{-1} \sum_{n=0}^{N} \epsilon^n (T_0^{-1})^{n+1}$, and it gives the desired inverse, as is easily verified.

Our last assertion connects the spectrum $\sigma(T)$ with the spectral resolution $\{E(\lambda)\}$.

Proposition 6.9 *For each $f \in \mathcal{H}$, the Lebesgue-Stieltjes measure corresponding to $F(\lambda) = (E(\lambda)f, f)$ is supported on $\sigma(T)$.*

To put it another way, $F(\lambda)$ is constant on each open interval of the complement of $\sigma(T)$.

To prove this, let J be one of the open intervals in the complement of $\sigma(T)$, $x_0 \in J$, and J_0 the sub-interval centered at x_0 of length 2ϵ, with $\epsilon < \|(T - x_0 I)^{-1}\|$. First note that if z has non-vanishing imaginary part then $(T - zI)^{-1}$ is given by $\Phi_z(T)$, with $\Phi_z(t) = (t - z)^{-1}$. Hence $(T - zI)^{-1}(T - \overline{z}I)^{-1}$ is given by $\Psi_z(T)$, with $\Psi_z(t) = 1/|t - z|^2$. Therefore by the estimate given in (37) and the representation (36) applied to $\Phi = \Psi_z$, we obtain

$$\int \frac{dF(\lambda)}{|\lambda - z|^2} \leq A',$$

as long as z is complex and $|x_0 - z| < \epsilon$. We can therefore obtain the same inequality for x real, $|x_0 - x| < \epsilon$. Now integration in $x \in J_0$ using the fact that $\int_{J_\epsilon} \frac{dx}{|\lambda - x|^2} = \infty$ for every $\lambda \in J_\epsilon$, gives $\int_{J_\epsilon} dF(\lambda) = 0$. Thus $F(\lambda)$ is constant in J_ϵ, but since x_0 was an arbitrary point of J the function $F(\lambda)$ is constant throughout J and the proposition is proved.

7 Exercises

1. Let X be a set and $\mathcal{M}$ a non-empty collection of subsets of X. Prove that if $\mathcal{M}$ is closed under complements and countable unions of disjoint sets, then $\mathcal{M}$ is a σ-algebra.

[Hint: Any countable union of sets can be written as a countable union of disjoint sets.]

2. Let $(X, \mathcal{M}, \mu)$ be a measure space. One can define the **completion** of this space as follows. Let $\overline{\mathcal{M}}$ be the collection of sets of the form $E \cup Z$, where $E \in \mathcal{M}$, and $Z \subset F$ with $F \in \mathcal{M}$ and $\mu(F) = 0$. Also, define $\overline{\mu}(E \cup Z) = \mu(E)$. Then:

(a) $\overline{\mathcal{M}}$ is the smallest σ-algebra containing $\mathcal{M}$ and all subsets of elements of $\mathcal{M}$ of measure zero.

(b) The function $\overline{\mu}$ is a measure on $\overline{\mathcal{M}}$, and this measure is complete.

[Hint: To prove $\overline{\mathcal{M}}$ is a σ-algebra it suffices to see that if $E_1 \subset \overline{\mathcal{M}}$, then $E_1^c \subset \overline{\mathcal{M}}$. Write $E_1 = E \cup Z$ with $Z \subset F$, E and F in $\mathcal{M}$. Then $E_1^c = (E \cup F)^c \cup (F - Z)$.]

3. Consider the exterior Lebesgue measure m_* introduced in Chapter 1. Prove that a set E in $\mathbb{R}^d$ is Carathéodory measurable if and only if E is Lebesgue measurable in the sense of Chapter 1.

[Hint: If E is Lebesgue measurable and A is any set, choose a G_δ set G such that $A \subset G$ and $m_*(A) = m(G)$. Conversely, if E is Carathéodory measurable and $m_*(E) < \infty$, choose a G_δ set G with $E \subset G$ and $m_*(E) = m_*(G)$. Then $G - E$ has exterior measure 0.]

4. Let r be a rotation of $\mathbb{R}^d$. Using the fact that the mapping $x \mapsto r(x)$ preserves Lebesgue measure (see Problem 4 in Chapter 2 and Exercise 26 in Chapter 3), show that it induces a measure-preserving map of the sphere S^{d-1} with its measure $d\sigma$. A converse is stated in Problem 4.

5. Use the polar coordinate formula to prove the following:

(a) $\int_{\mathbb{R}^d} e^{-\pi|x|^2}\, dx = 1$, when $d = 2$. Deduce from this that the same identity holds for all d.

(b) $\left(\int_0^\infty e^{-\pi r^2} r^{d-1}\, dr \right) \sigma(S^{d-1}) = 1$, and as a result, $\sigma(S^{d-1}) = 2\pi^{d/2}/\Gamma(d/2)$.

(c) If B is the unit ball, $v_d = m(B) = \pi^{d/2}/\Gamma(d/2 + 1)$, since this quantity equals $\left(\int_0^1 r^{d-1}\, dr \right) \sigma(S^{d-1})$. (See Exercise 14 in Chapter 2.)

6. A version of Green's formula for the unit ball B in $\mathbb{R}^d$ can be stated as follows. Suppose u and v are a pair of functions that are in $C^2(\overline{B})$. Then one has

$$\int_B (v\triangle u - u\triangle v)\, dx = \int_{S^{d-1}} \left(v \frac{\partial u}{\partial n} - u \frac{\partial v}{\partial n} \right) d\sigma.$$

Here S^{d-1} is the unit sphere with $d\sigma$ the measure defined in Section 3.2, and $\partial u/\partial n$, $\partial v/\partial n$ denote the directional derivatives of u and v (respectively) along the inner normals to S^{d-1}.

Show that the above can be derived from Lemma 4.5 of the previous chapter by taking $\eta = \eta_\epsilon^+$ and letting $\epsilon \to 0$.

7. There is an alternate version of the mean-value property given in (21) of Chapter 5. It can be stated as follows. Suppose u is harmonic in Ω, and the closure of the ball of center x_0 and radius r is contained in Ω. Then

$$u(x_0) = c \int_{S^{d-1}} u(x_0 + ry)\, d\sigma(y), \qquad \text{with } c^{-1} = \sigma(S^{d-1}).$$

Conversely, a continuous function satisfying this mean-value property is harmonic.

[Hint: This can be proved as a direct consequence of the corresponding result for averages over balls (Theorem 4.27 in Chapter 5), or can be deduced from Exercise 6.]

8. The fact that the Lebesgue measure is uniquely characterized by its translation invariance can be made precise by the following assertion: If μ is a Borel measure on $\mathbb{R}^d$ that is translation-invariant, and is finite on compact sets, then μ is a multiple of Lebesgue measure m. Prove this theorem by proceeding as follows.

(a) Suppose Q_a denotes a translate of the cube $\{x : 0 < x_j \leq a,\ j = 1, 2, \ldots, d\}$ of side length a. If we let $\mu(Q_1) = c$, then $\mu(Q_{1/n}) = cn^{-d}$ for each integer n.

(b) As a result μ is absolutely continuous with respect to m, and there is a locally integrable function f such that

$$\mu(E) = \int_E f \, dx.$$

(c) By the differentiation theorem (Corollary 1.7 in Chapter 3) it follows that $f(x) = c$ a.e., and hence $\mu = cm$.

[Hint: Q_1 can be written as a disjoint union of n^d translates of $Q_{1/n}$.]

9. Let $C([a, b])$ denote the vector space of continuous functions on the closed and bounded interval $[a, b]$. Suppose we are given a Borel measure μ on this interval, with $\mu([a, b]) < \infty$. Then

$$f \mapsto \ell(f) = \int_a^b f(x) \, d\mu(x)$$

is a linear functional on $C([a, b])$, with ℓ positive in the sense that $\ell(f) \geq 0$ if $f \geq 0$.

Prove that, conversely, for any linear functional ℓ on $C([a, b])$ that is positive in the above sense, there is a unique finite Borel measure μ so that $\ell(f) = \int_a^b f \, d\mu$ for $f \in C([a, b])$.

[Hint: Suppose $a = 0$ and $u \geq 0$. Define $F(u)$ by $F(u) = \lim_{\epsilon \to 0} \ell(f_\epsilon)$, where

$$f_\epsilon(x) = \begin{cases} 1 & \text{for } 0 \leq x \leq u, \\ 0 & \text{for } u + \epsilon \leq x, \end{cases}$$

and f_ϵ is linear between u and $u + \epsilon$. (See Figure 3.) Then F is increasing and right-continuous, and $\ell(f)$ can be written as $\int_a^b f(x) \, dF(x)$ via Theorem 3.5.]

The result also holds if $[a, b]$ is replaced by a closed infinite interval; we then assume that ℓ is defined on the continuous functions of bounded support, and obtain that the resulting μ is finite on all bounded intervals.

A generalization is given in Problem 5.

10. Suppose ν, ν_1, ν_2 are signed measures on $(X, \mathcal{M})$ and μ a (positive) measure on $\mathcal{M}$. Using the symbols $\perp$ and $\ll$ defined in Section 4.2, prove:

(a) If $\nu_1 \perp \mu$ and $\nu_2 \perp \mu$, then $\nu_1 + \nu_2 \perp \mu$.

(b) If $\nu_1 \ll \mu$ and $\nu_2 \ll \mu$, then $\nu_1 + \nu_2 \ll \mu$.

(c) $\nu_1 \perp \nu_2$ implies $|\nu_1| \perp |\nu_2|$.

(d) $\nu \ll |\nu|$.

(e) If $\nu \perp \mu$ and $\nu \ll \mu$, then $\nu = 0$.

11. Suppose that F is an increasing normalized function on $\mathbb{R}$, and let $F = F_A + F_C + F_J$ be the decomposition of F in Exercise 24 in Chapter 3; here F_A is

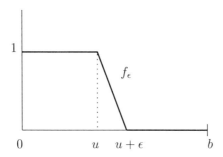

Figure 3. The function f_ϵ in Exercise 9

absolutely continuous, F_C is continuous with $F_C' = 0$ a.e, and F_J is a pure jump function. Let $\mu = \mu_A + \mu_C + \mu_J$ with μ, μ_A, μ_C, and μ_J the Borel measures associated to F, F_A, F_C, and F_J, respectively. Verify that:

(i) μ_A is absolutely continuous with respect to Lebesgue measure and $\mu_A(E) = \int_E F'(x)\,dx$ for every Lebesgue measurable set E.

(ii) As a result, if F is absolutely continuous, then $\int f\,d\mu = \int f\,dF = \int f(x)F'(x)\,dx$ whenever f and fF' are integrable.

(iii) $\mu_C + \mu_J$ and Lebesgue measure are mutually singular.

12. Suppose $\mathbb{R}^d - \{0\}$ is represented as $\mathbb{R}_+ \times S^{d-1}$, with $\mathbb{R}_+ = \{0 < r < \infty\}$. Then every open set in $\mathbb{R}^d - \{0\}$ can be written as a countable union of open rectangles of this product.

[Hint: Consider the countable collection of rectangles of the form

$$\{r_j < r < r_k'\} \times \{\gamma \in S^{d-1} : |\gamma - \gamma_\ell| < 1/n\}.$$

Here r_j and r_k' range over all positive rationals, and $\{\gamma_\ell\}$ is a countable dense set of S^{d-1}.]

13. Let m_j be the Lebesgue measure for the space $\mathbb{R}^{d_j}$, $j = 1, 2$. Consider the product $\mathbb{R}^d = \mathbb{R}^{d_1} \times \mathbb{R}^{d_2}$ ($d = d_1 + d_2$), with m the Lebesgue measure on $\mathbb{R}^d$. Show that m is the completion (in the sense of Exercise 2) of the product measure $m_1 \times m_2$.

14. Suppose $(X_j, \mathcal{M}_j, \mu_j)$, $1 \leq j \leq k$, is a finite collection of measure spaces. Show that parallel with the case $k = 2$ considered in Section 3 one can construct a product measure $\mu_1 \times \mu_2 \times \cdots \times \mu_k$ on $X = X_1 \times X_2 \times \cdots \times X_k$. In fact, for any set $E \subset X$ such that $E = E_1 \times E_2 \times \cdots \times E_k$, with $E_j \subset \mathcal{M}_j$ for all j, define $\mu_0(E) = \prod_{j=1}^{k} \mu_j(E_j)$. Verify that μ_0 extends to a premeasure on the algebra $\mathcal{A}$ of finite disjoint unions of such sets, and then apply Theorem 1.5.

15. The product theory extends to infinitely many factors, under the requisite assumptions. We consider measure spaces $(X_j, \mathcal{M}_j, \mu_j)$ with $\mu_j(X_j) = 1$ for all but finitely many j. Define a **cylinder set** E as

$$\{x = (x_j),\ x_j \in E_j,\ E_j \in \mathcal{M}_j, \text{but } E_j = X_j \text{ for all but finitely many } j\}.$$

For such a set define $\mu_0(E) = \prod_{j=1}^{\infty} \mu_j(E_j)$. If $\mathcal{A}$ is the algebra generated by the cylinder sets, μ_0 extends to a premeasure on $\mathcal{A}$, and we can apply Theorem 1.5 again.

16. Consider the d-dimensional torus $\mathbb{T}^d = \mathbb{R}^d/\mathbb{Z}^d$. Identify $\mathbb{T}^d$ as $\mathbb{T}^1 \times \cdots \times \mathbb{T}^1$ (d factors) and let μ be the product measure on $\mathbb{T}^d$ given by $\mu = \mu_1 \times \mu_2 \times \cdots \times \mu_d$, where μ_j is Lebesgue measure on X_j identified with the circle $\mathbb{T}$. That is, if we represent each point in X_j uniquely as x_j with $0 < x_j \leq 1$, then the measure μ_j is the induced Lebesgue measure on $\mathbb{R}^1$ restricted to $(0, 1]$.

(a) Check that the completion μ is Lebesgue measure induced on the cube $Q = \{x :\ 0 < x_j \leq 1,\ j = 1, \ldots, d\}$.

(b) For each function f on Q let $\tilde{f}$ be its extension to $\mathbb{R}^d$ which is periodic, that is, $\tilde{f}(x + z) = \tilde{f}(x)$ for every $z \in \mathbb{Z}^d$. Then f is measurable on $\mathbb{T}^d$ if and only if $\tilde{f}$ is measurable on $\mathbb{R}^d$, and f is continuous on $\mathbb{T}^d$ if and only if $\tilde{f}$ is continuous on $\mathbb{R}^d$.

(c) Suppose f and g are integrable on $\mathbb{T}^d$. Show that the integral defining $(f * g)(x) = \int_{\mathbb{T}^d} f(x - y)g(y)\,dy$ is finite for a.e. x, that $f * g$ is integrable over $\mathbb{T}^d$, and that $f * g = g * f$.

(d) For any integrable function f on $\mathbb{T}^d$, write

$$f \sim \sum_{n \in \mathbb{Z}^d} a_n e^{2\pi i n \cdot x}$$

to mean that $a_n = \int_{\mathbb{T}^d} f(x) e^{-2\pi i n \cdot x}\,dx$. Prove that if g is also integrable, and $g \sim \sum_{n \in \mathbb{Z}^d} b_n e^{2\pi i n \cdot x}$, then

$$f * g \sim \sum_{n \in \mathbb{Z}^d} a_n b_n e^{2\pi i n \cdot x}.$$

(e) Verify that $\{e^{2\pi i n \cdot x}\}_{n \in \mathbb{Z}^d}$ is an orthonormal basis for $L^2(\mathbb{T}^d)$. As a result $\|f\|_{L^2(\mathbb{T}^d)} = \sum_{n \in \mathbb{Z}^d} |a_n|^2$.

(f) Let f be any continuous periodic function on $\mathbb{T}^d$. Then f can be uniformly approximated by finite linear combinations of the exponentials $\{e^{2\pi i n \cdot x}\}_{n \in \mathbb{Z}^d}$.

[Hint: For (e), reduce to the case $d = 1$ by Fubini's theorem. To prove (f) let $g(x) = g_\epsilon(x) = \epsilon^{-d}$, if $0 < x_j \leq \epsilon, j = 1, \ldots, d$, and $g_\epsilon(x) = 0$ elsewhere in Q. Then $(f * g_\epsilon)(x) \to f(x)$ uniformly as $\epsilon \to 0$. However $(f * g_\epsilon)(x) = \sum a_n b_n e^{2\pi i n x}$ with $b_n = \int_{\mathbb{T}^d} g_\epsilon(x) e^{-2\pi i n \cdot x}\,dx$, and $\sum |a_n b_n| < \infty$.]

17. By reducing to the case $d = 1$, show that each "rotation" $x \mapsto x + \alpha$ of the torus $\mathbb{T}^d = \mathbb{R}^d / \mathbb{Z}^d$ is measure preserving, for any $\alpha \in \mathbb{R}^d$.

18. Suppose τ is a measure-preserving transformation on a measure space (X, μ) with $\mu(X) = 1$. Recall that a measurable set E is invariant if $\tau^{-1}(E)$ and E differ by a set of measure zero. A sharper notion is to require that $\tau^{-1}(E)$ equal E. Prove that if E is any invariant set, there is a set E' so that $E' = \tau^{-1}(E')$, and E and E' differ by a set of measure zero.

[Hint: Let $E' = \limsup_{n \to \infty} \{\tau^{-n}(E)\} = \bigcap_{n=0}^{\infty} \left(\bigcup_{k \geq n} \tau^{-k}(E) \right)$.]

19. Let τ be a measure-preserving transformation on (X, μ) with $\mu(X) = 1$. Then τ is ergodic if and only if whenever ν is absolutely continuous with respect to μ and ν is invariant (that is, $\nu(\tau^{-1}(E)) = \nu(E)$ for all measurable sets E), then $\nu = c\mu$, with c a constant.

20. Suppose τ is a measure-preserving transformation on (X, μ). If

$$\mu(\tau^{-n}(E) \cap F) \to \mu(E)\mu(F)$$

as $n \to \infty$ for all measurable sets E and F, then $(T^n f, g) \to (f, 1)(1, g)$ whenever $f, g \in L^2(X)$ with $(Tf)(x) = f(\tau(x))$. Thus τ is mixing.

[Hint: By linearity the hypothesis implies the conclusion whenever f and g are simple functions.]

21. Let $\mathbb{T}^d$ be the torus, and $\tau : x \mapsto x + \alpha$ the mapping arising in Exercise 17. Then τ is ergodic if and only if $\alpha = (\alpha_1, \ldots, \alpha_d)$ with $\alpha_1, \alpha_2, \ldots, \alpha_d$, and 1 are linearly independent over the rationals. To do this show that:

(a) $\dfrac{1}{m} \sum_{k=0}^{m-1} f(\tau^k(x)) \to \displaystyle\int_{\mathbb{T}^d} f(x)\, dx$ as $m \to \infty$, for each $x \in \mathbb{T}^d$, whenever f is continuous and periodic and α satisfies the hypothesis.

(b) Prove as a result that in this case τ is uniquely ergodic.

[Hint: Use (f) in Exercise 16.]

22. Let $X = \prod_{i=1}^{\infty} X_i$, where each (X_i, μ_i) is identical to (X_1, μ_1), with $\mu_1(X_1) = 1$, and let μ be the corresponding product measure defined in Exercise 15. Define the **shift** $\tau : X \to X$ by $\tau((x_1, x_2, \ldots)) = (x_2, x_3, \ldots)$ for $x = (x_i) \in \prod_{i=1}^{\infty} X_i$.

(a) Verify that τ is a measure-preserving transformation.

(b) Prove that τ is ergodic by showing that it is mixing.

(c) Note that in general τ is not uniquely ergodic.

If we define the corresponding shift on the two-sided infinite product, then τ is also a measure-preserving isomorphism.

[Hint: For (b) note that $\mu(\tau^{-n}(E \cap F)) = \mu(E)\mu(F)$ whenever E and F are cylinder sets and n is sufficiently large. For (c) note that, for example, if we fix a point $\overline{x} \in X_1$, the set $E = \{(x_i) : x_j = \overline{x} \text{ all } j\}$ is invariant.]

23. Let $X = \prod_{i=1}^{\infty} Z(2)$, where each factor is the two-point space $Z(2) = \{0, 1\}$ with $\mu_1(0) = \mu_1(1) = 1/2$, and suppose μ denotes the product measure on X. Consider the mapping $D : X \to [0, 1]$ given by $D(\{a_j\}) \to \sum_{j=1}^{\infty} \frac{a_j}{2^j}$. Then there are denumerable sets $Z_1 \subset X$ and $Z_2 \subset [0, 1]$, such that:

(a) D is a bijection from $X - Z_1$ to $[0, 1] - Z_2$.

(b) A set E in X is measurable if and only if $D(E)$ is measurable in $[0, 1]$, and $\mu(E) = m(D(E))$, where m is Lebesgue measure on $[0, 1]$.

(c) The shift map on $\prod_{i=1}^{\infty} Z(2)$ then becomes the doubling map of example (b) in Section 5.4.

24. Consider the following generalization of the doubling map. For each integer m, $m \geq 2$, we define the map τ_m of $(0, 1]$ by $\tau(x) = mx \mod 1$.

(a) Verify that τ is measure-preserving for Lebesgue measure.

(b) Show that τ is mixing, hence ergodic.

(c) Prove as a consequence that almost every number x is normal in the scale m, in the following sense. Consider the m-adic expansion of x,

$$x = \sum_{j=1}^{\infty} \frac{a_j}{m^j}, \qquad \text{where each } a_j \text{ is an integer } 0 \leq a_j \leq m - 1.$$

Then x is **normal** if for each integer k, $0 \leq k \leq m - 1$,

$$\frac{\#\{j : a_j = k, \ 1 \leq j \leq n\}}{N} \to \frac{1}{m} \qquad \text{as } N \to \infty.$$

Note the analogy with the equidistribution statements in Section 2, Chapter 4, of Book I.

25. Show that the mean ergodic theorem still holds if we replace the assumption that T is an isometry by the assumption that T is a **contraction**, that is, $\|Tf\| \leq \|f\|$ for all $f \in \mathcal{H}$.

[Hint: Prove that T is a contraction if and only if T^* is a contraction, and use the identity $(f, T^* f) = (Tf, f)$.]

26. There is an L^2 version of the maximal ergodic theorem. Suppose τ is a measure-preserving transformation on (X, μ). Here we do not assume that $\mu(X) < \infty$. Then

$$f^*(x) = \sup \frac{1}{m} \sum_{k=0}^{m-1} |f(\tau^k(x))|$$

satisfies

$$\|f^*\|_{L^2(X)} \le c\|f\|_{L^2(X)}, \qquad \text{whenever } f \in L^2(X).$$

The proof is the same as outlined in Problem 6, Chapter 5 for the maximal function on $\mathbb{R}^d$. With this, extend the pointwise ergodic theorem to the case where $\mu(X) = \infty$, as follows:

(a) Show that $\lim_{m\to\infty} \frac{1}{m} \sum_{k=0}^{m-1} f(\tau^k(x))$ converges for a.e. x to $P(f)(x)$ for every $f \in L^2(X)$, because this holds for a dense subspace of $L^2(X)$.

(b) Prove that the conclusion holds for every $f \in L^1(X)$, because it holds for the dense subspace $L^1(X) \cap L^2(X)$.

27. We saw that if $\|f_n\|_{L^2} \le 1$, then $\frac{f_n(x)}{n} \to 0$ as $n \to \infty$ for a.e. x. However, show that the analogue where one replaces the L^2-norm by the L^1-norm fails, by constructing a sequence $\{f_n\}$, $f_n \in L^1(X)$, $\|f_n\|_{L^1} \le 1$, but with $\limsup_{n\to\infty} \frac{f_n(x)}{n} = \infty$ for a.e. x.

[Hint: Find intervals $I_n \subset [0,1]$, so that $m(I_n) = 1/(n \log n)$ but $\limsup_{n\to\infty}\{I_n\} = [0,1]$. Then take $f_n(x) = n \log n \chi_{I_n}$.]

28. We know by the Borel-Cantelli lemma that if $\{E_n\}$ is a collection of measurable sets in a measure a space (X, μ) and $\sum_{n=1}^{\infty} \mu(E_n) < \infty$ then $E = \limsup_{n\to\infty}\{E_n\}$ has measure zero.

In the opposite direction, if τ is a mixing measure-preserving transformation on X (with $\mu(X) = 1$), then whenever $\sum_{n=1}^{\infty} \mu(E_n) = \infty$, there are integers $m = m_n$ so that if $E_n' = \tau^{-m_n}(E_n)$, then $\limsup_{n\to\infty}(E_n') = X$, except for a set of measure 0.

8 Problems

1. Suppose Φ is a C^1 bijection of an open set $\mathcal{O}$ in $\mathbb{R}^d$ onto another open set $\mathcal{O}'$ in $\mathbb{R}^d$.

(a) If E is a measurable subset of $\mathcal{O}$, then $\Phi(E)$ is also measurable.

(b) $m(\Phi(E)) = \int_E |\det \Phi'(x)|\, dx$, where Φ' is the Jacobian of Φ.

(c) $\int_{\mathcal{O}'} f(y)\, dy = \int_{\mathcal{O}} f(\Phi(x))\,|\det \Phi'(x)|\, dx$ whenever f is integrable on $\mathcal{O}'$.

[Hint: To prove (a) follow the argument in Exercise 8, Chapter 1. For (b) assume E is a bounded open set, and write E as $\bigcup_{j=1}^{\infty} Q_j$, where Q_j are cubes whose interiors are disjoint, and whose diameters are less than ϵ. Let z_k be the center of Q_k. Then if $x \in Q_k$,

$$\Phi(x) = \Phi(z_k) + \Phi'(z_k)(x - z_k) + o(\epsilon),$$

hence $\Phi(Q_k) = \Phi(z_k) + \Phi'(z_k)(Q_k - z_k) + o(\epsilon)$, and as a result $(1 - \eta(\epsilon))\Phi'(z_k)(Q_k - z_k) \subset \Phi(Q_k) - \Phi(z_k) \subset (1 + \eta(\epsilon))\Phi'(z_k)(Q_k - z_k)$, where $\eta(\epsilon) \to 0$ as $\epsilon \to 0$. This means that

$$m(\Phi(\mathcal{O})) = \sum_k m(\Phi(Q_k)) = \sum_k |\det(\Phi'(z_k))|\, m(Q_k) + o(1) \qquad \text{as } \epsilon \to 0$$

on account of the linear transformation property of the Lebesgue measure given in Problem 4 of Chapter 2. Note that (b) is (c) for $f(\Phi(x)) = \chi_E(x)$.]

2. Show as a consequence of the previous problem: the measure $d\mu = \frac{dx\,dy}{y^2}$ in the upper half-plane $\mathbb{R}_+^2 = \{z = x + iy,\ y > 0\}$ is preserved by any fractional linear transformation $z \mapsto \frac{az+b}{cz+d}$, where $\begin{pmatrix} a & b \\ c & d \end{pmatrix}$ belongs to $SL_2(\mathbb{R})$.

3. Let S be a hypersurface in $\mathbb{R}^d = \mathbb{R}^{d-1} \times \mathbb{R}$, given by

$$S = \{(x, y) \in \mathbb{R}^{d-1} \times \mathbb{R} : y = F(x)\},$$

with F a C^1 function defined on an open set Ω in $\mathbb{R}^{d-1}$. For each subset $E \subset \Omega$ we write $\widehat{E}$ for the corresponding subset of S given by $\widehat{E} = \{(x, F(x)) : x \in E\}$. We note that the Borel sets of S can be defined in terms of the metric on S (which is the restriction of the Euclidean metric on $\mathbb{R}^d$). Thus if E is a Borel set in Ω, then $\widehat{E}$ is a Borel subset of S.

(a) Let μ be the Borel measure on S given by

$$\mu(\widehat{E}) = \int_E \sqrt{1 + |\nabla F|^2}\, dx.$$

If B is a ball in Ω, let $\widehat{B}^\delta = \{(x, y) \in \mathbb{R}^d,\ d((x, y), \widehat{B}) < \delta\}$. Show that

$$\mu(\widehat{B}) = \lim_{\delta \to 0} \frac{1}{2\delta} m((\widehat{B})^\delta),$$

where m denotes the d-dimensional Lebesgue measure. This result is analogous to Theorem 4.4 in Chapter 3.

(b) One may apply (a) to the case when S is the (upper) half of the unit sphere in $\mathbb{R}^d$, given by $y = F(x)$, $F(x) = (1 - |x|^2)^{1/2}$, $|x| < 1$, $x \in \mathbb{R}^{d-1}$. Show that in this case $d\mu = d\sigma$, the measure on the sphere arising in the polar coordinate formula in Section 3.2.

(c) The above conclusion allows one to write an explicit formula for $d\sigma$ in terms of spherical coordinates. Take, for example, the case $d = 3$, and write $y = \cos\theta$, $x = (x_1, x_2) = (\sin\theta\cos\varphi, \sin\theta\sin\varphi)$ with $0 \le \theta < \pi/2$, $0 \le \varphi < 2\pi$. Then according to (a) and (b) the element of area $d\sigma$ equals $(1 - |x|^2)^{-1/2}\, dx$. Use the change of variable theorem in Problem 1 to deduce that in this case $d\sigma = \sin\theta\, d\theta\, d\varphi$. This may be generalized to d dimensions, $d \ge 2$, to obtain the formulas in Section 2.4 of the appendix in Book I.

4.* Let μ be a Borel measure on the sphere S^{d-1} which is rotation-invariant in the following sense: $\mu(r(E)) = \mu(E)$, for every rotation r of $\mathbb{R}^d$ and each Borel subset E of S^{d-1}. If $\mu(S^{d-1}) < \infty$, then μ is a constant multiple of the measure σ arising in the polar coordinate integration formula.

[Hint: Show that

$$\int_{S^{d-1}} Y_k(x) \, d\mu(x) = 0$$

for every surface spherical harmonic of degree $k \geq 1$. As a result, there is a constant c so that

$$\int_{S^{d-1}} f \, d\mu = c \int_{S^{d-1}} f \, d\sigma$$

for every continuous function f on S^{d-1}.]

5.* Suppose X is a metric space, and μ is a Borel measure on X with the property that $\mu(B) < \infty$ for every ball B. Define $C_0(X)$ to be the vector space of continuous functions on X that are each supported in some closed ball. Then $\ell(f) = \int_X f \, d\mu$ defines a linear functional on $C_0(X)$ that is positive, that is, $\ell(f) \geq 0$ if $f \geq 0$.

Conversely, for any positive linear functional ℓ on $C_0(X)$, there exists a unique Borel measure μ that is finite on all balls, such that $\ell(f) = \int f \, d\mu$.

6. Consider an automorphism A of $\mathbb{T}^d = \mathbb{R}^d/\mathbb{Z}^d$, that is, A is a linear isomorphism of $\mathbb{R}^d$ that preserves the lattice $\mathbb{Z}^d$. Note that A can be written as a $d \times d$ matrix whose entries are integers, with $\det A = \pm 1$. Define the mapping $\tau : \mathbb{T}^d \to \mathbb{T}^d$ by $\tau(x) = A(x)$.

(a) Observe that τ is a measure-preserving isomorphism of $\mathbb{T}^d$.

(b) Show that τ is ergodic (in fact, mixing) if and only if A has no eigenvalues of the form $e^{2\pi i p/q}$, where p and q are integers.

(c) Note that τ is never uniquely ergodic.

[Hint: The condition (b) is the same as $(A^t)^q$ has no invariant vectors, where A^t is the transpose of A. Note also that $f(\tau^k(x)) = e^{2\pi i (A^t)^k(n) \cdot x}$ where $f(x) = e^{2\pi i n \cdot x}$.]

7.* There is a version of the maximal ergodic theorem that is akin to the "rising sun lemma" and Exercise 6 in Chapter 3.

Suppose f is real-valued, and $f^{\#}(x) = \sup_m \frac{1}{m} \sum_{k=0}^{m-1} f(\tau^k(x))$. Let $E_0 = \{x : f^{\#}(x) > 0\}$. Then

$$\int_{E_0} f(x) \, dx \geq 0.$$

As a result (when we apply this to $f(x) - \alpha$), we get when $f \geq 0$ that

$$\mu\{x : f^*(x) > \alpha\} \leq \frac{1}{\alpha} \int_{\{f^*(x) > \alpha\}} f(x) \, dx.$$

In particular, the constant A in Theorem 5.3 can be taken to be 1.

8. Let $X = [0, 1)$, $\tau(x) = \langle 1/x \rangle$, $x \neq 0$, $\tau(0) = 0$. Here $\langle x \rangle$ denotes the fractional part of x. With the measure $d\mu = \frac{1}{\log 2} \frac{dx}{1+x}$, we have of course $\mu(X) = 1$.

Show that τ is a measure-preserving transformation.

[Hint: $\sum_{k=1}^{\infty} \frac{1}{(x+k)(x+k+1)} = \frac{1}{1+x}$.]

9.* The transformation τ in the previous problem is ergodic.

10.* The connection between continued fractions and the transformation $\tau(x) = \langle 1/x \rangle$ will now be described. A **continued fraction**, $a_0 + 1/(a_1 + 1/a_2) \cdots$, also written as $[a_0 a_1 a_2 \cdots]$, where the a_j are positive integers, can be assigned to any positive real number x in the following way. Starting with x, we successively transform it by two alternating operations: reducing it modulo 1 to lie in $[0, 1)$, and then taking the reciprocal of that number. The integers a_j that arise then define the continued fraction of x.

Thus we set $x = a_0 + r_0$, where $a_0 = [x] =$ the greatest integer in x, and $r_0 \in [0, 1)$. Next we write $1/r_0 = a_1 + r_1$, with $a_1 = [1/r_0]$, $r_1 \in [0, 1)$, to obtain successively $1/r_{n-1} = a_n + r_n$, where $a_n = [1/r_{n-1}]$, $r_n \in [0, 1)$. If $r_n = 0$ for some n, we write $a_k = 0$ for all $k > n$, and say that such a continued fraction **terminates**.

Note that if $0 \leq x < 1$, then $r_0 = x$ and $a_1 = [1/x]$, while $r_1 = \langle 1/x \rangle = \tau(x)$. More generally then, $a_k(x) = [1/\tau^{k-1}(x)] = a_1 \tau^{k-1}(x)$. The following properties of continued fractions of positive real numbers x are known:

(a) The continued fraction of x terminates if and only if x is rational.

(b) If $x = [a_0 a_1 \cdots a_n \cdots]$, and $x_N = [a_0 a_1 \cdots a_N 00 \cdots]$, then $x_N \to x$ as $N \to \infty$. The sequence $\{x_N\}$ gives essentially an optimal approximation of x by rationals.

(c) The continued fraction is periodic, that is, $a_{k+N} = a_k$ for some $N \geq 1$, and all sufficiently large k, if and only if x is an algebraic number of degree ≤ 2 over the rationals.

(d) One can conclude that $\frac{a_1 + a_2 + \cdots + a_n}{n} \to \infty$ as $n \to \infty$ for almost every x. In particular, the set of numbers x whose continued fractions $[a_0 a_1 \cdots a_n \cdots]$ are bounded has measure zero.

[Hint: For (d) apply a consequence of the pointwise ergodic theorem, which is as follows: Suppose $f \geq 0$, and $\int f \, d\mu = \infty$. If τ is ergodic, then $\frac{1}{m} \sum_{k=0}^{m-1} f(\tau^k(x)) \to \infty$ for a.e. x as $m \to \infty$. In the present case take $f(x) = [1/x]$.]

7 Hausdorff Measure and Fractals

Carathéodory developed a remarkably simple generalization of Lebesgue's measure theory which in particular allowed him to define the p-dimensional measure of
a set in q-dimensional space. In what follows, I present
a small addition.... a clarification of p-dimensional
measure that leads immediately to an extension to
non-integral p, and thus gives rise to sets of fractional
dimension.

F. Hausdorff, 1919

I coined *fractal* from the Latin adjective *fractus*. The
corresponding Latin verb *frangere* means to "break":
to create irregular fragments.

B. Mandelbrot, 1977

The deeper study of the geometric properties of sets often requires
an analysis of their extent or "mass" that goes beyond what can be
expressed in terms of Lebesgue measure. It is here that the notions
of the dimension of a set (which can be fractional) and an associated
measure play a crucial role.

Two initial ideas may help to provide an intuitive grasp of the concept
of the dimension of a set. The first can be understood in terms of how
the set replicates under scalings. Given the set E, let us suppose that
for some positive number n we have that $nE = E_1 \cup \cdots \cup E_m$, where the
sets E_j are m essentially disjoint congruent copies of E. Note that if
E were a line segment this would hold with $m = n$; if E were a square,
we would have $m = n^2$; if E were a cube, then $m = n^3$; etc. Thus, more
generally, we might be tempted to say that E has dimension α if $m = n^\alpha$.
Observe that if E is the Cantor set $\mathcal{C}$ in $[0, 1]$, then $3\mathcal{C}$ consists of 2 copies
of $\mathcal{C}$, one in $[0, 1]$ and the other in $[2, 3]$. Here $n = 3$, $m = 2$, and we would
be led to conclude that $\log 2 / \log 3$ is the dimension of the Cantor set.

Another approach is relevant for curves that are not necessarily rectifiable. Start with a curve $\Gamma = \{\gamma(t) : a \leq t \leq b\}$, and for each $\epsilon > 0$ consider polygonal lines joining $\gamma(a)$ to $\gamma(b)$, whose vertices lie on successive points of Γ, with each segment not exceeding ϵ in length. Denote by $\#(\epsilon)$ the least number of segments that arise for such polygonal lines. If $\#(\epsilon) \approx \epsilon^{-1}$ as $\epsilon \to 0$, then Γ is rectifiable. However, $\#(\epsilon)$ may well grow more rapidly than ϵ^{-1} as $\epsilon \to 0$. If we had $\#(\epsilon) \approx \epsilon^{-\alpha}$, $1 < \alpha$, then, in the spirit of the previous example, it would be natural to say that Γ has dimension α. These considerations have even an interest in other parts of science. For instance, in studying the question of determining the length of the border of a country or its coastline, L.F. Richardson found that the length of the west coast of Britain obeyed the empirical law $\#(\epsilon) \approx \epsilon^{-\alpha}$, with α approximately 1.5. Thus one might conclude that the coast has fractional dimension!

While there are a number of different ways to make some of these heuristic notions precise, the theory that has the widest scope and greatest flexibility is the one involving Hausdorff measure and Hausdorff dimension. Probably the most elegant and simplest illustration of this theory can be seen in terms of its application to a general class of self-similar sets, and this is what we consider first. Among these are the curves of von Koch type, and these can have any dimension between 1 and 2.

Next, we turn to an example of a space-filling curve, which, broadly speaking, falls under the scope of self-replicating constructions. Not only does this curve have an intrinsic interest, but its nature reveals the important fact that from the point of view of measure theory the unit interval and the unit square are the same.

Our final topic is of a somewhat different nature. It begins with the realization of an unexpected regularity that all subsets of $\mathbb{R}^d$ (of finite Lebesgue measure) enjoy, when $d \geq 3$. This property fails in two dimensions, and the key counter-example is the Besicovitch set. This set appears also in a number of other problems. While it has measure zero, this is barely so, since its Hausdorff dimension is necessarily 2.

1 Hausdorff measure

The theory begins with the introduction of a new notion of volume or mass. This "measure" is closely tied with the idea of dimension which prevails throughout the subject. More precisely, following Hausdorff, one considers for each appropriate set E and each $\alpha > 0$ the quantity $m_\alpha(E)$, which can be interpreted as the α-dimensional mass of E among sets of dimension α, where the word "dimension" carries (for now) only an intuitive meaning. Then, if α is larger than the dimension of the set E, the set has a negligible mass, and we have $m_\alpha(E) = 0$. If α is smaller than the dimension of E, then E is very large (comparatively), hence $m_\alpha(E) = \infty$. For the critical case when α is the dimension of E, the quantity $m_\alpha(E)$ describes the actual α-dimensional size of the set.

Two examples, to which we shall return in more detail later, illustrate this circle of ideas.

First, recall that the standard Cantor set $\mathcal{C}$ in $[0,1]$ has zero Lebesgue measure. This statement expresses the fact that $\mathcal{C}$ has one-dimensional mass or length equal to zero. However, we shall prove that $\mathcal{C}$ has a well-defined fractional Hausdorff dimension of $\log 2 / \log 3$, and that the corresponding Hausdorff measure of the Cantor set is positive and finite.

Another illustration of the theory developed below consists of starting with Γ, a rectifiable curve in the plane. Then Γ has zero two-dimensional Lebesgue measure. This is intuitively clear, since Γ is a one-dimensional object in a two-dimensional space. This is where the Hausdorff measure comes into play: the quantity $m_1(\Gamma)$ is not only finite, but precisely equal to the length of Γ as we defined it in Section 3.1 of Chapter 3.

We first consider the relevant exterior measure, defined in terms of coverings, whose restriction to the Borel sets is the desired Hausdorff measure.

For any subset E of $\mathbb{R}^d$, we define the **exterior α-dimensional Hausdorff measure** of E by

$$m_\alpha^*(E) = \lim_{\delta \to 0} \inf \left\{ \sum_k (\text{diam } F_k)^\alpha : E \subset \bigcup_{k=1}^\infty F_k, \text{ diam } F_k \leq \delta \text{ all } k \right\},$$

where diam S denotes the diameter of the set S, that is, diam $S = \sup\{|x - y| : x, y \in S\}$. In other words, for each $\delta > 0$ we consider covers of E by countable families of (arbitrary) sets with diameter less than δ, and take the infimum of the sum $\sum_k (\text{diam } F_k)^\alpha$. We then define $m_\alpha^*(E)$ as the limit of these infimums as δ tends to 0. We note that the quantity

$$\mathcal{H}_\alpha^\delta(E) = \inf \left\{ \sum_k (\text{diam } F_k)^\alpha : E \subset \bigcup_{k=1}^\infty F_k, \text{ diam } F_k \leq \delta \text{ all } k \right\}$$

is increasing as δ decreases, so that the limit

$$m_\alpha^*(E) = \lim_{\delta \to 0} \mathcal{H}_\alpha^\delta(E)$$

exists, although $m_\alpha^*(E)$ could be infinite. We note that in particular, one has $\mathcal{H}_\alpha^\delta(E) \leq m_\alpha^*(E)$ for all $\delta > 0$. When defining the exterior measure $m_\alpha^*(E)$ it is important to require that the coverings be of sets of arbitrarily small diameters; this is the thrust of the definition $m_\alpha^*(E) = \lim_{\delta \to 0} \mathcal{H}_\alpha^\delta(E)$. This requirement, which is not relevant for Lebesgue measure, is needed to ensure the basic additive feature stated in Property 3 below. (See also Exercise 12.)

Scaling is the key notion that appears at the heart of the definition of the exterior Hausdorff measure. Loosely speaking, the measure of a set scales according to its dimension. For instance, if Γ is a one-dimensional subset of $\mathbb{R}^d$, say a smooth curve of length L, then $r\Gamma$ has total length rL. If Q is a cube in $\mathbb{R}^d$, the volume of rQ is $r^d|Q|$. This feature is captured in the definition of exterior Hausdorff measure by the fact that if the set F is scaled by r, then $(\text{diam } F)^\alpha$ scales by r^α. This key idea reappears in the study of self-similar sets in Section 2.2.

We begin with a list of properties satisfied by the Hausdorff exterior measure.

Property 1 (Monotonicity) *If $E_1 \subset E_2$, then $m_\alpha^*(E_1) \leq m_\alpha^*(E_2)$.*

This is straightforward, since any cover of E_2 is also a cover of E_1.

Property 2 (Sub-additivity) $m_\alpha^*(\bigcup_{j=1}^\infty E_j) \leq \sum_{j=1}^\infty m_\alpha^*(E_j)$ *for any countable family $\{E_j\}$ of sets in $\mathbb{R}^d$.*

For the proof, fix δ, and choose for each j a cover $\{F_{j,k}\}_{k=1}^\infty$ of E_j by sets of diameter less than δ such that $\sum_k (\text{diam } F_{j,k})^\alpha \leq \mathcal{H}_\alpha^\delta(E_j) + \epsilon/2^j$. Since $\bigcup_{j,k} F_{j,k}$ is a cover of E by sets of diameter less than δ, we must have

$$\mathcal{H}_\alpha^\delta(E) \leq \sum_{j=1}^\infty \mathcal{H}_\alpha^\delta(E_j) + \epsilon$$

$$\leq \sum_{j=1}^\infty m_\alpha^*(E_j) + \epsilon.$$

Since ϵ is arbitrary, the inequality $\mathcal{H}_\alpha^\delta(E) \leq \sum m_\alpha^*(E_j)$ holds, and we let δ tend to 0 to prove the countable sub-additivity of m_α^*.

Property 3 *If $d(E_1, E_2) > 0$, then $m_\alpha^*(E_1 \cup E_2) = m_\alpha^*(E_1) + m_\alpha^*(E_2)$.*

It suffices to prove that $m_\alpha^*(E_1 \cup E_2) \geq m_\alpha^*(E_1) + m_\alpha^*(E_2)$ since the reverse inequality is guaranteed by sub-additivity. Fix $\epsilon > 0$ with $\epsilon < d(E_1, E_2)$. Given any cover of $E_1 \cup E_2$ with sets $F_1, F_2 \ldots$, of diameter less than δ, where $\delta < \epsilon$, we let

$$F_j' = E_1 \cap F_j \quad \text{and} \quad F_j'' = E_2 \cap F_j.$$

Then $\{F'_j\}$ and $\{F''_j\}$ are covers for E_1 and E_2, respectively, and are disjoint. Hence,

$$\sum_j (\text{diam } F'_j)^\alpha + \sum_i (\text{diam } F''_i)^\alpha \leq \sum_k (\text{diam } F_k)^\alpha.$$

Taking the infimum over the coverings, and then letting δ tend to zero yields the desired inequality.

At this point, we note that m^*_α satisfies all the properties of a metric Carathéodory exterior measure as discussed in Chapter 6. Thus m^*_α is a countably additive measure when restricted to the Borel sets. We shall therefore restrict ourselves to Borel sets and write $m_\alpha(E)$ instead of $m^*_\alpha(E)$. The measure m_α is called the α-**dimensional Hausdorff measure**.

Property 4 *If $\{E_j\}$ is a countable family of disjoint Borel sets, and $E = \bigcup_{j=1}^\infty E_j$, then*

$$m_\alpha(E) = \sum_{j=1}^\infty m_\alpha(E_j).$$

For what follows in this chapter, the full additivity in the above property is not needed, and we can manage with a weaker form whose proof is elementary and not dependent on the developments of Chapter 6. (See Exercise 2.)

Property 5 *Hausdorff measure is invariant under translations*

$$m_\alpha(E + h) = m_\alpha(E) \quad \text{for all } h \in \mathbb{R}^d,$$

and rotations

$$m_\alpha(rE) = m_\alpha(E),$$

where r is a rotation in $\mathbb{R}^d$.
Moreover, it scales as follows:

$$m_\alpha(\lambda E) = \lambda^\alpha m_\alpha(E) \quad \text{for all } \lambda > 0.$$

These conclusions follow once we observe that the diameter of a set S is invariant under translations and rotations, and satisfies $\text{diam}(\lambda S) = \lambda \, \text{diam}(S)$ for $\lambda > 0$.

We describe next a series of properties of Hausdorff measure, the first of which is immediate from the definitions.

Property 6 *The quantity $m_0(E)$ counts the number of points in E, while $m_1(E) = m(E)$ for all Borel sets $E \subset \mathbb{R}$. (Here m denotes the Lebesgue measure on $\mathbb{R}$.)*

In fact, note that in one dimension every set of diameter δ is contained in an interval of length δ (and for an interval its length equals its Lebesgue measure).

In general, d-dimensional Hausdorff measure in $\mathbb{R}^d$ is, up to a constant factor, equal to Lebesgue measure.

Property 7 *If E is a Borel subset of $\mathbb{R}^d$, then $c_d m_d(E) = m(E)$ for some constant c_d that depends only on the dimension d.*

The constant c_d equals $m(B)/(\text{diam } B)^d$, for the unit ball B; note that this ratio is the same for all balls B in $\mathbb{R}^d$, and so $c_d = v_d/2^d$ (where v_d denotes the volume of the unit ball). The proof of this property relies on the so-called iso-diametric inequality, which states that among all sets of a given diameter, the ball has largest volume. (See Problem 2.) Without using this geometric fact one can prove the following substitute.

Property 7′ *If E is a Borel subset of $\mathbb{R}^d$ and $m(E)$ is its Lebesgue measure, then $m_d(E) \approx m(E)$, in the sense that*

$$c_d m_d(E) \leq m(E) \leq 2^d c_d m_d(E).$$

Using Exercise 26 in Chapter 3 we can find for every $\epsilon, \delta > 0$, a covering of E by balls $\{B_j\}$, such that diam $B_j < \delta$, while $\sum_j m(B_j) \leq m(E) + \epsilon$. Now,

$$\mathcal{H}_d^\delta(E) \leq \sum_j (\text{diam } B_j)^d = c_d^{-1} \sum_j m(B_j) \leq c_d^{-1}(m(E) + \epsilon).$$

Letting δ and ϵ tend to 0, we get $m_d(E) \leq c_d^{-1} m(E)$. For the reverse direction, let $E \subset \bigcup_j F_j$ be a covering with $\sum_j (\text{diam } F_j)^d \leq m_d(E) + \epsilon$. We can always find closed balls B_j centered at a point of F_j so that $B_j \supset F_j$ and diam $B_j = 2 \text{ diam } F_j$. However, $m(E) \leq \sum_j m(B_j)$, since $E \subset \bigcup_j B_j$, and the last sum equals

$$\sum c_d(\text{diam } B_j)^d = 2^d c_d \sum (\text{diam } F_j)^d \leq 2^d c_d \left(m_d(E) + \epsilon\right).$$

Letting $\epsilon \to 0$ gives $m(E) \leq 2^d c_d m_d(E)$.

Property 8 *If $m_\alpha^*(E) < \infty$ and $\beta > \alpha$, then $m_\beta^*(E) = 0$. Also, if $m_\alpha^*(E) > 0$ and $\beta < \alpha$, then $m_\beta^*(E) = \infty$.*

Indeed, if diam $F \leq \delta$, and $\beta > \alpha$, then

$$(\text{diam } F)^\beta = (\text{diam } F)^{\beta-\alpha}(\text{diam } F)^\alpha \leq \delta^{\beta-\alpha}(\text{diam } F)^\alpha.$$

Consequently

$$\mathcal{H}_\beta^\delta(E) \leq \delta^{\beta-\alpha}\mathcal{H}_\alpha^\delta(E) \leq \delta^{\beta-\alpha}m_\alpha^*(E).$$

Since $m_\alpha^*(E) < \infty$ and $\beta - \alpha > 0$, we find in the limit as δ tends to 0, that $m_\beta^*(E) = 0$.

The contrapositive gives $m_\beta^*(E) = \infty$ whenever $m_\alpha^*(E) > 0$ and $\beta < \alpha$.

We now make some easy observations that are consequences of the above properties.

1. If I is a finite line segment in $\mathbb{R}^d$, then $0 < m_1(I) < \infty$.

2. More generally, if Q is a k-cube in $\mathbb{R}^d$ (that is, Q is the product of k non-trivial intervals and $d - k$ points), then $0 < m_k(Q) < \infty$.

3. If $\mathcal{O}$ is a non-empty open set in $\mathbb{R}^d$, then $m_\alpha(\mathcal{O}) = \infty$ whenever $\alpha < d$. Indeed, this follows because $m_d(\mathcal{O}) > 0$.

4. Note that we can always take $\alpha \leq d$. This is because when $\alpha > d$, m_α vanishes on every ball, and hence on all of $\mathbb{R}^d$.

2 Hausdorff dimension

Given a Borel subset E of $\mathbb{R}^d$, we deduce from Property 8 that there exists a unique α such that

$$m_\beta(E) = \begin{cases} \infty & \text{if } \beta < \alpha, \\ 0 & \text{if } \alpha < \beta. \end{cases}$$

In other words, α is given by

$$\alpha = \sup\{\beta : m_\beta(E) = \infty\} = \inf\{\beta : m_\beta(E) = 0\}.$$

We say that E has **Hausdorff dimension** α, or more succinctly, that E has dimension α. We shall write $\alpha = \dim E$. At the critical value α we can say no more than that in general the quantity $m_\alpha(E)$ satisfies $0 \leq m_\alpha(E) \leq \infty$. If E is bounded and the inequalities are strict, that is, $0 < m_\alpha(E) < \infty$, we say that E has **strict Hausdorff dimension** α. The term **fractal** is commonly applied to sets of fractional dimension.

In general, calculating the Hausdorff measure of a set is a difficult problem. However, it is possible in some cases to bound this measure from above and below, and hence determine the dimension of the set in question. A few examples will illustrate these new concepts.

2.1 Examples

The Cantor set

The first striking example consists of the Cantor set $\mathcal{C}$, which was constructed in Chapter 1 by successively removing the middle-third intervals in $[0, 1]$.

Theorem 2.1 *The Cantor set $\mathcal{C}$ has strict Hausdorff dimension $\alpha = \log 2/\log 3$.*

The inequality

$$m_\alpha(\mathcal{C}) \leq 1$$

follows from the construction of $\mathcal{C}$ and the definitions. Indeed, recall from Chapter 1 that $\mathcal{C} = \bigcap C_k$, where each C_k is a finite union of 2^k intervals of length 3^{-k}. Given $\delta > 0$, we first choose K so large that $3^{-K} < \delta$. Since the set C_K covers $\mathcal{C}$ and consists of 2^K intervals of diameter $3^{-K} < \delta$, we must have

$$\mathcal{H}_\alpha^\delta(\mathcal{C}) \leq 2^K(3^{-K})^\alpha.$$

However, α satisfies precisely $3^\alpha = 2$, hence $2^K(3^{-K})^\alpha = 1$, and therefore $m_\alpha(\mathcal{C}) \leq 1$.

The reverse inequality, which consists of proving that $0 < m_\alpha(\mathcal{C})$, requires a further idea. Here we rely on the Cantor-Lebesgue function, which maps $\mathcal{C}$ *surjectively* onto $[0, 1]$. The key fact we shall use about this function is that it satisfies a precise continuity condition that reflects the dimension of the Cantor set.

A function f defined on a subset E of $\mathbb{R}^d$ satisfies a **Lipschitz condition** on E if there exists $M > 0$ such that

$$|f(x) - f(y)| \leq M|x - y| \quad \text{for all } x, y \in E.$$

More generally, a function f satisfies a **Lipschitz condition with exponent** γ (or is **Hölder** γ) if

$$|f(x) - f(y)| \leq M|x - y|^\gamma \quad \text{for all } x, y \in E.$$

The only interesting case is when $0 < \gamma \leq 1$. (See Exercise 3.)

Lemma 2.2 *Suppose a function f defined on a compact set E satisfies a Lipschitz condition with exponent γ. Then*

(i) $m_\beta(f(E)) \leq M^\beta m_\alpha(E)$ *if $\beta = \alpha/\gamma$.*

(ii) $\dim f(E) \leq \frac{1}{\gamma} \dim E$.

Proof. Suppose $\{F_k\}$ is a countable family of sets that covers E. Then $\{f(E \cap F_k)\}$ covers $f(E)$ and, moreover, $f(E \cap F_k)$ has diameter less than $M(\operatorname{diam} F_k)^\gamma$. Hence

$$\sum_k (\operatorname{diam} f(E \cap F_k))^{\alpha/\gamma} \leq M^{\alpha/\gamma} \sum_k (\operatorname{diam} F_k)^\alpha,$$

and part (i) follows. This result now immediately implies conclusion (ii).

Lemma 2.3 *The Cantor-Lebesgue function F on $\mathcal{C}$ satisfies a Lipschitz condition with exponent $\gamma = \log 2/\log 3$.*

Proof. The function F was constructed in Section 3.1 of Chapter 3 as the limit of a sequence $\{F_n\}$ of piecewise linear functions. The function F_n increases by at most 2^{-n} on each interval of length 3^{-n}. So the slope of F_n is always bounded by $(3/2)^n$, and hence

$$|F_n(x) - F_n(y)| \leq \left(\frac{3}{2}\right)^n |x - y|.$$

Moreover, the approximating sequence also satisfies $|F(x) - F_n(x)| \leq 1/2^n$. These two estimates together with an application of the triangle inequality give

$$|F(x) - F(y)| \leq |F_n(x) - F_n(y)| + |F(x) - F_n(x)| + |F(y) - F_n(y)|$$
$$\leq \left(\frac{3}{2}\right)^n |x - y| + \frac{2}{2^n}.$$

Having fixed x and y, we then minimize the right hand side by choosing n so that both terms have the same order of magnitude. This is achieved by taking n so that $3^n|x - y|$ is between 1 and 3. Then, we see that

$$|F(x) - F(y)| \leq c2^{-n} = c(3^{-n})^\gamma \leq M|x - y|^\gamma,$$

since $3^\gamma = 2$ and 3^{-n} is not greater than $|x - y|$. This argument is repeated in Lemma 2.8 below.

With $E = \mathcal{C}$, f the Cantor-Lebesgue function, and $\alpha = \gamma = \log 2/\log 3$, the two lemmas give

$$m_1([0,1]) \leq M^\beta m_\alpha(\mathcal{C}).$$

Thus $m_\alpha(\mathcal{C}) > 0$, and we find that $\dim \mathcal{C} = \log 2 / \log 3$.

The proof of this example is typical in the sense that the inequality $m_\alpha(\mathcal{C}) < \infty$ is usually easier to obtain than $0 < m_\alpha(\mathcal{C})$. Also, with some extra effort, it is possible to show that the $\log 2 / \log 3$-dimensional Hausdorff measure of $\mathcal{C}$ is precisely 1. (See Exercise 7.)

Rectifiable curves

A further example of the role of dimension comes from looking at continuous curves in $\mathbb{R}^d$. Recall that a continuous curve $\gamma : [a, b] \to \mathbb{R}^d$ is said to be **simple** if $\gamma(t_1) \neq \gamma(t_2)$ whenever $t_1 \neq t_2$, and **quasi-simple** if the mapping $t \mapsto z(t)$ is injective for t in the complement of finitely many points.

Theorem 2.4 *Suppose the curve γ is continuous and quasi-simple. Then γ is rectifiable if and only if $\Gamma = \{\gamma(t) : a \leq t \leq b\}$ has strict Hausdorff dimension one. Moreover, in this case the length of the curve is precisely its one-dimensional measure $m_1(\Gamma)$.*

Proof. Suppose to begin with that Γ is a rectifiable curve of length L, and consider an arc-length parametrization $\tilde{\gamma}$ such that $\Gamma = \{\tilde{\gamma}(t) : 0 \leq t \leq L\}$. This parametrization satisfies the Lipschitz condition

$$|\tilde{\gamma}(t_1) - \tilde{\gamma}(t_2)| \leq |t_1 - t_2|.$$

This follows since $|t_1 - t_2|$ is the length of the curve between t_1 and t_2, which is greater than the distance from $\tilde{\gamma}(t_1)$ to $\tilde{\gamma}(t_2)$. Since $\tilde{\gamma}$ satisfies the conditions of Lemma 2.2 with exponent 1 and $M = 1$, we find that

$$m_1(\Gamma) \leq L.$$

To prove the reverse inequality, we let $a = t_0 < t_1 < \cdots < t_N = b$ denote a partition of $[a, b]$ and let

$$\Gamma_j = \{\gamma(t) : t_j \leq t \leq t_{j+1}\},$$

so that $\Gamma = \bigcup_{j=0}^{N-1} \Gamma_j$, and hence

$$m_1(\Gamma) = \sum_{j=0}^{N-1} m_1(\Gamma_j)$$

by an application of Property 4 of the Hausdorff measure and the fact that Γ is quasi-simple. Indeed, by removing finitely many points the union $\bigcup_{j=0}^{N-1} \Gamma_j$ becomes disjoint, while the points removed clearly have zero m_1-measure. We next claim that $m_1(\Gamma_j) \geq \ell_j$, where ℓ_j is the distance from $\gamma(t_j)$ to $\gamma(t_{j+1})$, that is, $\ell_j = |\gamma(t_{j+1}) - \gamma(t_j)|$. To see this, recall that Hausdorff measure is rotation-invariant, and introduce new orthogonal coordinates x and y such that $[\gamma(t_j), \gamma(t_j + 1)]$ is the segment $[0, \ell_j]$ on the x-axis. The projection $\pi(x, y) = x$ satisfies the Lipschitz condition

$$|\pi(P) - \pi(Q)| \leq |P - Q|,$$

and clearly the segment $[0, \ell_j]$ on the x-axis is contained in the image $\pi(\Gamma_j)$. Therefore, Lemma 2.2 guarantees

$$\ell_j \leq m_1(\Gamma_j),$$

and thus $m_1(\Gamma) \geq \sum \ell_j$. Since by definition the length L of Γ is the supremum of the sums $\sum \ell_j$ over all partitions of $[a, b]$, we find that $m_1(\Gamma) \geq L$, as desired.

Conversely, if Γ has strict Hausdorff dimension 1, then $m_1(\Gamma) < \infty$, and the above argument shows that Γ is rectifiable.

The reader may note the resemblance of this characterization of rectifiability and an earlier one in terms of Minkowski content, given in Chapter 3. In this connection we point out that there is a different notion of dimension that is sometimes used instead of Hausdorff dimension. For a compact set E, this dimension is given in terms of the size of $E^\delta = \{x \in \mathbb{R}^d : d(x, E) < \delta\}$ as $\delta \to 0$. One observes that if E is a k-dimensional cube in $\mathbb{R}^d$, then $m(E^\delta) \leq c\delta^{d-k}$ as $\delta \to 0$, with m the Lebesgue measure of $\mathbb{R}^d$. With this in mind, the **Minkowski dimension** of E is defined by

$$\inf \{\beta : m(E^\delta) = O(\delta^{d-\beta}) \text{ as } \delta \to 0\}.$$

One can show that the Hausdorff dimension of a set does not exceed its Minkowski dimension, but that equality does not hold in general. More details may be found in Exercises 17 and 18.

The Sierpinski triangle

A Cantor-like set can be constructed in the plane as follows. We begin with a (solid) closed equilateral triangle S_0, whose sides have unit length. Then, as a first step we remove the shaded open equilateral triangle pictured in Figure 1.

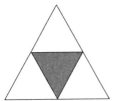 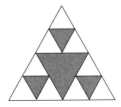 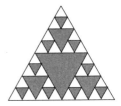

Figure 1. Construction of the Sierpinski triangle

This leaves three closed triangles whose union we denote by S_1. Each triangle is half the size of the original (or parent) triangle S_0, and these smaller closed triangles are said to be of the first **generation**: the triangles in S_1 are the children of the parent S_0. In the second step, we repeat the process in each triangle of the first generation. Each such triangle has three children of the second generation. We denote by S_2 the union of the three triangles in the second generation. We then repeat this process to find a sequence S_k of compact sets which satisfy the following properties:

(a) Each S_k is a union of 3^k closed equilateral triangles of side length 2^{-k}. (These are the triangles of the kth generation.)

(b) $\{S_k\}$ is a decreasing sequence of compact sets; that is, $S_{k+1} \subset S_k$ for all $k \geq 0$.

The **Sierpinski triangle** is the compact set defined by

$$S = \bigcap_{k=0}^{\infty} S_k.$$

Theorem 2.5 *The Sierpinski triangle S has strict Hausdorff dimension $\alpha = \log 3/\log 2$.*

The inequality $m_\alpha(S) \leq 1$ follows immediately from the construction. Given $\delta > 0$, choose K so that $2^{-K} < \delta$. Since the set S_K covers S and consists of 3^K triangles each of diameter $2^{-K} < \delta$, we must have

$$\mathcal{H}_\alpha^\delta(S) \leq 3^K (2^{-K})^\alpha.$$

But since $2^\alpha = 3$, we find $\mathcal{H}_\alpha^\delta(S) \leq 1$, hence $m_\alpha(S) \leq 1$.

The inequality $m_\alpha(S) > 0$ is more subtle. For its proof we need to fix a special point in each triangle that appears in the construction of S.

We choose to call the lower left vertex of a triangle *the* vertex of that triangle. With this choice there are 3^k vertices of the kth generation. The argument that follows is based on the important fact that all these vertices belong to $\mathcal{S}$.

Suppose $\mathcal{S} \subset \bigcup_{j=1}^{\infty} F_j$, with diam $F_j < \delta$. We wish to prove that

$$\sum_j (\text{diam } F_j)^\alpha \geq c > 0$$

for some constant c. Clearly, each F_j is contained in a ball of twice the diameter of F_j, so upon replacing 2δ by δ and noting that $\mathcal{S}$ is compact, it suffices to show that if $\mathcal{S} \subset \bigcup_{j=1}^{N} B_j$, where $\mathcal{B} = \{B_j\}_{j=1}^{N}$ is a finite collection of balls whose diameters are less than δ, then

$$\sum_{j=1}^{N} (\text{diam } B_j)^\alpha \geq c > 0.$$

Suppose we have such a covering by balls. Consider the minimum diameter of the B_j, and choose k so that

$$2^{-k} \leq \min_{1 \leq j \leq N} \text{diam } B_j < 2^{-k+1}.$$

Lemma 2.6 *Suppose B is a ball in the covering $\mathcal{B}$ that satisfies*

$$2^{-\ell} \leq \text{diam } B < 2^{-\ell+1} \qquad \text{for some } \ell \leq k.$$

Then B contains at most $c3^{k-\ell}$ vertices of the kth generation.

In this chapter, we shall continue use the common practice of denoting by $c, c', \ldots$ generic constants whose values are unimportant and may change from one usage to another. We also use $A \approx B$ to denote that the quantities A and B are **comparable**, that is, $cB \leq A \leq c'B$, for appropriate constants c and c'.

Proof of Lemma 2.6. Let B^* denote the ball with same center as B but three times its diameter, and let $\triangle_k$ be a triangle of the kth generation whose vertex v lies in B. If $\triangle_\ell'$ denotes the triangle of the ℓth generation that contains $\triangle_k$, then since diam $B \geq 2^{-\ell}$,

$$v \in \triangle_k \subset \triangle_\ell' \subset B^*,$$

as shown in Figure 2.

Next, there is a positive constant c such that B^* can contain at most c distinct triangles of the ℓth generation. This is because triangles of the

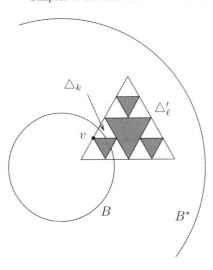

Figure 2. The setting in Lemma 2.6

ℓ^{th} generation have disjoint interiors and area equal to $c'4^{-\ell}$, while B^* has area at most equal to $c''4^{-\ell}$. Finally, each $\triangle'_\ell$ contains $3^{k-\ell}$ triangles of the kth generation, hence B can contain at most $c3^{k-\ell}$ vertices of triangles of the kth generation.

To complete the proof that $\sum_{j=1}^{N}(\text{diam } B_j)^\alpha \geq c > 0$, note that

$$\sum_{j=1}^{N}(\text{diam } B_j)^\alpha \geq \sum_{\ell} N_\ell 2^{-\ell\alpha},$$

where N_ℓ denotes the number of balls in $\mathcal{B}$ that satisfy $2^{-\ell} \leq \text{diam } B_j \leq 2^{-\ell+1}$. By the lemma, we see that the total number of vertices of triangles in the kth generation that can be covered by the collection $\mathcal{B}$ can be no more than $c\sum_{\ell} N_\ell 3^{k-\ell}$. Since all 3^k vertices of triangles in the kth generation belong to $\mathcal{S}$, and all vertices of the kth generation must be covered, we must have $c\sum_{\ell} N_\ell 3^{k-\ell} \geq 3^k$. Hence

$$\sum_{\ell} N_\ell 3^{-\ell} \geq c.$$

It now suffices to recall the definition of α which guarantees $2^{-\ell\alpha} = 3^{-\ell}$, and therefore

$$\sum_{j=1}^{N}(\text{diam } B_j)^\alpha \geq c,$$

as desired.

We give a final example that exhibits properties similar to the Cantor set and Sierpinski triangle. It is the curve discovered by von Koch in 1904.

The von Koch curve

Consider the unit interval $K_0 = [0, 1]$, which we may think of as lying on the x-axis in the xy-plane. Then consider the polygonal path K_1 illustrated in Figure 3, which consists of four equal line segments of length $1/3$.

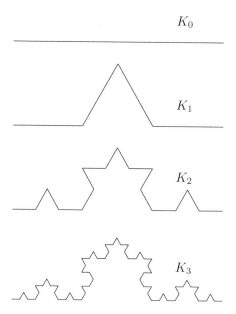

K_0

K_1

K_2

K_3

Figure 3. The first few stages in the construction of the von Koch curve

Let $K_1(t)$, for $0 \le t \le 1$, denote the parametrization of K_1 that has constant speed. In other words, as t travels from 0 to $1/4$, the point $K_1(t)$ travels on the first line segment. As t travels from $1/4$ to $1/2$, the point $K_1(t)$ travels on the second line segment, and so on. In particular, we see that $K_1(\ell/4)$ for $0 \le \ell \le 4$ correspond to the five vertices of K_1.

At the second stage of the construction we repeat the process of replacing each line segment in stage one by the corresponding polygonal line. We then obtain the polygonal curve K_2 illustrated in Figure 3. It has $16 = 4^2$ segments of length $1/9 = 3^{-2}$. We choose a parametrization

$K_2(t)$ $(0 \leq t \leq 1)$ of K_2 that has constant speed. Observe that $K_2(\ell/4^2)$ for $0 \leq \ell \leq 4^2$ gives all vertices of K_2, and that the vertices of K_1 belong to K_2, with

$$K_2(\ell/4) = K_1(\ell/4) \quad \text{for } 0 \leq \ell \leq 4.$$

Repeating this process indefinitely, we obtain a sequence of continuous polygonal curves $\{K_j\}$, where K_j consists of 4^j segments of length 3^{-j} each. If $K_j(t)$ $(0 \leq t \leq 1)$ is the parametrization of K_j that has constant speed, then the vertices are precisely at the points $K_j(\ell/4^j)$, and

$$K_{j'}(\ell/4^j) = K_j(\ell/4^j) \quad \text{for } 0 \leq \ell \leq 4^j$$

whenever $j' \geq j$.

In the limit as j tends to infinity, the polygonal lines K_j tend to the **von Koch curve** $\mathcal{K}$. Indeed, we have

$$|K_{j+1}(t) - K_j(t)| \leq 3^{-j} \quad \text{for all } 0 \leq t \leq 1 \text{ and } j \geq 0.$$

This is clear when $j = 0$, and follows by induction in j when we consider the nature of the construction of the jth stage. Since we may write

$$K_J(t) = K_1(t) + \sum_{j=1}^{J-1} (K_{j+1}(t) - K_j(t)),$$

the above estimate proves that the series

$$K_1(t) + \sum_{j=1}^{\infty} (K_{j+1}(t) - K_j(t))$$

converges absolutely and uniformly to a continuous function $\mathcal{K}(t)$ that is a parametrization of $\mathcal{K}$. Besides continuity, the function $\mathcal{K}(t)$ satisfies a regularity assumption that takes the form of a Lipschitz condition, as in the case of the Cantor-Lebesgue function.

Theorem 2.7 *The function $\mathcal{K}(t)$ satisfies a Lipschitz condition of exponent $\gamma = \log 3 / \log 4$, that is:*

$$|\mathcal{K}(t) - \mathcal{K}(s)| \leq M|t - s|^{\gamma} \quad \text{for all } t, s \in [0, 1].$$

We have already observed that $|K_{j+1}(t) - K_j(t)| \leq 3^{-j}$. Since K_j travels a distance of 3^{-j} in 4^{-j} units of time, we see that

$$|K_j'(t)| \leq \left(\frac{4}{3}\right)^j \quad \text{except when } t = \ell/4^j.$$

Consequently we must have

$$|K_j(t) - K_j(s)| \leq \left(\frac{4}{3}\right)^j |t - s|.$$

Moreover, $\mathcal{K}(t) = K_1(t) + \sum_{j=1}^{\infty}(K_{j+1}(t) - K_j(t))$. We now find ourselves in precisely the same situation as in the proof that the Cantor-Lebesgue function satisfies a Lipschitz condition with exponent $\log 2/\log 3$. We generalize that argument in the following lemma.

Lemma 2.8 *Suppose $\{f_j\}$ is a sequence of continuous functions on the interval $[0,1]$ that satisfy*

$$|f_j(t) - f_j(s)| \leq A^j|t - s| \quad \text{for some } A > 1,$$

and

$$|f_j(t) - f_{j+1}(t)| \leq B^{-j} \quad \text{for some } B > 1.$$

Then the limit $f(t) = \lim_{j\to\infty} f_j(t)$ exists and satisfies

$$|f(t) - f(s)| \leq M|t - s|^{\gamma},$$

where $\gamma = \log B/\log(AB)$.

Proof. The continuous limit f is given by the uniformly convergent series

$$f(t) = f_1(t) + \sum_{k=1}^{\infty}(f_{k+1}(t) - f_k(t)),$$

and therefore

$$|f(t) - f_j(t)| \leq \sum_{k=j}^{\infty}|f_{k+1}(t) - f_k(t)| \leq \sum_{k=j}^{\infty} B^{-k} \leq cB^{-j}.$$

The triangle inequality, an application of the inequality just obtained, and the inequality in the statement of the lemma give

$$|f(t) - f(s)| \leq |f_j(t) - f_j(s)| + |(f - f_j)(t)| + |(f - f_j)(s)|$$
$$\leq c(A^j|t - s| + B^{-j}).$$

For a fixed pair of numbers t and s with $t \neq s$, we choose j to minimize the sum $A^j|t - s| + B^{-j}$. This is essentially achieved by picking j so that two terms $A^j|t - s|$ and B^{-j} are comparable. More precisely, we choose a j that satisfies

$$(AB)^j|t - s| \leq 1 \quad \text{and} \quad 1 \leq (AB)^{j+1}|t - s|.$$

Since $|t - s| \leq 2$ and $AB > 1$, such a j must exist. The first inequality then gives

$$A^j|t - s| \leq B^{-j},$$

while raising the second inequality to the power γ, and using the fact that $(AB)^\gamma = B$ gives

$$1 \leq B^j|t - s|^\gamma.$$

Thus $B^{-j} \leq |t - s|^\gamma$, and consequently

$$|f(t) - f(s)| \leq c(A^j|t - s| + B^{-j}) \leq M|t - s|^\gamma,$$

as was to be shown.

In particular, this result with Lemma 2.2 implies that

$$\dim \mathcal{K} \leq \frac{1}{\gamma} = \frac{\log 4}{\log 3}.$$

To prove that $m_\gamma(\mathcal{K}) > 0$ and hence $\dim \mathcal{K} = \log 4 / \log 3$ requires an argument similar to the one given for the Sierpinski triangle. In fact, this argument generalizes to cover a general family of sets that have a self-similarity property. We therefore turn our attention to this general theory next.

Remarks. We mention some further facts about the von Koch curve. More details can be found in Exercises 13, 14, and 15 below.

1. The curve $\mathcal{K}$ is one in a family of similarly constructed curves. For each ℓ, $1/4 < \ell < 1/2$, consider at the first stage the curve $K_1^\ell(t)$ given by four line segments each of length ℓ, the first and last on the x-axis, and the second and third forming two sides of an isoceles triangle whose base lies on the x-axis. (See Figure 4.) The case $\ell = 1/3$ corresponds to the previously defined von Koch curve.

   Proceeding as in the case $\ell = 1/3$, one obtains a curve $\mathcal{K}^\ell$, and it can be seen that

   $$\dim(\mathcal{K}^\ell) = \frac{\log 4}{\log 1/\ell}.$$

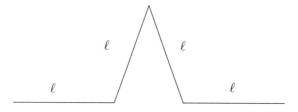

Figure 4. The curve $K_1^\ell(t)$

Thus for every α, $1 < \alpha < 2$, we have a curve of this kind of dimension α. Note that when $\ell \to 1/4$ the limiting curve is a straight line segment, which has dimension 1. When $\ell \to 1/2$, the limit can be seen to correspond to a "space-filling" curve.

2. The curves $t \mapsto K^\ell(t)$, $1/4 < \ell \le 1/2$, are each nowhere differentiable. One can also show that each curve is simple when $1/4 \le \ell < 1/2$.

2.2 Self-similarity

The Cantor set C, the Sierpinski triangle S, and von Koch curve K all share an important property: each of these sets contains scaled copies of itself. Moreover, each of these examples was constructed by iterating a process closely tied to its scaling. For instance, the interval $[0, 1/3]$ contains a copy of the Cantor set scaled by a factor of $1/3$. The same is true for the interval $[2/3, 1]$, and therefore

$$C = C_1 \cup C_2,$$

where C_1 and C_2 are scaled versions of C. Also, each interval $[0, 1/9]$, $[2/9, 3/9]$, $[6/9, 7/9]$ and $[8/9, 1]$ contains a copy of C scaled by a factor of $1/9$, and so on.

In the case of the Sierpinski triangle, each of the three triangles in the first generation contains a copy of S scaled by the factor of $1/2$. Hence

$$S = S_1 \cup S_2 \cup S_3,$$

where each S_j, $j = 1, 2, 3$, is obtained by scaling and translating the original Sierpinski triangle. More generally, every triangle in the kth generation is a copy of S scaled by the factor of $1/2^k$.

Finally, each line segment in the initial stage of the construction of the von Koch curve gives rise to a scaled and possibly rotated copy of the von Koch curve. In fact

$$\mathcal{K} = \mathcal{K}_1 \cup \mathcal{K}_2 \cup \mathcal{K}_3 \cup \mathcal{K}_4,$$

where $\mathcal{K}_j$, $j = 1, 2, 3, 4$, is obtained by scaling $\mathcal{K}$ by the factor of $1/3$ and translating and rotating it.

Thus these examples each contain replicas of themselves, but on a smaller scale. In this section, we give a precise definition of the resulting notion of self-similarity and prove a theorem determining the Hausdorff dimension of these sets.

A mapping $S : \mathbb{R}^d \to \mathbb{R}^d$ is said to be a **similarity** with **ratio** $r > 0$ if

$$|S(x) - S(y)| = r|x - y|.$$

It can be shown that every similarity of $\mathbb{R}^d$ is the composition of a translation, a rotation, and a dilation by r. (See Problem 3.)

Given finitely many similarities $S_1, \ldots, S_m$ with the same ratio r, we say that the set $F \subset \mathbb{R}^d$ is **self-similar** if

$$F = S_1(F) \cup \cdots \cup S_m(F).$$

We point out the relevance of the various examples we have already seen.

When $F = \mathcal{C}$ is the Cantor set, there are two similarities given by

$$S_1(x) = x/3 \quad \text{and} \quad S_2(x) = x/3 + 2/3$$

of ratio $1/3$. So $m = 2$ and $r = 1/3$.

In the case of $F = \mathcal{S}$, the Sierpinski triangle, the ratio is $r = 1/2$ and there are $m = 3$ similarities given by

$$S_1(x) = \frac{x}{2}, \quad S_2(x) = \frac{x}{2} + \alpha \quad \text{and} \quad S_3(x) = \frac{x}{2} + \beta.$$

Here, α and β are the points drawn in the first diagram in Figure 5.

If $F = \mathcal{K}$, the von Koch curve, we have

$$S_1(x) = \frac{x}{3}, \quad S_2(x) = \rho\frac{x}{3} + \alpha, \quad S_3(x) = \rho^{-1}\frac{x}{3} + \beta,$$

and

$$S_4(x) = \frac{x}{3} + \gamma,$$

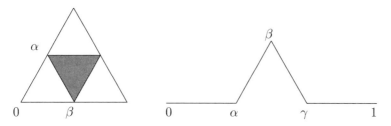

Figure 5. Similarities of the Sierpinski triangle and von Koch curve

where ρ is the rotation centered at the origin and of angle $\pi/3$. There are $m = 4$ similarities which have ratio $r = 1/3$. The points α, β, and γ are shown in the second diagram in Figure 5.

Another example, sometimes called the **Cantor dust** $\mathcal{D}$, is another two-dimensional version of the standard Cantor set. For each fixed $0 < \mu < 1/2$, the set $\mathcal{D}$ may be constructed by starting with the unit square $Q = [0,1] \times [0,1]$. At the first stage we remove everything but the four open squares in the corners of Q that have side length μ. This yields a union D_1 of four squares, as illustrated in Figure 6.

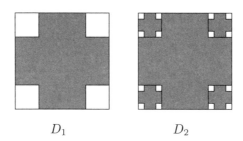

D_1 D_2

Figure 6. Construction of the Cantor dust

We repeat this process in each sub-square of D_1; that is, we remove everything but the four squares in the corner, each of side length μ^2. This gives a union D_2 of 16 squares. Repeating this process, we obtain a family $D_1 \supset D_2 \supset \cdots \supset D_k \supset \cdots$ of compact sets whose intersection defines the Cantor dust corresponding to the parameter μ.

There are here $m = 4$ similarities of ratio μ given by

$$
\begin{aligned}
S_1(x) &= \mu x, \\
S_2(x) &= \mu x + (0, 1 - \mu), \\
S_3(x) &= \mu x + (1 - \mu, 1 - \mu), \\
S_4(x) &= \mu x + (1 - \mu, 0).
\end{aligned}
$$

It is to be noted that $\mathcal{D}$ is the product $\mathcal{C}_\xi \times \mathcal{C}_\xi$, with $\mathcal{C}_\xi$ the Cantor set of constant dissection ξ, as defined in Exercise 3, of Chapter 1. Here $\xi = 1 - 2\mu$.

The first result we prove guarantees the existence of self-similar sets under the assumption that the similarities are contracting, that is, that their ratio satisfies $r < 1$.

Theorem 2.9 *Suppose $S_1, S_2, \ldots, S_m$ are m similartities, each with the same ratio r that satisfies $0 < r < 1$. Then there exists a unique non-empty compact set F such that*

$$ F = S_1(F) \cup \cdots \cup S_m(F). $$

The proof of this theorem is in the nature of a fixed point argument. We shall begin with some large ball B and iteratively apply the mappings $S_1, \ldots, S_m$. The fact that the similarities have ratio $r < 1$ will suffice to imply that this process contracts to a unique set F with the desired property.

Lemma 2.10 *There exists a closed ball B so that $S_j(B) \subset B$ for all $j = 1, \ldots, m$.*

Proof. Indeed, we note that if S is a similarity with ratio r, then

$$
\begin{aligned}
|S(x)| &\leq |S(x) - S(0)| + |S(0)| \\
&\leq r|x| + |S(0)|.
\end{aligned}
$$

If we require that $|x| \leq R$ implies $|S(x)| \leq R$, it suffices to choose R so that $rR + |S(0)| \leq R$, that is, $R \geq |S(0)|/(1 - r)$. In this fashion, we obtain for each S_j a ball B_j centered at the origin that satisfies $S_j(B_j) \subset B_j$. If B denotes the ball among the B_j with the largest radius, then the above shows that $S_j(B) \subset B$ for all j.

Now for any set A, let $\tilde{S}(A)$ denote the set given by

$$ \tilde{S}(A) = S_1(A) \cup \cdots \cup S_m(A). $$

Note that if $A \subset A'$, then $\tilde{S}(A) \subset \tilde{S}(A')$.

Also observe that while each S_j is a mapping from $\mathbb{R}^d$ to $\mathbb{R}^d$, the mapping $\tilde{S}$ is not a point mapping, but takes subsets of $\mathbb{R}^d$ to subsets of $\mathbb{R}^d$.

To exploit the notion of contraction with a ratio less than 1, we introduce the distance between two compact sets as follows. For each $\delta > 0$ and set A, we let

$$A^\delta = \{x : d(x, A) < \delta\}.$$

Hence A^δ is a set that contains A but which is slightly larger in terms of δ. If A and B are two compact sets, we define the **Hausdorff distance** as

$$\mathrm{dist}(A, B) = \inf\{\delta : \ B \subset A^\delta \text{ and } A \subset B^\delta\}.$$

Lemma 2.11 *The distance function* dist *defined on compact subsets of $\mathbb{R}^d$ satisfies*

(i) $\mathrm{dist}(A, B) = 0$ *if and only if $A = B$.*

(ii) $\mathrm{dist}(A, B) = \mathrm{dist}(B, A)$.

(iii) $\mathrm{dist}(A, B) \leq \mathrm{dist}(A, C) + \mathrm{dist}(C, B)$.

If $S_1, \ldots, S_m$ are similarities with ratio r, then

(iv) $\mathrm{dist}(\tilde{S}(A), \tilde{S}(B)) \leq r \, \mathrm{dist}(A, B)$.

The proof of the lemma is simple and may be left to the reader.

Using both lemmas we may now prove Theorem 2.9. We first choose B as in Lemma 2.10, and let $F_k = \tilde{S}^k(B)$, where $\tilde{S}^k$ denotes the kth composition of $\tilde{S}$, that is, $\tilde{S}^k = \tilde{S}^{k-1} \circ \tilde{S}$ with $\tilde{S}^1 = \tilde{S}$. Each F_k is compact, non-empty, and $F_k \subset F_{k-1}$, since $\tilde{S}(B) \subset B$. If we let

$$F = \bigcap_{k=1}^{\infty} F_k,$$

then F is compact, non-empty, and clearly $\tilde{S}(F) = F$, since applying $\tilde{S}$ to $\bigcap_{k=1}^{\infty} F_k$ yields $\bigcap_{k=2}^{\infty} F_k$, which also equals F.

Uniqueness of the set F is proved as follows. Suppose G is another compact set so that $\tilde{S}(G) = G$. Then, an application of part (iv) in Lemma 2.11 yields $\mathrm{dist}(F, G) \leq r \, \mathrm{dist}(F, G)$. Since $r < 1$, this forces $\mathrm{dist}(F, G) = 0$, so that $F = G$, and the proof of Theorem 2.9 is complete.

Under an additional technical condition, one can calculate the precise Hausdorff dimension of the self-similar set F. Loosely speaking, the restriction holds if the sets $S_1(F), \ldots, S_m(F)$ do not overlap too much. Indeed, if these sets were disjoint, then we could argue that

$$m_\alpha(F) = \sum_{j=1}^m m_\alpha(S_j(F)).$$

Since each S_j scales by r, we would then have $m_\alpha(S_j(F)) = r^\alpha m_\alpha(F)$. Hence

$$m_\alpha(F) = mr^\alpha m_\alpha(F).$$

If $m_\alpha(F)$ were finite, then we would have that $mr^\alpha = 1$; thus

$$\alpha = \frac{\log m}{\log 1/r}.$$

The restriction we impose is as follows. We say that the similarities $S_1, \ldots, S_m$ are **separated** if there is an bounded open set $\mathcal{O}$ so that

$$\mathcal{O} \supset S_1(\mathcal{O}) \cup \cdots \cup S_m(\mathcal{O}),$$

and the $S_j(\mathcal{O})$ are disjoint. It is not assumed that $\mathcal{O}$ contains F.

Theorem 2.12 *Suppose $S_1, S_2, \ldots, S_m$ are m separated similarities with the common ratio r that satisfies $0 < r < 1$. Then the set F has Hausdorff dimension equal to $\log m / \log(1/r)$.*

Observe first that when F is the Cantor set we may take $\mathcal{O}$ to be the open unit interval, and note that we have already proved that its dimension is $\log 2 / \log 3$. For the Sierpinski triangle the open unit triangle will do, and $\dim \mathcal{S} = \log 3 / \log 2$. In the example of the Cantor dust the open unit square works, and $\dim \mathcal{D} = \log m / \log \mu^{-1}$. Finally, for the von Koch curve we may take the interior of the triangle pictured in Figure 7, and we will have $\dim \mathcal{K} = \log 4 / \log 3$.

We now turn to the proof of Theorem 2.12, which will follow the same approach used in the case of the Sierpinski triangle. If $\alpha = \log m / \log(1/r)$, we claim that $m_\alpha(F) < \infty$, hence $\dim F \le \alpha$. Moreover, this inequality holds even without the separation assumption. Indeed, recall that

$$F_k = \tilde{S}^k(B),$$

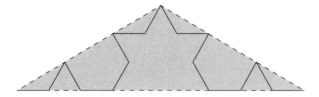

Figure 7. Open set in the separation of the von Koch similarities

and $\tilde{S}^k(B)$ is the union of m^k sets of diameter less than cr^k (with $c =$ diam B), each of the form

$$S_{n_1} \circ S_{n_2} \circ \cdots \circ S_{n_k}(B), \qquad \text{where } 1 \le n_i \le m \text{ and } 1 \le i \le k.$$

Consequently, if $cr^k \le \delta$, then

$$\mathcal{H}_\alpha^\delta(F) \le \sum_{n_1,\ldots,n_k} (\text{diam } S_{n_1} \circ \cdots \circ S_{n_k}(B))^\alpha$$
$$\le c'm^k r^{\alpha k}$$
$$\le c',$$

since $mr^\alpha = 1$, because $\alpha = \log m / \log(1/r)$. Since c' is independent of δ, we get $m_\alpha(F) \le c'$.

To prove $m_\alpha(F) > 0$, we now use the separation condition. We argue in parallel with the earlier calculation of the Hausdorff dimension of the Sierpinski triangle.

Fix a point $\bar{x}$ in F. We define *the* "vertices" of the kth generation as the m^k points that lie in F and are given by

$$S_{n_1} \circ \cdots \circ S_{n_k}(\bar{x}), \qquad \text{where } 1 \le n_1 \le m, \ldots, 1 \le n_k \le m.$$

Each vertex is labeled by $(n_1, \ldots, n_k)$. Vertices need not be distinct, so they are counted with their multiplicities.

Similarly, we define *the* "open sets" of the kth generation to be the m^k sets given by

$$S_{n_1} \circ \cdots \circ S_{n_k}(\mathcal{O}), \qquad \text{where } 1 \le n_1 \le m, \ldots, 1 \le n_k \le m,$$

and where $\mathcal{O}$ is fixed and chosen to satisfy the separation condition. Such open sets are again labeled by multi-indices $(n_1, n_2, \ldots, n_k)$ with $1 \le n_j \le m$, $1 \le j \le k$.

Then the open sets of the kth generation are disjoint, since those of the first generation are disjoint. Moreover if $k \geq \ell$, each open set of the ℓth generation contains $m^{k-\ell}$ open sets of the kth generation.

Suppose v is a vertex of the kth generation, and let $\mathcal{O}(v)$ denote the open set in the kth generation which is associated to v, that is, v and $\mathcal{O}(v)$ carry the same label $(n_1, n_2, \ldots, n_k)$. Since $\overline{x}$ is at a fixed distance from the original open set $\mathcal{O}$, and $\mathcal{O}$ has a finite diameter, we find that

(a) $d(v, \mathcal{O}(v)) \leq cr^k$.

(b) $c'r^k \leq \operatorname{diam} \mathcal{O}(v) \leq cr^k$.

As in the case of the Sierpinski triangle, it suffices to prove that if $\mathcal{B} = \{B_j\}_{j=1}^N$ is a finite collection of balls whose diameters are less than δ and whose union covers F, then

$$\sum_{j=1}^{N} (\operatorname{diam} B_j)^\alpha \geq c > 0.$$

Suppose we have such a covering by balls, and choose k so that

$$r^k \leq \min_{1 \leq j \leq N} \operatorname{diam} B_j < r^{k-1}.$$

Lemma 2.13 *Suppose B is a ball in the covering $\mathcal{B}$ that satisfies*

$$r^\ell \leq \operatorname{diam} B < r^{\ell-1} \quad \text{for some } \ell \leq k.$$

Then B contains at most $cm^{k-\ell}$ vertices of the kth generation.

Proof. If v is a vertex of the kth generation with $v \in B$, and $\mathcal{O}(v)$ denotes the corresponding open set of the kth generation, then, for some fixed dilate B^* of B, properties (a) and (b) above guarantee that $\mathcal{O}(v) \subset B^*$, and B^* also contains the open set of generation ℓ that contains $\mathcal{O}(v)$.

Since B^* has volume $cr^{d\ell}$, and each open set in the ℓth generation has volume $\approx r^{d\ell}$ (by property (b) above), B^* can contain at most c open sets of generation ℓ. Hence B^* contains at most $cm^{k-\ell}$ open sets of the kth generation. Consequently, B can contain at most $cm^{k-\ell}$ vertices of the kth generation, and the lemma is proved.

For the final argument, let N_ℓ denote the number of balls in $\mathcal{B}$ so that

$$r^\ell \leq \operatorname{diam} B_j \leq r^{\ell-1}.$$

By the lemma, we see that the total number of vertices of the kth generation that can be covered by the collection $\mathcal{B}$ can be no more than

$c \sum_\ell N_\ell m^{k-\ell}$. Since all m^k vertices of the kth generation belong to F, we must have $c \sum_\ell N_\ell m^{k-\ell} \geq m^k$, and hence

$$\sum_\ell N_\ell m^{-\ell} \geq c.$$

The definition of α gives $r^{\ell\alpha} = m^{-\ell}$, and therefore

$$\sum_{j=1}^N (\text{diam } B_j)^\alpha \geq \sum_\ell N_\ell r^{\ell\alpha} \geq c,$$

and the proof of Theorem 2.12 is complete.

3 Space-filling curves

The year 1890 heralded an important discovery: Peano constructed a continuous curve that filled an entire square in the plane. Since then, many variants of his construction have been given. We shall describe here a construction that has the feature of elucidating an additional significant fact. It is that from the point of measure theory, speaking broadly, the unit interval and unit square are "isomorphic."

Theorem 3.1 *There exists a curve $t \mapsto \mathcal{P}(t)$ from the unit interval to the unit square with the following properties:*

(i) *$\mathcal{P}$ maps $[0, 1]$ to $[0, 1] \times [0, 1]$ continuously and surjectively.*

(ii) *$\mathcal{P}$ satisfies a Lipschitz condition of exponent $1/2$, that is,*

$$|\mathcal{P}(t) - \mathcal{P}(s)| \leq M|t - s|^{1/2}.$$

(iii) *The image under $\mathcal{P}$ of any sub-interval $[a, b]$ is a compact subset of the square of (two-dimensional) Lebesgue measure exactly $b - a$.*

The third conclusion can be elaborated further.

Corollary 3.2 *There are subsets $Z_1 \subset [0, 1]$ and $Z_2 \subset [0, 1] \times [0, 1]$, each of measure zero, such that $\mathcal{P}$ is bijective from*

$$[0, 1] - Z_1 \quad to \quad [0, 1] \times [0, 1] - Z_2$$

and measure preserving. In other words, E is measurable if and only if $\mathcal{P}(E)$ is measurable, and

$$m_1(E) = m_2(\mathcal{P}(E)).$$

Here m_1 and m_2 denote the Lebesgue measures in $\mathbb{R}^1$ and $\mathbb{R}^2$, respectively.

We shall call the function $t \mapsto \mathcal{P}(t)$ the **Peano mapping**. Its image is called the **Peano curve**.

Several observations help clarify the nature of the conclusions of the theorem. Suppose that $F : [0,1] \to [0,1] \times [0,1]$ is continuous and surjective. Then:

(a) F cannot be Lipschitz of exponent $\gamma > 1/2$. This follows at once from Lemma 2.2, which states that

$$\dim \ F([0,1]) \leq \frac{1}{\gamma} \dim[0,1],$$

so that $2 \leq 1/\gamma$ as desired.

(b) F cannot be injective. Indeed, if this were the case, then the inverse G of F would exist and would be continuous. Given any two points $a \neq b$ in $[0,1]$, we would get a contradiction by looking at two distinct curves in the square that join $F(a)$ and $F(b)$, since the image of these two curves under G would have to intersect at points between a and b. In fact, given any open disc D in the square, there always exists $x \in D$ so that $F(t) = F(s) = x$ yet $t \neq s$.

The proof of Theorem 3.1 will follow from a careful study of a natural class of mappings that associate sub-squares in $[0,1] \times [0,1]$ to sub-intervals in $[0,1]$. This implements the approach implicit in Hilbert's iterative procedure, which he set forth in the first three stages in Figure 8.

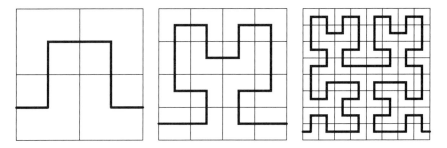

Figure 8. Construction of the Peano curve

We turn now to the study of the general class of mappings.

3.1 Quartic intervals and dyadic squares

The **quartic intervals** arise when $[0, 1]$ is successively sub-divided by powers of 4. For instance, the first generation quartic intervals are the closed intervals

$$I_1 = [0, 1/4], \quad I_2 = [1/4, 1/2], \quad I_3 = [1/2, 3/4], \quad I_4 = [3/4, 1].$$

The second generation quartic intervals are obtained by sub-dividing each interval of the first generation by 4. Hence there are $16 = 4^2$ quartic intervals of the second generation. In general, there are 4^k quartic intervals of the kth generation, each of the form $[\frac{\ell}{4^k}, \frac{\ell+1}{4^k}]$, where ℓ is integral with $0 \leq \ell < 4^k$.

A **chain** of quartic intervals is a decreasing sequence of intervals

$$I^1 \supset I^2 \supset \cdots \supset I^k \supset \cdots,$$

where I^k is a quartic interval of the kth generation (hence $|I^k| = 4^{-k}$).

Proposition 3.3 *Chains of quartic intervals satisfy the following properties:*

(i) *If $\{I^k\}$ is a chain of quartic intervals, then there exists a unique $t \in [0, 1]$ such that $t \in \bigcap_k I^k$.*

(ii) *Conversely, given $t \in [0, 1]$, there is a chain $\{I^k\}$ of quartic intervals such that $t \in \bigcap_k I^k$.*

(iii) *The set of t for which the chain in part (ii) is not unique is a set of measure zero (in fact, this set is countable).*

Proof. Part (i) follows from the fact that $\{I^k\}$ is a decreasing sequence of compact sets whose diameters go to 0.

For part (ii), we fix t and note that for each k there exists at least one quartic interval I^k with $t \in I^k$. If t is of the form $\ell/4^k$, where $0 < \ell < 4^k$, then there are exactly two quartic intervals of the kth generation that contain t. Hence, the set of points for which the chain is not unique is precisely the set of **dyadic rationals**

$$\frac{\ell}{4^k}, \quad \text{where } 1 \leq k, \text{ and } 0 < \ell < 4^k.$$

Note that of course, these fractions are the same as those of the form $\ell'/2^{k'}$ with $0 < \ell' < 2^{k'}$. This set is countable, hence has measure 0.

It is clear that each chain $\{I^k\}$ of quartic intervals can be represented naturally by a string $.a_1 a_2 \cdots a_k \cdots$, where each a_k is either $0, 1, 2$, or 3. Then the point t corresponding to this chain is given by

$$t = \sum_{k=1}^{\infty} \frac{a_k}{4^k}.$$

The points where ambiguity occurs are precisely those where $a_k = 3$ for all sufficiently large k, or equivalently where $a_k = 0$ for all sufficiently large k.

Part of our description of the Peano mapping will follow from associating to each quartic interval a dyadic square. These **dyadic squares** are obtained by sub-dividing the unit square $[0, 1] \times [0, 1]$ in the plane by successively bisecting the sides.

For instance, dyadic squares of the first generation arise from bisecting the sides of the unit square. This yields four closed squares S_1, S_2, S_3 and S_4, each of side length $1/2$ and area $|S_i| = 1/4$, for $i = 1, \ldots, 4$.

The dyadic squares of the second generation are obtained by bisecting each dyadic square of the first generation, and so on. In general, there are 4^k squares of the kth generation, each of side length $1/2^k$ and area $1/4^k$.

A **chain** of dyadic squares is a decreasing sequence of squares

$$S^1 \supset S^2 \supset \cdots \supset S^k \supset \cdots,$$

where S^k is a dyadic square of the kth generation.

Proposition 3.4 *Chains of dyadic squares have the following properties:*

(i) *If $\{S^k\}$ is a chain of dyadic squares, then there exists a unique $x \in [0, 1] \times [0, 1]$ such that $x \in \bigcap_k S^k$.*

(ii) *Conversely, given $x \in [0, 1] \times [0, 1]$, there is a chain $\{S^k\}$ of dyadic squares such that $x \in \bigcap_k S^k$.*

(iii) *The set of x for which the chain in part (ii) is not unique is a set of measure zero.*

In this case, the set of ambiguities consists of all points (x_1, x_2) where one of the coordinates is a dyadic rational. Geometrically, this set is the (countable) union of vertical and horizontal segments in $[0, 1] \times [0, 1]$ determined by the grid of dyadic rationals. This set has measure zero.

Moreover, each chain of dyadic squares can be represented by a string $.b_1 b_2 \cdots$, where each b_k is either $0, 1, 2$ or 3. Then

(1)
$$x = \sum_{k=1}^{\infty} \frac{\bar{b}_k}{2^k},$$

where

$$\begin{aligned}
\bar{b}_k &= (0,0) \quad \text{if } b_k = 0, \\
\bar{b}_k &= (0,1) \quad \text{if } b_k = 1, \\
\bar{b}_k &= (1,0) \quad \text{if } b_k = 2, \\
\bar{b}_k &= (1,1) \quad \text{if } b_k = 3.
\end{aligned}$$

3.2 Dyadic correspondence

A **dyadic correspondence** is a mapping Φ from quartic intervals to dyadic squares that satisfies:

(1) Φ is bijective.

(2) Φ respects generations.

(3) Φ respects inclusion.

By (2), we mean that if I is a quartic interval of the kth generation, then $\Phi(I)$ is a dyadic square of the kth generation. By (3), we mean that if $I \subset J$, then $\Phi(I) \subset \Phi(J)$.

For example, the trivial, or standard correspondence assigns to the string $.a_1 a_2 \cdots$ the string $.b_1 b_2 \cdots$ with $b_k = a_k$.

Given a dyadic correspondence Φ, the **induced mapping** Φ^* maps $[0,1]$ to $[0,1] \times [0,1]$ and is given as follows. If $\{t\} = \bigcap I^k$ where $\{I^k\}$ is a chain of quartic intervals, then, since $\{\Phi(I^k)\}$ is a chain of dyadic squares, we may let

$$\Phi^*(t) = x = \bigcap \Phi(I^k).$$

We note that Φ^* is well-defined except on a (countable) set of measure zero, (those points t that are represented by more than one quartic chain.)

A moment's reflection will show that if I' is a quartic interval of the kth generation, then the images $\Phi^*(I') = \{\Phi^*(t), \ t \in I'\}$, comprise the dyadic square of the kth generation $\Phi(I')$. Thus $\Phi^*(I') = \Phi(I')$, and hence $m_1(I') = m_2(\Phi^*(I'))$.

Theorem 3.5 *Given a dyadic correspondence Φ, there exist sets $Z_1 \subset [0,1]$ and $Z_2 \subset [0,1] \times [0,1]$, each of measure zero, so that:*

(i) Φ^* *is a bijection on* $[0, 1] - Z_1$ *to* $[0, 1] \times [0, 1] - Z_2$.

(ii) E *is measurable if and only if* $\Phi^*(E)$ *is measurable.*

(iii) $m_1(E) = m_2(\Phi^*(E))$.

Proof. First, let $\mathcal{N}_1$ denote the collection of chains of those quartic intervals arising in (iii) of Proposition 3.3, those for which the points in $I = [0, 1]$ are not uniquely representable. Similarly, let $\mathcal{N}_2$ denote the collection of chains of those dyadic squares for which the corresponding points in the square $I \times I$ are not uniquely representable.

Since Φ is a bijection from chains of quartic intervals to chains of dyadic squares, it is also a bijection from $\mathcal{N}_1 \cup \Phi^{-1}(\mathcal{N}_2)$ to $\Phi(\mathcal{N}_1) \cup \mathcal{N}_2$, and hence also of their complements. Let Z_1 be the subset of I consisting of all points in I that can be represented (according to (i) of Proposition 3.3) by the chains in $\mathcal{N}_1 \cup \Phi^{-1}(\mathcal{N}_2)$, and let Z_2 be the set of points in the square that can be represented by dyadic squares in $\Phi(\mathcal{N}_1) \cup \mathcal{N}_2$. Then Φ^*, the induced mapping, is well-defined on $I - Z_1$, and gives a bijection of $I - Z_1$ to $(I \times I) - Z_2$. To prove that both Z_1 and Z_2 have measure zero, we invoke the following lemma. We suppose $\{f_k\}_{k=1}^{\infty}$ is a fixed given sequence, with each f_k either $0, 1, 2$, or 3.

Lemma 3.6 *Let*

$$E_0 = \{x = \sum_{k=1}^{\infty} a_k/4^k, \quad \text{where } a_k \neq f_k \text{ for all sufficiently large } k\}.$$

Then $m(E_0) = 0$.

Indeed, if we fix r, then $m(\{x : a_r \neq f_r\}) = 3/4$, and

$$m(\{x : a_r \neq f_r \text{ and } a_{r+1} \neq f_{r+1}\}) = (3/4)^2, \quad \text{etc.}$$

Thus $m(\{x : a_k \neq f_k, \text{ all } k \geq r\}) = 0$, and E_0 is a countable union of such sets, from which the lemma follows.

There is a similar statement for points in the square $S = I \times I$ in terms of the representation (1).

Note that as a result the set of points in I corresponding to chains in $\mathcal{N}_1$ form a set of measure zero. In fact, we may use the lemma for the sequence for which $f_k = 1$, for all k, since the elements of $\mathcal{N}_1$ correspond to sequences $\{a_k\}$ with $a_k = 0$ for all sufficiently large k, or $a_k = 3$ for all sufficiently large k.

Similarly, the points in the square S corresponding to $\mathcal{N}_2$ form a set of measure zero. To see this, take for example $f_k = 1$ for k odd, and $f_k = 2$

for k even, and note that $\mathcal{N}_2$ corresponds to all sequences $\{a_k\}$ where one of the following four exclusive alternatives holds for all sufficiently large k: either a_k is 0 or 1; or a_k is 2 or 3; or a_k is 0 or 2; or a_k is 1 or 3. By similar reasoning the points $\Phi^{-1}(\mathcal{N}_2)$ and $\Phi(\mathcal{N}_1)$ form sets of measure zero in I and $I \times I$ respectively.

We now turn to the proof that Φ^* (which is a bijection from $I - Z_1$ to $(I \times I) - Z_2$) is measure preserving. For this it is useful to recall Theorem 1.4 in Chapter 1, whereby any open set $\mathcal{O}$ in the unit interval I can be realized as a countable union $\bigcup_{j=1}^{\infty} I_j$, where each I_j is a closed interval and the I_j have disjoint interiors. Moreover, an examination of the proof shows that the intervals can be taken to be dyadic, that is, of the form $[\ell/2^j, (\ell+1)/2^j]$, for appropriate integers ℓ and j. Further, such an interval is itself a quartic interval if j is even, $j = 2k$, or the union of two quartic intervals $[(2\ell)/2^{2k}, (2\ell+1)/2^{2k}]$ and $[(2\ell+1)/2^{2k}, (2\ell+2)/2^{2k}]$, if j is odd, $j = 2k - 1$. Thus any open set in I can be given as a union of quartic intervals whose interiors are disjoint. Similarly, any open set in the square $I \times I$ is a union of dyadic squares whose interiors are disjoint.

Now let E be any set of measure zero in $I - Z_1$ and $\epsilon > 0$. Then we can cover $E \subset \bigcup_j I_j$, where I_j are quartic intervals and $\sum_j m_1(I_j) < \epsilon$. Because $\Phi^*(E) \subset \bigcup_j \Phi^*(I_j)$, then

$$m_2(\Phi^*(E)) \le \sum m_2(\Phi^*(I_j)) = \sum m_1(I_j) < \epsilon.$$

Thus $\Phi^*(E)$ is measurable and $m_2(\Phi^*(E)) = 0$. Similarly, $(\Phi^*)^{-1}$ maps sets of measure zero in $(I \times I) - Z_2$ to sets of measure zero in I.

Now the argument above also shows that if $\mathcal{O}$ is any open set in I, then $\Phi^*(\mathcal{O} - Z_1)$ is measurable, and $m_2(\Phi^*(\mathcal{O} - Z_1)) = m_1(\mathcal{O})$. Thus this identity goes over to G_δ sets in I. Since any measurable set differs from a G_δ set by a set of measure zero, we see that we have established that $m_2(\Phi^*(E)) = m_1(E)$ for any measurable subset of E of $I - Z_1$. The same argument can be applied to $(\Phi^*)^{-1}$, and this completes the proof of the theorem.

The Peano mapping will be obtained as Φ^* for a special correspondence Φ.

3.3 Construction of the Peano mapping

The particular dyadic correspondence we now present provides us with the steps to follow when tracing the approximations of the Peano curve. The main idea behind its construction is that as we go from one quartic interval in the kth generation to the next quartic interval in the same generation, we move from a dyadic square of the kth generation to another square of the kth generation that shares a common side.

More precisely, we say that two quartic intervals in the same generation are **adjacent** if they share a point in common. Also, two squares in the same generation are **adjacent** if they share a side in common.

Lemma 3.7 *There is a unique dyadic correspondence Φ so that:*

(i) *If I and J are two adjacent intervals of the same generation, then $\Phi(I)$ and $\Phi(J)$ are two adjacent squares (of the same generation).*

(ii) *In generation k, if I_- is the left-most interval and I_+ the right-most interval, then $\Phi(I_-)$ is the left-lower square and $\Phi(I_+)$ is the right-lower square.*

Part (ii) of the lemma is illustrated in Figure 9.

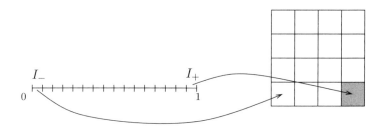

Figure 9. Special dyadic correspondence

Given a square S and its four immediate sub-squares, an acceptable **traverse** is an ordering of the sub-squares S_1, S_2, S_3, and S_4, so that S_j and S_{j+1} are adjacent for $j = 1, 2, 3$. With such an ordering, we note that if we color S_1 white, and then alternate black and white, the square S_3 is also white, while S_2 and S_4 are black. The important point to remember is that if the first square in a traverse is white, then the last square is black.

The key observation is the following. Suppose we are given a square S, and a side σ of S. If S_1 is any of the immediate four sub-squares in S, then there exists a unique traverse S_1, S_2, S_3, and S_4 so that the last square S_4 has a side in common with σ. With the initial square S_1 in the lower-left corner of S, the four possibilities which correspond to the four choices of σ, are illustrated in Figure 10.

We may now begin the inductive description of the dyadic correspondence satisfying the conditions in the lemma. On quartic intervals of the first generation we assign the square $S_j = \Phi(I_j)$, as pictured in Figure 11.

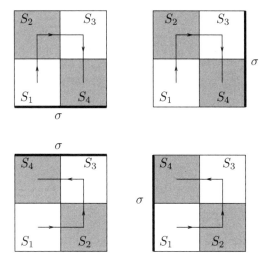

Figure 10. Traverses

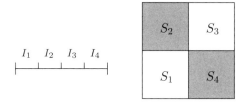

Figure 11. Initial step of the correspondence

Now suppose Φ has been defined for all quartic intervals of generation less than or equal to k. We now write the intervals in generation k in increasing order as $I_1, \ldots, I_{4^k}$, and let $S_j = \Phi(I_j)$. We then divide I_1 into four quartic intervals of generation $k+1$ and denote them by $I_{1,1}$, $I_{1,2}$, $I_{1,3}$, and $I_{1,4}$, where the intervals are chosen in increasing order.

Then, we assign to each interval $I_{1,j}$ a dyadic square $\Phi(I_{1,j}) = S_j$ of generation $k+1$ contained in S_1 so that:

(a) $S_{1,1}$ is the lower-left sub-square of S_1,

(b) $S_{1,4}$ touches the side that S_1 shares with S_2,

(c) $S_{1,1}$, $S_{1,2}$, $S_{1,3}$, and $S_{1,4}$ is a traverse.

This is possible, since the induction hypothesis guarantees that S_2 is adjacent to S_1.

This settles the assignments for the sub-squares of S_1, so we now turn our attention to S_2. Let $I_{2,1}$, $I_{2,2}$, $I_{2,3}$, and $I_{2,4}$ denote the quartic intervals of generation $k+1$ in I_2, written in increasing order. First, we take $S_{2,1} = \Phi(I_{2,1})$ to be the sub-square of S_2 which is adjacent to $S_{1,4}$. This can be done because $S_{1,4}$ touches S_2 by construction. Note that we leave S_1 from a black square ($S_{1,4}$), and enter S_2 in a white square ($S_{2,1}$). Since S_3 is adjacent to S_2, we may now find a traverse $S_{2,1}$, $S_{2,2}$, $S_{2,3}$ and $S_{2,4}$ so that $S_{2,4}$ touches S_3.

We may then repeat this process in each interval I_j and square S_j, $j = 3, \ldots, 4^k$. Note that at each stage the square $S_{j,1}$ (the "entering" square) is white, while $S_{j,4}$ (the "exiting" square) is black.

In the final step, the induction hypothesis guarantees that S_{4^k} is the lower-right corner square. Moreover, since S_{4^k-1} must be adjacent to S_{4^k} it must be either above it, or to the left of it, so we enter a square of the $(k+1)^{\text{st}}$ generation along an upper or left side. The entering square is a white square, and we traverse to the lower right corner sub-square of S_{4^k}, which is a black square.

This concludes the inductive step, hence the proof of Lemma 3.7.

We may now begin the actual description of the Peano curve. For each generation k we construct a polygonal line which consists of vertical and horizontal line segments connecting the centers of consecutive squares. More precisely, let Φ denote the dyadic correspondence in Lemma 3.7, and let $S_1, \ldots, S_{4^k}$ be the squares of the kth generation ordered according to Φ, that is, $\Phi(I_j) = S_j$. Let t_j denote the middle point of I_j,

$$t_j = \frac{j - \frac{1}{2}}{4^k} \quad \text{for } j = 1, \ldots, 4^k.$$

Let x_j be the center of the square S_j, and define

$$\mathcal{P}_k(t_j) = x_j.$$

Also set

$$\mathcal{P}_k(0) = (0, 1/2^{k+1}) = x_0 \quad \text{where } t_0 = 0,$$

and

$$\mathcal{P}_k(1) = (1, 1/2^{k+1}) = x_{4^k+1} \quad \text{where } t_{4^k+1} = 1.$$

Then, we extend $\mathcal{P}_k(t)$ to the unit interval $0 \le t \le 1$ by linearity along the sub-intervals determined by the division points $t_0, \ldots, t_{4^k+1}$.

Note that the distance $|x_j - x_{j+1}| = 1/2^k$, while $|t_j - t_{j+1}| = 1/4^k$ for $0 \leq j \leq 4^k$. Also

$$|x_1 - x_0| = |x_{4^k} - x_{4^{k+1}}| = \frac{1}{2 \cdot 2^k},$$

while

$$|t_1 - t_0| = |t_{4^k} - t_{4^{k+1}}| = \frac{1}{2 \cdot 4^k}.$$

Therefore $\mathcal{P}'_k(t) = 4^k 2^{-k} = 2^k$ except when $t = t_j$.

As a result,

$$|\mathcal{P}_k(t) - \mathcal{P}_k(s)| \leq 2^k |t - s|.$$

However,

$$|\mathcal{P}_{k+1}(t) - \mathcal{P}_k(t)| \leq \sqrt{2}\, 2^{-k},$$

because when $\ell/4^k \leq t \leq (\ell+1)/4^k$, then $\mathcal{P}_{k+1}(t)$ and $\mathcal{P}_k(t)$ belong to the same dyadic square of generation k.

Therefore the limit

$$\mathcal{P}(t) = \lim_{k \to \infty} \mathcal{P}_k(t) = \mathcal{P}_1(t) + \sum_{j=1}^{\infty} \mathcal{P}_{j+1}(t) - \mathcal{P}_j(t)$$

exists, and defines a continuous function in view of the uniform convergence. By Lemma 2.8 we conclude that

$$|\mathcal{P}(t) - \mathcal{P}(s)| \leq M |t - s|^{1/2},$$

and $\mathcal{P}$ satisfies a Lipschitz condition of exponent of $1/2$.

Moreover, each $\mathcal{P}_k(t)$ visits each dyadic square of generation k as t ranges in $[0, 1]$. Hence $\mathcal{P}$ is dense in the unit square, and by continuity we find that $t \mapsto \mathcal{P}(t)$ is a surjection.

Finally, to prove the measure preserving property of $\mathcal{P}$, it suffices to establish $\mathcal{P} = \Phi^*$.

Lemma 3.8 *If Φ is the dyadic correspondence in Lemma 3.7, then $\Phi^*(t) = \mathcal{P}(t)$ for every $0 \leq t \leq 1$.*

Proof. First, we observe that $\Phi^*(t)$ is unambiguously defined for every t. Indeed, suppose $t \in \bigcap_k I^k$ and $t \in \bigcap_k J^k$ are two chains of quartic intervals; then I^k and J^k must be adjacent for sufficiently large

k. Thus $\Phi(I^k)$ and $\Phi(J^k)$ must be adjacent squares for all sufficiently large k. Hence

$$\bigcap_k \Phi(I^k) = \bigcap_k \Phi(J^k).$$

Next, directly from our construction we have

$$\bigcap_k \Phi(I^k) = \lim \mathcal{P}_k(t) = \mathcal{P}(t).$$

This gives the desired conclusion.

The argument also shows that $\mathcal{P}(I) = \Phi(I)$ for any quartic interval I. Now recall that any interval (a, b) can be written as $\bigcup_j I_j$, where the I_j are quartic intervals with disjoint interiors. Because $\mathcal{P}(I_j) = \Phi(I_j)$, these are then dyadic squares with disjoint interiors. Since $\mathcal{P}(a, b) = \bigcup_k \mathcal{P}(I_j)$, we have

$$m_2(\mathcal{P}(a, b)) = \sum_{j=1}^\infty m_2(\mathcal{P}(I_j)) = \sum_{j=1}^\infty m_2(\Phi(I_j)) = \sum_{j=1}^\infty m_1(I_j) = m_1(a, b).$$

This proves conclusion (iii) of Theorem 3.1. The other conclusions having already been established, we need only note that the corollary is contained in Theorem 3.5.

As a result, we conclude that $t \mapsto \mathcal{P}(t)$ also induces a measure preserving mapping from $[0, 1]$ to $[0, 1] \times [0, 1]$. This concludes the proof of Theorem 3.1.

4* Besicovitch sets and regularity

We begin by presenting a surprising regularity property enjoyed by all measurable subsets (of finite measure) of $\mathbb{R}^d$ when $d \geq 3$. As we shall see, the fact that the corresponding phenomenon does not hold for $d = 2$ is due to the existence of a remarkable set that was discovered by Besicovitch. A construction of a set of this kind will be detailed in Section 4.4.

We first fix some notation. For each unit vector γ on the sphere, $\gamma \in S^{d-1}$, and each $t \in \mathbb{R}$ we consider the **plane** $\mathcal{P}_{t,\gamma}$, which is defined as the $(d-1)$-dimensional affine hyperplane perpendicular to γ and of "signed distance" t from the origin.[1] The plane $\mathcal{P}_{t,\gamma}$ is given by

$$\mathcal{P}_{t,\gamma} = \{x \in \mathbb{R}^d : x \cdot \gamma = t\}.$$

---

[1]Note that there are two planes perpendicular to γ and of distance $|t|$ from the origin; this accounts for the fact that t may be either positive or negative.

We observe that each $\mathcal{P}_{t,\gamma}$ carries a natural $(d-1)$ Lebesgue measure, denoted by m_{d-1}. In fact, if we complete γ to an orthonormal basis $e_1, e_2, \ldots, e_{d-1}, \gamma$ of $\mathbb{R}^d$, then we can write any $x \in \mathbb{R}^d$ in terms of the corresponding coordinates as $x = x_1 e_1 + x_2 e_2 + \cdots + x_d \gamma$. When we set $x \in \mathbb{R}^d = \mathbb{R}^{d-1} \times \mathbb{R}$ with $(x_1, \ldots, x_{d-1}) \in \mathbb{R}^{d-1}$, $x_d \in \mathbb{R}$, then the measure m_{d-1} on $\mathcal{P}_{t,\gamma}$ is the Lebesgue measure on $\mathbb{R}^{d-1}$. This definition of m_{d-1} is independent of the choice of orthonormal vectors $e_1, e_2, \ldots, e_{d-1}$, since Lebesgue measure is invariant under rotations. (See Problem 4, Chapter 2, or Exercise 26, Chapter 3.)

With these preliminaries out of the way, we define for each subset $E \subset \mathbb{R}^d$ the **slice** of E cut out by the plane $\mathcal{P}_{t,\gamma}$ as

$$E_{t,\gamma} = E \cap \mathcal{P}_{t,\gamma}.$$

We now consider the slices $E_{t,\gamma}$ as t varies, where E is measurable and γ is fixed. (See Figure 12.)

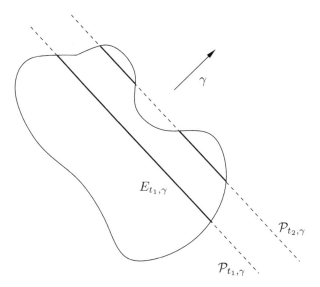

Figure 12. The slices $E \cap \mathcal{P}_{t,\gamma}$ as t varies

We observe that for almost every t the set $E_{t,\gamma}$ is m_{d-1} measurable and, moreover, $m_{d-1}(E_{t,\gamma})$ is a measurable function of t. This is a direct consequence of Fubini's theorem and the above decomposition, $\mathbb{R}^d = \mathbb{R}^{d-1} \times \mathbb{R}$. In fact, so long as the direction γ is pre-assigned, not much more can be said in general about the function $t \mapsto m_{d-1}(E_{t,\gamma})$.

However, when $d \geq 3$ the nature of the function is dramatically different for "most" γ. This is contained in the following theorem.

Theorem 4.1 *Suppose E is of finite measure in $\mathbb{R}^d$, with $d \geq 3$. Then for almost every $\gamma \in S^{d-1}$:*

(i) *$E_{t,\gamma}$ is measurable for all $t \in \mathbb{R}$.*

(ii) *$m_{d-1}(E_{t,\gamma})$ is continuous in $t \in \mathbb{R}$.*

Moreover, the function of t defined by $\mu(t, \gamma) = m_{d-1}(E_{t,\gamma})$ satisfies a Lipschitz condition with exponent α for any α with $0 < \alpha < 1/2$.

The almost everywhere assertion is with respect to the natural measure $d\sigma$ on S^{d-1} that arises in the polar coordinate formula in Section 3.2 of the previous chapter.

We recall that a function f is Lipschitz with exponent α if

$$|f(t_1) - f(t_2)| \leq A|t_1 - t_2|^\alpha \quad \text{for some } A.$$

A significant part of (i) is that for a.e. γ, the slice $E_{t,\gamma}$ is measurable for *all* values of the parameter t. In particular, one has the following.

Corollary 4.2 *Suppose E is a set of measure zero in $\mathbb{R}^d$ with $d \geq 3$. Then, for almost every $\gamma \in S^{d-1}$, the slice $E_{t,\gamma}$ has zero measure for all $t \in \mathbb{R}$.*

The fact that there is no analogue of this when $d = 2$ is a consequence of the existence of a **Besicovitch set**, (also called a "Kakeya set"), which is defined as a set that satisfies the three conditions in the theorem below.

Theorem 4.3 *There exists a set $\mathcal{B}$ in $\mathbb{R}^2$ that:*

(i) *is compact,*

(ii) *has Lebesgue measure zero,*

(iii) *contains a translate of every unit line segment.*

Note that with $F = \mathcal{B}$ and $\gamma \in S^1$ one has $m_1(F \cap \mathcal{P}_{t_0,\gamma}) \geq 1$ for some t_0. If $m_1(F \cap \mathcal{P}_{t,\gamma})$ were continuous in t, then this measure would be strictly positive for an interval in t containing t_0, and thus we would have $m_2(F) > 0$, by Fubini's theorem. This contradiction shows that the analogue of Theorem 4.1 cannot hold for $d = 2$.

While the set $\mathcal{B}$ has zero two-dimensional measure, this assertion cannot be improved by replacing this measure by α-dimensional Hausdorff measure, with $\alpha < 2$.

Theorem 4.4 *Suppose F is any set that satisfies the conclusions (i) and (iii) of Theorem 4.3. Then F has Hausdorff dimension 2.*

4.1 The Radon transform

Theorems 4.1 and 4.4 will be derived by an analysis of the regularity properties of the Radon transform $\mathcal{R}$. The operator $\mathcal{R}$ arises in a number of problems in analysis, and was already considered in Chapter 6 of Book I.

For an appropriate function f on $\mathbb{R}^d$, the **Radon transform** of f is defined by

$$\mathcal{R}(f)(t, \gamma) = \int_{\mathcal{P}_{t,\gamma}} f.$$

The integration is performed over the plane $\mathcal{P}_{t,\gamma}$ with respect to the measure m_{d-1} discussed above. We first make the following simple observation:

1. If f is continuous and has compact support, then f is of course integrable on *every* plane $\mathcal{P}_{t,\gamma}$, and so $\mathcal{R}(f)(t, \gamma)$ is defined for all $(t, \gamma) \in \mathbb{R} \times S^{d-1}$. Moreover it is a continuous function of the pair (t, γ) and has compact support in the t-variable.

2. If f is merely Lebesgue integrable, then f may fail to be measurable or integrable on $\mathcal{P}_{t,\gamma}$ for some (t, γ), and thus $\mathcal{R}(f)(t, \gamma)$ is not defined for those (t, γ).

3. Suppose f is the characteristic function of the set E, that is, $f = \chi_E$. Then $\mathcal{R}(f)(t, \gamma) = m_{d-1}(E_{t,\gamma})$ if $E_{t,\gamma}$ is measurable.

It is this last property that links the Radon transform to our problem. Key estimates in this conclusion involve a maximal "Radon transform" defined by

$$\mathcal{R}^*(f)(\gamma) = \sup_{t \in \mathbb{R}} |\mathcal{R}(f)(t, \gamma)|,$$

as well as corresponding expressions controlling the Lipschitz character of $\mathcal{R}(f)(t, \gamma)$ as a function of t. A basic fact inherent in our analysis is that the regularity of the Radon transform actually improves as the dimension of the underlying space increases.

Theorem 4.5 *Suppose f is continuous and has compact support in $\mathbb{R}^d$ with $d \geq 3$. Then*

$$(2) \qquad \int_{S^{d-1}} \mathcal{R}^*(f)(\gamma)\, d\sigma(\gamma) \leq c \left[\|f\|_{L^1(\mathbb{R}^d)} + \|f\|_{L^2(\mathbb{R}^d)} \right]$$

for some constant $c > 0$ that does not depend on f.

An inequality of this type is a typical "a priori" estimate. It is obtained first under some regularity assumption on the function f, and then a limiting argument allows one to pass to the more general case when f belongs to $L^1 \cap L^2$.

We make some comments about the appearance of both the L^1-norm and L^2-norm in (2). The L^2-norm imposes a crucial local control of the kind that is necessary for the desired regularity. (See Exercise 27.) However, without some restriction on f of a global nature, the function f might fail to be integrable on *any* plane $\mathcal{P}_{t,\gamma}$, as the example $f(x) = 1/(1+|x|^{d-1})$ shows. Note that this function belongs to $L^2(\mathbb{R}^d)$ if $d \geq 3$, but not to $L^1(\mathbb{R}^d)$.

The proof of Theorem 4.5 actually gives an essentially stronger result, which we state as a corollary.

Corollary 4.6 *Suppose f is continuous and has compact support in $\mathbb{R}^d$, $d \geq 3$. Then for any α, $0 < \alpha < 1/2$, the inequality (2) holds with $\mathcal{R}^*(f)(\gamma)$ replaced by*

$$(3) \qquad \sup_{t_1 \neq t_2} \frac{|\mathcal{R}(f)(t_1, \gamma) - \mathcal{R}(f)(t_2, \gamma)|}{|t_1 - t_2|^\alpha}.$$

The proof of the theorem relies on the interplay between the Radon transform and the Fourier transform.

For fixed $\gamma \in S^{d-1}$, we let $\hat{\mathcal{R}}(f)(\lambda, \gamma)$ denote the Fourier transform of $\mathcal{R}(f)(t, \gamma)$ in the t-variable

$$\hat{\mathcal{R}}(f)(\lambda, \gamma) = \int_{-\infty}^{\infty} \mathcal{R}(f)(t, \gamma) e^{-2\pi i \lambda t} \, dt.$$

In particular, we use $\lambda \in \mathbb{R}$ to denote the dual variable of t.

We also write $\hat{f}$ for the Fourier transform of f as a function on $\mathbb{R}^d$, namely

$$\hat{f}(\xi) = \int_{\mathbb{R}^d} f(x) e^{-2\pi i x \cdot \xi} \, dx.$$

Lemma 4.7 *If f is continuous with compact support, then for every $\gamma \in S^{d-1}$ we have*

$$\hat{\mathcal{R}}(f)(\lambda, \gamma) = \hat{f}(\lambda \gamma).$$

The right-hand side is just the Fourier transform of f evaluated at the point $\lambda \gamma$.

Proof. For each unit vector γ we use the adapted coordinate system described above: $x = (x_1, \ldots, x_d)$ where γ coincides with the x_d direction. We can then write each $x \in \mathbb{R}^d$ as $x = (u, t)$ with $u \in \mathbb{R}^{d-1}$, $t \in \mathbb{R}$, where $x \cdot \gamma = t = x_d$ and $u = (x_1, \ldots, x_{d-1})$. Moreover

$$\int_{\mathcal{P}_{t,\gamma}} f = \int_{\mathbb{R}^{d-1}} f(u, t) \, du,$$

and Fubini's theorem shows that $\int_{\mathbb{R}^d} f(x) \, dx = \int_{-\infty}^{\infty} \left(\int_{\mathcal{P}_{t,\gamma}} f \right) dt$. Applying this to $f(x) e^{-2\pi i x \cdot (\lambda \gamma)}$ in place of $f(x)$ gives

$$\hat{f}(\lambda \gamma) = \int_{\mathbb{R}^d} f(x) e^{-2\pi i x \cdot (\lambda \gamma)} \, dx = \int_{-\infty}^{\infty} \left(\int_{\mathbb{R}^{d-1}} f(u, t) \, du \right) e^{-2\pi i \lambda t} \, dt$$

$$= \int_{-\infty}^{\infty} \left(\int_{\mathcal{P}_{t,\gamma}} f \right) e^{-2\pi i \lambda t} \, dt.$$

Therefore $\hat{f}(\lambda \gamma) = \hat{\mathcal{R}}(f)(\lambda, \gamma)$, and the lemma is proved.

Lemma 4.8 *If f is continuous with compact support, then*

$$\int_{S^{d-1}} \left(\int_{-\infty}^{\infty} |\hat{\mathcal{R}}(f)(\lambda, \gamma)|^2 |\lambda|^{d-1} d\lambda \right) d\sigma(\gamma) = 2 \int_{\mathbb{R}^d} |f(x)|^2 dx.$$

Let us observe the crucial point that the greater the dimension d, the larger the factor $|\lambda|^{d-1}$ as $|\lambda|$ tends to infinity. Hence the greater the dimension, the better the decay of the Fourier transform $\hat{\mathcal{R}}(f)(\lambda, \gamma)$, and so the better the regularity of the Radon transform $\mathcal{R}(f)(t, \gamma)$ as a function of t.

Proof. The Plancherel formula in Chapter 5 guarantees that

$$2 \int_{\mathbb{R}^d} |f(x)|^2 \, dx = 2 \int_{\mathbb{R}^d} |\hat{f}(\xi)|^2 \, d\xi.$$

Changing to polar coordinates $\xi = \lambda \gamma$ where $\lambda > 0$ and $\gamma \in S^{d-1}$, we obtain

$$2 \int_{\mathbb{R}^d} |\hat{f}(\xi)|^2 \, d\xi = 2 \int_{S^{d-1}} \int_0^{\infty} |\hat{f}(\lambda \gamma)|^2 \lambda^{d-1} \, d\lambda \, d\sigma(\gamma).$$

We now observe that a simple change of variables provides

$$\int_{S^{d-1}} \int_0^{\infty} |\hat{f}(\lambda \gamma)|^2 \lambda^{d-1} \, d\lambda \, d\sigma(\gamma) = \int_{S^{d-1}} \int_{-\infty}^0 |\hat{f}(\lambda \gamma)|^2 |\lambda|^{d-1} \, d\lambda \, d\sigma(\gamma),$$

and the proof is complete once we invoke the result of Lemma 4.7.

The final ingredient in the proof of Theorem 4.5 consists of the following:

Lemma 4.9 *Suppose*

$$F(t) = \int_{-\infty}^{\infty} \hat{F}(\lambda) e^{2\pi i \lambda t} \, d\lambda,$$

where

$$\sup_{\lambda \in \mathbb{R}} |\hat{F}(\lambda)| \leq A \quad and \quad \int_{-\infty}^{\infty} |\hat{F}(\lambda)|^2 |\lambda|^{d-1} d\lambda \leq B^2.$$

Then

(4) $$\sup_{t \in \mathbb{R}} |F(t)| \leq c(A + B).$$

Moreover, if $0 < \alpha < 1/2$, then

(5) $$|F(t_1) - F(t_2)| \leq c_\alpha |t_1 - t_2|^\alpha (A + B) \quad for \ all \ t_1, \ t_2.$$

Proof. The first inequality is obtained by considering separately the two cases $|\lambda| \leq 1$ and $|\lambda| > 1$. We write

$$F(t) = \int_{|\lambda| \leq 1} \hat{F}(\lambda) e^{2\pi i \lambda t} \, d\lambda + \int_{|\lambda| > 1} \hat{F}(\lambda) e^{2\pi i \lambda t} \, d\lambda.$$

Clearly, the first integral is bounded by cA. To estimate the second integral it suffices to bound $\int_{|\lambda| > 1} |\hat{F}(\lambda)| \, d\lambda$. An application of the Cauchy-Schwarz inequality gives

$$\int_{|\lambda| > 1} |\hat{F}(\lambda)| d\lambda \leq \left(\int_{|\lambda| > 1} |\hat{F}(\lambda)|^2 |\lambda|^{d-1} d\lambda \right)^{1/2} \left(\int_{|\lambda| > 1} |\lambda|^{-d+1} d\lambda \right)^{1/2}.$$

This last integral is convergent precisely when $-d + 1 < -1$, which is equivalent to $d > 2$, namely $d \geq 3$, which we assume. Hence $|F(t)| \leq c(A + B)$ as desired.

To establish Lipschitz continuity, we first note that

$$F(t_1) - F(t_2) = \int_{-\infty}^{\infty} \hat{F}(\lambda) \left[e^{2\pi i \lambda t_1} - e^{2\pi i \lambda t_2} \right] d\lambda.$$

Since one has the inequality[2] $|e^{ix} - 1| \le |x|$, we immediately see that

$$|e^{2\pi i \lambda t_1} - e^{2\pi i \lambda t_2}| \le c|t_1 - t_2|^\alpha \lambda^\alpha \quad \text{if } 0 \le \alpha < 1.$$

We may then write the difference $F(t_1) - F(t_2)$ as a sum of two integrals. The integral over $|\lambda| \le 1$ is clearly bounded by $cA|t_1 - t_2|^\alpha$. The second integral, the one over $|\lambda| > 1$, can be estimated from above by

$$|t_1 - t_2|^\alpha \int_{|\lambda|>1} |\hat{F}(\lambda)||\lambda|^\alpha \, d\lambda.$$

An application of the Cauchy-Schwarz inequality show that this last integral is majorized by

$$\left(\int_{|\lambda|>1} |\hat{F}(\lambda)|^2 |\lambda|^{d-1} \, d\lambda \right)^{1/2} \left(\int_{|\lambda|>1} |\lambda|^{-d+1+2\alpha} \, d\lambda \right)^{1/2} \le c_\alpha B,$$

since the second integral is finite if $-d + 1 + 2\alpha < -1$, and in particular this holds if $\alpha < 1/2$ when $d \ge 3$. This concludes the proof of the lemma.

We now gather these results to prove the theorem. For each $\gamma \in S^{d-1}$ let

$$F(t) = \mathcal{R}(f)(t, \gamma).$$

Note that with this definition we have

$$\sup_{t \in \mathbb{R}} |F(t)| = \mathcal{R}^*(f)(\gamma).$$

Let

$$A(\gamma) = \sup_\lambda |\hat{F}(\lambda)| \quad \text{and} \quad B^2(\gamma) = \int_{-\infty}^{\infty} |\hat{F}(\lambda)|^2 |\lambda|^{d-1} \, d\lambda.$$

Then by (4)

$$\sup_{t \in \mathbb{R}} |F(t)| \le c(A(\gamma) + B(\gamma)).$$

However, we observed that $\hat{F}(\lambda) = \hat{f}(\lambda\gamma)$, and hence

$$A(\gamma) \le \|f\|_{L^1(\mathbb{R}^d)}.$$

---

[2] The distance in the plane from the point e^{ix} to the point 1 is shorter than the length of the arc on the unit circle joining them.

Therefore,

$$|\mathcal{R}^*(f)(\gamma)|^2 \leq c(A(\gamma)^2 + B(\gamma)^2),$$

and thus

$$\int_{S^{d-1}} |\mathcal{R}^*(f)(\gamma)|^2 \, d\sigma(\gamma) \leq c(\|f\|^2_{L^1(\mathbb{R}^d)} + \|f\|^2_{L^2(\mathbb{R}^d)}),$$

since $\int B^2(\gamma) \, d\sigma(\gamma) = 2\|f\|^2_{L^2}$ by Lemma 4.8. Consequently,

$$\int_{S^{d-1}} \mathcal{R}^*(f)(\gamma) \, d\sigma(\gamma) \leq c(\|f\|_{L^1(\mathbb{R}^d)} + \|f\|_{L^2(\mathbb{R}^d)}).$$

Note that the identity we have used,

$$\mathcal{R}(f)(t, \gamma) = \int_{-\infty}^{\infty} \hat{F}(\lambda) e^{2\pi i \lambda t} \, d\lambda,$$

with $F(t) = \mathcal{R}(f)(t, \gamma)$, is justified for almost every $\gamma \in S^{d-1}$ by the Fourier inversion result in Theorem 4.2 of Chapter 2. Indeed, we have seen that $A(\gamma)$ and $B(\gamma)$ are finite for almost every γ, and thus $\hat{F}$ is integrable for those γ. This completes the proof of the theorem. The corollary follows the same way if we use (5) instead of (4).

We now return to the situation in the plane to see what information we may deduce from the above analysis. The inequality (2) as it stands does not hold when $d = 2$. However, a modification of it does hold, and this will be used in the proof of Theorem 4.4.

If $f \in L^1(\mathbb{R}^d)$ we define

$$\mathcal{R}_\delta(f)(t, \gamma) = \frac{1}{2\delta} \int_{t-\delta}^{t+\delta} \mathcal{R}(f)(s, \gamma) \, ds$$

$$= \frac{1}{2\delta} \int_{t-\delta \leq x \cdot \gamma \leq t+\delta} f(x) \, dx.$$

In this definition of $\mathcal{R}_\delta(f)(t, \gamma)$ we integrate the function f in a small "strip" of width 2δ around the plane $\mathcal{P}_{t,\gamma}$. Thus $\mathcal{R}_\delta$ is an average of Radon transforms.

We let

$$\mathcal{R}_\delta^*(f)(\gamma) = \sup_{t \in \mathbb{R}} |\mathcal{R}_\delta(f)(t, \gamma)|.$$

Theorem 4.10 *If f is continuous with compact support, then*

$$\int_{S^1} \mathcal{R}_\delta^*(f)(\gamma)\, d\sigma(\gamma) \leq c(\log 1/\delta)^{1/2} \left(\|f\|_{L^1(\mathbb{R}^2)} + \|f\|_{L^2(\mathbb{R}^2)} \right)$$

when $0 < \delta \leq 1/2$.

The same argument as in the proof of Theorem 4.5 applies here, except that we need a modified version of Lemma 4.9. More precisely, let us set

$$F_\delta(t) = \int_{-\infty}^{\infty} \hat{F}(\lambda) \left(\frac{e^{2\pi i(t+\delta)\lambda} - e^{2\pi i(t-\delta)\lambda}}{2\pi i\lambda(2\delta)} \right) d\lambda,$$

and suppose that

$$\sup_\lambda |\hat{F}(\lambda)| \leq A \quad \text{and} \quad \int_{-\infty}^{\infty} |\hat{F}(\lambda)|^2 |\lambda|\, d\lambda \leq B.$$

Then we claim that

(6) $$\sup_t |F_\delta(t)| \leq c(\log 1/\delta)^{1/2}(A + B).$$

Indeed, we use the fact that $|(\sin x)/x| \leq 1$ to see that, in the definition of $F_\delta(t)$, the integral over $|\lambda| \leq 1$ gives the cA. Also, the integral over $|\lambda| > 1$ can be split and is bounded by the sum

$$\int_{1<|\lambda|\leq 1/\delta} |\hat{F}(\lambda)|\, d\lambda + \frac{c}{\delta} \int_{1/\delta \leq |\lambda|} |\hat{F}(\lambda)||\lambda|^{-1}\, d\lambda.$$

The first integral above can be estimated by the Cauchy-Schwarz inequality, as follows

$$\int_{1<|\lambda|\leq 1/\delta} |\hat{F}(\lambda)|\, d\lambda \leq c \left(\int_{1<|\lambda|\leq 1/\delta} |\hat{F}(\lambda)|^2 |\lambda|\, d\lambda \right)^{1/2} \left(\int_{1<|\lambda|\leq 1/\delta} |\lambda|^{-1}\, d\lambda \right)^{1/2}$$
$$\leq cB(\log 1/\delta)^{1/2}.$$

Finally, we also note that

$$\frac{c}{\delta} \int_{1/\delta \leq |\lambda|} |\hat{F}(\lambda)||\lambda|^{-1}\, d\lambda \leq c \left(\int_{1/\delta \leq |\lambda|} |\hat{F}(\lambda)|^2 |\lambda|\, d\lambda \right)^{1/2} \frac{1}{\delta} \left(\int_{1/\delta \leq |\lambda|} |\lambda|^{-3}\, d\lambda \right)^{1/2}$$
$$\leq cB$$

and this establishes (6), and hence the theorem.

4.2 Regularity of sets when $d \geq 3$

We now extend to the general context the basic estimates for the Radon transform, proved for continuous functions of compact support. This will yield the regularity result formulated in Theorem 4.1.

Proposition 4.11 *Suppose $d \geq 3$, and let f belong to $L^1(\mathbb{R}^d) \cap L^2(\mathbb{R}^d)$. Then for a.e. $\gamma \in S^{d-1}$ we can assert the following:*

(a) *f is measurable and integrable on the plane $\mathcal{P}_{t,\gamma}$, for every $t \in \mathbb{R}$.*

(b) *The function $\mathcal{R}(f)(t, \gamma)$ is continuous in t and satisfies a Lipschitz condition with exponent α for each $\alpha < 1/2$. Moreover, the inequality (2) of Theorem 4.5 and its variant with (3) hold for f.*

We prove this in a series of steps.

Step 1. We consider $f = \chi_\mathcal{O}$, the characteristic function of a bounded open set $\mathcal{O}$. Here the assertion (a) is evident since $\mathcal{O} \cap \mathcal{P}_{t,\gamma}$ is an open and bounded set in $\mathcal{P}_{t,\gamma}$ and is bounded. Thus $\mathcal{R}(f)(t, \gamma)$ is defined for all (t, γ).

Next we can find a sequence $\{f_n\}$ of non-negative continuous functions of compact support so that for every x, $f_n(x)$ increases to $f(x)$ as $n \to \infty$. Thus $\mathcal{R}(f_n)(t, \gamma) \to \mathcal{R}(f)(t, \gamma)$ for every (t, γ) by the monotone convergence theorem, and also $\mathcal{R}^*(f_n)(\gamma) \to \mathcal{R}^*(f)(\gamma)$ for each $\gamma \in S^{d-1}$. As a result we see that the inequality (2) is valid for $f = \chi_\mathcal{O}$, with $\mathcal{O}$ open and bounded.

Step 2. We now consider $f = \chi_E$, where E is a set of measure zero, and take first the case when the set E is bounded. Then we can find a decreasing sequence $\{\mathcal{O}_n\}$ of open and bounded sets, such that $E \subset \mathcal{O}_n$, while $m(\mathcal{O}_n) \to 0$ as $n \to \infty$.

Let $\tilde{E} = \bigcap \mathcal{O}_n$. Since $\tilde{E} \cap \mathcal{P}_{t,\gamma}$ is measurable for every (t, γ), the functions $\mathcal{R}(\chi_{\tilde{E}})(t, \gamma)$ and $\mathcal{R}^*(\chi_{\tilde{E}})(\gamma)$ are well-defined. However, $\mathcal{R}^*(\chi_{\tilde{E}})(\gamma) \leq \mathcal{R}^*(\chi_{\mathcal{O}_n})(\gamma)$, while the $\mathcal{R}^*(\chi_{\mathcal{O}_n})$ decrease. Thus the inequality (2) we have just proved for $f = \chi_{\mathcal{O}_n}$ shows that $\mathcal{R}^*(\chi_{\tilde{E}})(\gamma) = 0$ for a.e. γ. The fact that $E \subset \tilde{E}$ then implies that for a.e. γ, the set $E \cap \mathcal{P}_{t,\gamma}$ has $(d-1)$-dimensional measure zero for every $t \in \mathbb{R}$. This conclusion immediately extends to the case when E is not necessarily bounded, by writing E as a countable union of bounded sets of measure zero. Therefore Corollary 4.2 is proved.

Step 3. Here we assume that f is a bounded measurable function supported on a bounded set. Then by familiar arguments we can find a sequence $\{f_n\}$ of continuous functions that are uniformly bounded, supported in a fixed compact set, and so that $f_n(x) \to f(x)$ a.e. By the bounded convergence theorem, $\|f_n - f\|_{L^1}$ and $\|f_n - f\|_{L^2}$ both tend to zero as $n \to \infty$, and upon selecting a subsequence if necessary, we can suppose that $\|f_n - f\|_{L^1} + \|f_n - f\|_{L^2} \leq 2^{-n}$. By what we have just proved in Step 2 we have, for a.e. $\gamma \in S^{d-1}$, that $f_n(x) \to f(x)$ on $\mathcal{P}_{t,\gamma}$ a.e. with respect to the measure m_{d-1}, for each $t \in \mathbb{R}$. Thus again by the bounded convergence theorem for those (t, γ), we see that $\mathcal{R}(f_n)(t, \gamma) \to \mathcal{R}(f)(t, \gamma)$, and this limit defines $\mathcal{R}(f)$. Now applying Theorem 4.5 to $f_n - f_{n-1}$ gives

$$\sum_{n=1}^{\infty} \int_{S^{d-1}} \mathcal{R}^*(f_n - f_{n-1})(\gamma) \, d\sigma(\gamma) \leq c \sum_{n=1}^{\infty} 2^{-n} < \infty.$$

This means that

$$\sum_n \sup_t |\mathcal{R}(f_n)(t, \gamma) - \mathcal{R}(f_{n-1})(t, \gamma)| < \infty,$$

for a.e. $\gamma \in S^{d-1}$, and hence for those γ the sequence of functions $\mathcal{R}(f_n)(t, \gamma)$ converges uniformly. As a consequence, for those γ the function $\mathcal{R}(f)(t, \gamma)$ is continuous in t, and the inequality (2) is valid for this f. The inequality with (3) is deduced in the same way.

Finally, we deal with the general f in $L^1 \cap L^2$ by approximating it by a sequence of bounded functions each with bounded support. The details of the argument are similar to the case treated above and are left to the reader.

Observe that the special case $f = \chi_E$ of the proposition gives us Theorem 4.1.

4.3 Besicovitch sets have dimension 2

Here we prove Theorem 4.4, that any Besicovitch set necessarily has Hausdorff dimension 2. We use Theorem 4.10, namely, the inequality

$$\int_{S^1} \mathcal{R}^*_\delta(f)(\gamma) \, d\sigma(\gamma) \leq c(\log 1/\delta)^{1/2} \left(\|f\|_{L^1(\mathbb{R}^2)} + \|f\|_{L^2(\mathbb{R}^2)} \right).$$

This inequality was proved under the assumption that f was continuous and had compact support. In the present situation it goes over without difficulty to the general case where $f \in L^1 \cap L^2$, by an easy limiting argument, since it is clear that $\mathcal{R}^*_\delta(f_n)(\gamma)$ converges to $\mathcal{R}^*_\delta(f)(\gamma)$ for all γ if $f_n \to f$ in the L^1-norm.

Now suppose F is a Besicovitch set and α is fixed with $0 < \alpha < 2$. Assume that $F \subset \bigcup_{i=1}^{\infty} B_i$ is a covering, where B_i are balls with diameter less than a given number. We must show that

$$\sum_i (\operatorname{diam} B_i)^\alpha \geq c_\alpha > 0.$$

We proceed in two steps, considering first a simple situation that will make clear the idea of the proof.

Case 1. We suppose first that all the balls B_i have the same diameter δ (with $\delta \leq 1/2$) and also that there are only a finite number, say N, of balls in the covering. We must prove that $N\delta^\alpha \geq c_\alpha$.

Let B_i^* denote the double of B_i and $F^* = \bigcup_i B_i^*$. Then, we clearly have

$$m(F^*) \leq cN\delta^2.$$

Since F is a Besicovitch set, for each $\gamma \in S^1$ there is a segment s_γ of unit length, perpendicular to γ, and which is contained in F. Also, by construction, any translate by less than δ of a point in s_γ must belong to F^*. Hence

$$\mathcal{R}_\delta^*(\chi_{F^*})(\gamma) \geq 1 \quad \text{for every } \gamma.$$

If we take $f = \chi_{F^*}$ in the inequality (6), and note that the Cauchy-Schwarz inequality implies

$$\|\chi_{F^*}\|_{L^1(\mathbb{R}^2)} \leq c\|\chi_{F^*}\|_{L^2(\mathbb{R}^2)} \leq c(m(F^*))^{1/2},$$

then we obtain

$$c \leq N^{1/2}\delta(\log 1/\delta)^{1/2}.$$

This implies $N\delta^\alpha \geq c$ for $\alpha < 2$.

Case 2. We now treat the general case. Suppose $F \subset \bigcup_{i=1}^{\infty} B_i$, where the balls B_i each have diameter less than 1. For each integer k let N_k be the number of balls in the collection $\{B_i\}$ for which

$$2^{-k-1} \leq \operatorname{diam} B_i \leq 2^{-k}.$$

We need to show that $\sum_{k=0}^{\infty} N_k 2^{-k\alpha} \geq c_\alpha$. In fact, we shall prove the stronger result that there exists a positive integer k' such that $N_{k'} 2^{-k'\alpha} \geq c_\alpha$.

Let

$$F_k = F \cap \left(\bigcup_{2^{-k-1} \leq \text{diam } B_i \leq 2^{-k}} B_i \right),$$

and let

$$F_k^* = \bigcup_{2^{-k-1} \leq \text{diam } B_i \leq 2^{-k}} B_i^*,$$

where B_i^* denotes the double of B_i. Then we note that

$$m_1(F_k^*) \leq cN_k 2^{-2k} \quad \text{for all } k.$$

Since F is a Besicovitch set, for each $\gamma \in S^1$ there is a segment s_γ of unit length entirely contained in F. We now make precise the fact that for some k, a large proportion of s_γ belongs to F_k.

We pick a sequence of real numbers $\{a_k\}_{k=0}^\infty$ such that $0 \leq a_k \leq 1$, $\sum a_k = 1$, but a_k does not tend to zero too quickly. For instance, we may choose $a_k = c_\epsilon 2^{-\epsilon k}$ with $c_\epsilon = 1 - 2^{-\epsilon}$, and $\epsilon > 0$ but ϵ sufficiently small.

Then, for some k we must have

$$m_1(s_\gamma \cap F_k) \geq a_k.$$

Otherwise, since $F = \bigcup F_k$, we would have

$$m_1(s_\gamma \cap F) < \sum a_k = 1,$$

and this contradicts the fact that $m_1(s_\gamma \cap F) = 1$, since s_γ is entirely contained in F.

Therefore, with this k, we must have

$$\mathcal{R}_{2^{-k}}^*(\chi_{F_k^*})(\gamma) \geq a_k,$$

because any point of distance less than 2^{-k} from F_k must belong to F_k^*. Since the choice of k may depend on γ, we let

$$E_k = \{\gamma : \mathcal{R}_{2^{-k}}^*(\chi_{F_k^*})(\gamma) \geq a_k\}.$$

By our previous observations, we have

$$S^1 = \bigcup_{k=1}^\infty E_k,$$

and so for at least one k, which we denote by k', we have

$$m(E_{k'}) \geq 2\pi a_{k'},$$

for otherwise $m(S_1) < 2\pi \sum a_k = 2\pi$. As a result

$$2\pi a_{k'}^2 = 2\pi a_{k'} a_{k'}$$

$$\leq \int_{E_{k'}} a_{k'} \, d\sigma(\gamma)$$

$$\leq \int_{S_1} \mathcal{R}_{2^{-k'}}^* (\chi_{F_{k'}^*})(\gamma) \, d\sigma(\gamma).$$

By the fundamental inequality (6) we get

$$a_{k'}^2 \leq c(\log 2^{k'})^{1/2} \|\chi_{F_{k'}^*}\|_{L^2(\mathbb{R}^2)}.$$

Recalling that by our choice $a_k \approx 2^{-\epsilon k}$, and noting that $\|\chi_{F_{k'}^*}\|_{L^2} \leq c N_{k'}^{1/2} 2^{-k'}$, we obtain

$$2^{(1-2\epsilon)k'} \leq c(\log 2^{k'})^{1/2} N_{k'}^{1/2}.$$

Finally, this last inequality guarantees that $N_{k'} 2^{-\alpha k'} \geq c_\alpha$ as long as $4\epsilon < 2 - \alpha$.

This concludes the proof of the theorem.

4.4 Construction of a Besicovitch set

There are a number of different constructions of Besicovitch sets. The one we have chosen to describe here involves the concept of self-replicating sets, an idea that permeates much of the discussion of this chapter.

We consider the Cantor set of constant dissection $\mathcal{C}_{1/2}$, which for simplicity we shall write as $\mathcal{C}$, and which is defined in Exercise 3, Chapter 1. Note that $\mathcal{C} = \bigcap_{k=0}^{\infty} C_k$, where $C_0 = [0,1]$, and C_k is the union of 2^k closed intervals of length 4^{-k} obtained by removing from C_{k-1} the 2^{k-1} centrally situated open intervals of length $\frac{1}{2} \cdot 4^{-k+1}$. The set $\mathcal{C}$ can also be represented as the set of points $x \in [0,1]$ of the form $x = \sum_{k=1}^{\infty} \epsilon_k/4^k$, with ϵ_k either 0 or 3.

We now place a copy of $\mathcal{C}$ on the x-axis of the plane $\mathbb{R}^2 = \{(x,y)\}$, and a copy of $\frac{1}{2}\mathcal{C}$ on the line $y = 1$. That is, we put $E_0 = \{(x,y) : x \in \mathcal{C}, \ y = 0\}$ and $E_1 = \{(x,y) : 2x \in \mathcal{C}, \ y = 1\}$. The set F that will play the central role is defined as the union of all line segments that join a point of E_0 with a point of E_1. (See Figure 13.)

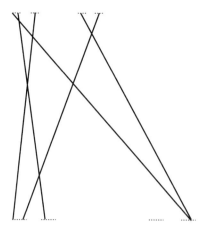

Figure 13. Several line segments joining E_0 with E_1

Theorem 4.12 *The set F is compact and of two-dimensional measure zero. It contains a translate of any unit line segment whose slope is a number s that lies outside the intervals $(-1, 2)$.*

Once the theorem is proved, our job is done. Indeed, a finite union of rotations of the set F contains unit segments of any slope, and that set is therefore a Besicovitch set.

The proof of the required properties of the set F amounts to showing the following paradoxical facts about the set $\mathcal{C} + \lambda\mathcal{C}$, for $\lambda > 0$. Here $\mathcal{C} + \lambda\mathcal{C} = \{x_1 + \lambda x_2 : x_1 \in \mathcal{C}, x_2 \in \mathcal{C}\}$:

- $\mathcal{C} + \lambda\mathcal{C}$ has one-dimensional measure zero, for a.e. λ.

- $\mathcal{C} + \frac{1}{2}\mathcal{C}$ is the interval $[0, 3/2]$.

Let us see how these two assertions imply the theorem. First, we note that the set F is closed (and hence compact), because both E_0 and E_1 are closed. Next observe that with $0 < y < 1$, the slice F^y of the set F is exactly $(1 - y)\mathcal{C} + \frac{y}{2}\mathcal{C}$. This set is obtained from the set $\mathcal{C} + \lambda\mathcal{C}$, where $\lambda = y/(2(1 - y))$, by scaling with the factor $1 - y$. Hence F^y is of measure zero whenever $\mathcal{C} + \lambda\mathcal{C}$ is also of measure zero. Moreover, under the mapping $y \mapsto \lambda$, sets of measure zero in $(0, \infty)$ correspond to sets of measure zero in $(0, 1)$. (For this see, for example, Exercise 8 in Chapter 1, or Problem 1 in Chapter 6.) Therefore, the first assertion and Fubini's theorem prove that the (two-dimensional) measure of F is zero.

Finally the slope s of the segment joining the point $(x_0, 0)$, with the point $(x_1, 1)$ is $s = 1/(x_1 - x_0)$. Thus the quantity s can be realized if

$x_1 \in \mathcal{C}/2$ and $x_0 \in \mathcal{C}$, that is, if $1/s \in \mathcal{C}/2 - \mathcal{C}$. However, by an obvious symmetry $\mathcal{C} = 1 - \mathcal{C}$, and so the condition becomes $1/s \in \mathcal{C}/2 + \mathcal{C} - 1$, which by the second assertion is $1/s \in [-1, 1/2]$. This last is equivalent with $s \notin (-1, 2)$.

Our task therefore remains the proof of the two assertions above. The proof of the second is nearly trivial. In fact,

$$\frac{2}{3}\left(\mathcal{C} + \frac{1}{2}\mathcal{C}\right) = \frac{2}{3}\mathcal{C} + \frac{1}{3}\mathcal{C},$$

and this set consists of all x of the form $x = \sum_{k=1}^{\infty}\left(\frac{2\epsilon_k}{3} + \frac{\epsilon_k'}{3}\right)4^{-k}$, where ϵ_k and ϵ_k' are independently 0 or 3. Since then $\frac{2\epsilon_k}{3} + \frac{\epsilon_k'}{3}$ can take any of the values 0, 1, 2, or 3, we have that $\frac{2}{3}\left(\mathcal{C} + \frac{1}{2}\mathcal{C}\right) = [0, 1]$, and hence $\mathcal{C} + \frac{1}{2}\mathcal{C} = [0, 3/2]$.

The proof that $m(\mathcal{C} + \lambda\mathcal{C}) = 0$ for a.e. λ

We come to the main point: that $\mathcal{C} + \lambda\mathcal{C}$ has measure zero for almost all λ. We show this by examining the self-replicating properties of the sets $\mathcal{C}$ and $\mathcal{C} + \lambda\mathcal{C}$.

We know that $\mathcal{C} = \mathcal{C}_1 \cup \mathcal{C}_2$, where $\mathcal{C}_1$ and $\mathcal{C}_2$ are two similar copies of $\mathcal{C}$, obtained with similarity ratio $1/4$, and given by $\mathcal{C}_1 = \frac{1}{4}\mathcal{C}$ and $\mathcal{C}_2 = \frac{1}{4}\mathcal{C} + \frac{3}{4}$. Thus $\mathcal{C}_1 \subset [0, 1/4]$ and $\mathcal{C}_2 \subset [3/4, 1]$. Iterating ℓ times this decomposition of $\mathcal{C}$, that is, reaching the ℓth "generation," we can write

$$(7) \qquad \mathcal{C} = \bigcup_{1 \le j \le 2^\ell} \mathcal{C}_j^\ell,$$

with $\mathcal{C}_1^\ell = (1/4)^\ell \mathcal{C}$ and each $\mathcal{C}_j^\ell$ a translate of $\mathcal{C}_1^\ell$.

We consider in the same way the set

$$\mathcal{K}(\lambda) = \mathcal{C} + \lambda\mathcal{C},$$

and we shall sometimes omit the λ and write $\mathcal{K}(\lambda) = \mathcal{K}$, when this causes no confusion. By its definition we have

$$\mathcal{K} = \mathcal{K}_1 \cup \mathcal{K}_2 \cup \mathcal{K}_3 \cup \mathcal{K}_4,$$

where $\mathcal{K}_1 = \mathcal{C}_1 + \lambda\mathcal{C}_1$, $\mathcal{K}_2 = \mathcal{C}_1 + \lambda\mathcal{C}_2$, $\mathcal{K}_3 = \mathcal{C}_2 + \lambda\mathcal{C}_1$, and $\mathcal{K}_4 = \mathcal{C}_2 + \lambda\mathcal{C}_2$. An iteration of this decomposition using (7) gives

$$(8) \qquad \mathcal{K} = \bigcup_{1 \le i \le 4^\ell} \mathcal{K}_i^\ell,$$

where each $\mathcal{K}_i^\ell$ equals $\mathcal{C}_{j_1}^\ell + \lambda \mathcal{C}_{j_2}^\ell$ for a pair of indices j_1, j_2. In fact, this relation among the indices sets up a bijection between the i with $1 \le i \le 4^\ell$, and the pair j_1, j_2 with $1 \le j_1 \le 2^\ell$ and $1 \le j_2 \le 2^\ell$. Note that each $\mathcal{K}_i^\ell$ is a translate of $\mathcal{K}_1^\ell$, and each $\mathcal{K}_i^\ell$ is also obtained from $\mathcal{K}$ by a similarity of ratio $4^{-\ell}$. Now note that $\mathcal{C} = \mathcal{C}/4 \bigcup (\mathcal{C}/4 + 3/4)$ implies that

$$\mathcal{K}(\lambda) = \mathcal{C} + \lambda \mathcal{C} = (\mathcal{C} + \frac{\lambda}{4}\mathcal{C}) \cup (\mathcal{C} + \frac{\lambda}{4}\mathcal{C} + \frac{3\lambda}{4})$$
$$= \mathcal{K}(\lambda/4) \cup (\mathcal{K}(\lambda/4) + \frac{3\lambda}{4}).$$

Thus $\mathcal{K}(\lambda)$ has measure zero if and only if $\mathcal{K}(\lambda/4)$ has measure zero. Hence it suffices to prove that $\mathcal{K}(\lambda)$ has measure zero for a.e. $\lambda \in [1, 4]$.

After these preliminaries let us observe that we immediately obtain that $m(\mathcal{K}(\lambda)) = 0$ for some special λ's, those for which the following **coincidence** takes place: for some ℓ and a pair i and i' with $i \neq i'$,

$$\mathcal{K}_i^\ell(\lambda) = \mathcal{K}_{i'}^\ell(\lambda).$$

Indeed, if we have this coincidence, then (8) gives

$$m(\mathcal{K}(\lambda)) \le \sum_{i=1,\ i \neq i'}^{4^\ell} m(\mathcal{K}_i^\ell(\lambda)) = (4^\ell - 1)4^{-\ell} m(\mathcal{K}(\lambda)),$$

and this implies $m(\mathcal{K}(\lambda)) = 0$.

The key insight below is that, in a quantitative sense, the λ's for which this coincidence takes place are "dense" relative to the size of ℓ. More precisely, we have the following.

Proposition 4.13 *Suppose λ_0 and ℓ are given, with $1 \le \lambda_0 \le 4$ and ℓ a positive integer. Then, there exist a $\bar{\lambda}$ and a pair i, i' with $i \neq i'$ such that*

(9) $$\mathcal{K}_i^\ell(\bar{\lambda}) = \mathcal{K}_{i'}^\ell(\bar{\lambda}) \quad \text{and} \quad |\bar{\lambda} - \lambda_0| \le c4^{-\ell}.$$

Here c is a constant independent of λ_0 and ℓ.

This is proved on the basis of the following observation.

Lemma 4.14 *For every λ_0 there is a pair $1 \le i_1, i_2 \le 4$, with $i_1 \neq i_2$ such that $\mathcal{K}_{i_1}(\lambda_0)$ and $\mathcal{K}_{i_2}(\lambda_0)$ intersect.*

Proof. Indeed, if the $\mathcal{K}_i$ are disjoint for $1 \leq i \leq 4$ then for sufficiently small δ the $\mathcal{K}_i^\delta$ are also disjoint. Here we have used the notation that F^δ denotes the set of points of distance less than δ from F. (See Lemma 3.1 in Chapter 1.) However, $\mathcal{K}^\delta = \bigcup_{i=1}^4 \mathcal{K}_i^\delta$, and by similarity $m(\mathcal{K}^{4\delta}) = 4m(\mathcal{K}_i^\delta)$. Thus by the disjointness of the $\mathcal{K}_i^\delta$ we have $m(\mathcal{K}^\delta) = m(\mathcal{K}^{4\delta})$, which is a contradiction, since $\mathcal{K}^{4\delta} - \mathcal{K}^\delta$ contains an open ball (of radius $3\delta/2$). The lemma is therefore proved.

Now apply the lemma for our given λ_0 and write $\mathcal{K}_{i_1} = \mathcal{C}_{\mu_1} + \lambda_0 \mathcal{C}_{\nu_1}$, $\mathcal{K}_{i_2} = \mathcal{C}_{\mu_2} + \lambda_0 \mathcal{C}_{\nu_2}$, where the μ's and ν's are either 1 or 2. However, since $i_1 \neq i_2$ we have $\mu_1 \neq \mu_2$ or $\nu_1 \neq \nu_2$ (or both). Assume for the moment that $\nu_1 \neq \nu_2$.

The fact that $\mathcal{K}_{i_1}(\lambda_0)$ and $\mathcal{K}_{i_2}(\lambda_0)$ intersect means that there are pairs of numbers (a, b) and (a', b'), with $a \in \mathcal{C}_{\mu_1}$, $b \in \mathcal{C}_{\nu_1}$, $a' \in \mathcal{C}_{\mu_2}$, and $b' \in \mathcal{C}_{\nu_2}$ such that

$$(10) \qquad\qquad a + \lambda_0 b = a' + \lambda_0 b'.$$

Note that the fact that $\nu_1 \neq \nu_2$ means that $|b - b'| \geq 1/2$. Next, looking at the ℓth generation we find via (7) that there are indices $1 \leq j_1, j_2, j_1', j_2' \leq 2^\ell$, so that $a \in \mathcal{C}_{j_1}^\ell \subset \mathcal{C}_{\mu_1}$, $b \in \mathcal{C}_{j_2}^\ell \subset \mathcal{C}_{\nu_1}$, $a' \in \mathcal{C}_{j_1'}^\ell \subset \mathcal{C}_{\mu_2}$, $b' \in \mathcal{C}_{j_2'}^\ell \subset \mathcal{C}_{\nu_2}$. We also observe that the above sets are translates of each other, that is, $\mathcal{C}_{j_1}^\ell = \mathcal{C}_{j_1'}^\ell + \tau_1$ and $\mathcal{C}_{j_2}^\ell = \mathcal{C}_{j_2'}^\ell + \tau_2$, with $|\tau_k| \leq 1$. Hence if i and i' correspond to the pairs (j_1, j_2) and (j_1', j_2'), respectively, we have

$$(11) \qquad \mathcal{K}_i^\ell(\lambda) = \mathcal{K}_{i'}^\ell(\lambda) + \tau(\lambda) \quad \text{with } \tau(\lambda) = \tau_1 + \lambda \tau_2.$$

Now let (A, B) be the pair that corresponds to (a', b') under the above translations, namely

$$(12) \qquad\qquad A = a' + \tau_1, \qquad B = b' + \tau_2.$$

We claim there is a $\overline{\lambda}$ such that

$$(13) \qquad\qquad A + \overline{\lambda} B = a' + \overline{\lambda} b'.$$

In fact, by (12) we have put B in $\mathcal{C}_{j_2}^\ell \subset \mathcal{C}_{\nu_1}$, while b' is in $\mathcal{C}_{j_2'}^\ell \subset \mathcal{C}_{\nu_2}$. Thus $|B - b'| \geq 1/2$, since $\nu_1 \neq \nu_2$. We can therefore solve (13) by taking $\overline{\lambda} = (A - a')/(b' - B)$. Now we compare this with (10), and get $\lambda_0 = (a - a')/(b' - b)$. Moreover, $|A - a| \leq 4^{-\ell}$ and $|B - b| \leq 4^{-\ell}$, since A and a both lie in $\mathcal{C}_{j_1}^\ell$, and B and b lie in $\mathcal{C}_{j_2}^\ell$. This yields the inequality

$$(14) \qquad\qquad |\overline{\lambda} - \lambda_0| \leq c4^{-\ell}.$$

Also, (12) and (13) clearly imply $\tau(\overline{\lambda}) = \tau_1 + \overline{\lambda}\tau_2 = 0$, and this together with (11) proves the coincidence.

Therefore our proposition is proved under the restriction we made earlier that $\nu_1 \neq \nu_2$. The situation where instead $\mu_1 \neq \mu_2$ is obtained from the case $\nu_1 \neq \nu_2$ if we replace λ_0 by λ_0^{-1}. Note that $\mathcal{K}_i^\ell(\lambda_0) = \mathcal{K}_{i'}^\ell(\lambda_0)$ if and only if $\mathcal{C}_{j_1}^\ell + \lambda_0 \mathcal{C}_{j_2}^\ell = \mathcal{C}_{j_1'}^\ell + \lambda_0 \mathcal{C}_{j_2'}^\ell$ and this is the same as $\mathcal{C}_{j_2}^\ell + \lambda_0^{-1} \mathcal{C}_{j_1}^\ell = \mathcal{C}_{j_2'}^\ell + \lambda_0^{-1} \mathcal{C}_{j_1'}^\ell$. This allows us to reduce to the case $\mu_1 \neq \mu_2$, since $\mathcal{C}_{j_1}^\ell \subset \mathcal{C}_{\mu_1}$ and $\mathcal{C}_{j_1'}^\ell \subset \mathcal{C}_{\mu_2}$. Here the fact that $1 \leq \lambda_0 \leq 4$ gives $\lambda_0^{-1} \leq 1$ and guarantees that the constant c in (9) can be taken to be independent of λ_0. The proposition is therefore established.

Note that as a consequence, the following holds near the points $\overline{\lambda}$ where the coincidence (9) takes place: If $|\lambda - \overline{\lambda}| \leq \epsilon 4^{-\ell}$, then

$$(15) \qquad \mathcal{K}_i^\ell(\lambda) = \mathcal{K}_{i'}^\ell(\lambda) + \tau(\lambda) \quad \text{with } |\tau(\lambda)| \leq \epsilon 4^{-\ell}.$$

In fact, this is (11) together with the observation that

$$|\tau(\lambda)| = |\tau(\lambda) - \tau(\overline{\lambda})| \leq |\lambda - \overline{\lambda}|,$$

since $|\tau(\lambda)| = \tau_1 + \lambda \tau_2$ and $|\tau_2| \leq 1$.

The assertion (15) leads to the following more elaborate version of itself:

> There is a set Λ of full measure such that whenever $\lambda \in \Lambda$ and $\epsilon > 0$ are given, there are ℓ and a pair i, i' so that (15) holds.[3]

Indeed, for fixed $\epsilon > 0$, let Λ_ϵ denote the set of λ that satisfies (15) for some ℓ, i and i'. For any interval I of length not exceeding 1, we have

$$m(\Lambda_\epsilon \cap I) \geq \epsilon 4^{-\ell} \geq c^{-1} \epsilon m(I),$$

because of (9) and (15). Thus Λ_ϵ^c has no points of Lebesgue density, hence Λ_ϵ^c has measure zero, and thus Λ_ϵ is a set of full measure. (See Corollary 1.5 in Chapter 3.) Since $\Lambda = \bigcap_\epsilon \Lambda_\epsilon$, and Λ_ϵ decreases with ϵ, we see that Λ also has full measure and our assertion is proved.

Finally, our theorem will be established once we show that $m(\mathcal{K}(\lambda)) = 0$ whenever $\lambda \in \Lambda$. To prove this, we assume contrariwise that $m(\mathcal{K}(\lambda)) > 0$. Using again the point of density argument, there must be for any

---

[3]The terminology that Λ has "full measure" means that its complement has measure zero.

$0 < \delta < 1$, a non-empty open interval I with $m(\mathcal{K}(\lambda) \cap I) \geq \delta m(I)$. We then fix δ with $1/2 < \delta < 1$ and proceed. With this fixed δ, we select ϵ used below as $\epsilon = m(I)(1 - \delta)$. Next, find ℓ, i, and i' for which (15) holds. The existence of such indices is guaranteed by the hypothesis that $\lambda \in \Lambda$.

We then consider the two similarities (of ratio $4^{-\ell}$) that map $\mathcal{K}(\lambda)$ to $\mathcal{K}_i^\ell(\lambda)$ and $\mathcal{K}_{i'}^\ell(\lambda)$, respectively. These take the interval I to corresponding intervals I_i and $I_{i'}$, respectively, with $m(I_i) = m(I_{i'}) = 4^{-\ell}m(I)$. Moreover,

$$m(\mathcal{K}_i^\ell \cap I_i) \geq \delta m(I_i) \quad \text{and} \quad m(\mathcal{K}_{i'}^\ell \cap I_{i'}) \geq \delta m(I_{i'}).$$

Also, as in (15), $I_{i'} = I_i + \tau(\lambda)$, with $|\tau(\lambda)| \leq \epsilon 4^{-\ell}$. This shows that

$$m(I_i \cap I_{i'}) \geq m(I_i) - \tau(\lambda) \geq 4^{-\ell}m(I) - \epsilon 4^{-\ell} \geq \delta m(I_i),$$

since $\epsilon 4^{-\ell} = (1 - \delta)m(I_i)$. Thus $m(I_i - I_i \cap I_{i'}) \leq (1 - \delta)m(I_i)$, and

$$\begin{aligned} m(\mathcal{K}_i^\ell \cap I_i \cap I_{i'}) &\geq m(\mathcal{K}_i^\ell \cap I_i) - m(I_i - I_i \cap I_{i'}) \\ &\geq (2\delta - 1)m(I_i) \\ &> \frac{1}{2}m(I_i) \geq \frac{1}{2}m(I_i \cap I_{i'}). \end{aligned}$$

So $m(\mathcal{K}_i^\ell \cap I_i \cap I_{i'}) > \frac{1}{2}m(I_i \cap I_{i'})$ and the same holds for i' in place of i. Hence $m(\mathcal{K}_i^\ell \cap \mathcal{K}_{i'}^\ell) > 0$, and this contradicts the decomposition (8) and the fact that $m(\mathcal{K}_i^\ell) = 4^{-\ell}m(\mathcal{K})$ for every i. Therefore we obtain that $m(\mathcal{K}(\lambda)) = 0$ for every $\lambda \in \Lambda$, and the proof of Theorem 4.12 is now complete.

5 Exercises

1. Show that the measure m_α is not σ-finite on $\mathbb{R}^d$ if $\alpha < d$.

2. Suppose E_1 and E_2 are two compact subsets of $\mathbb{R}^d$ such that $E_1 \cap E_2$ contains at most one point. Show directly from the definition of the exterior measure that if $0 < \alpha \leq d$, and $E = E_1 \cup E_2$, then

$$m_\alpha^*(E) = m_\alpha^*(E_1) + m_\alpha^*(E_2).$$

[Hint: Suppose $E_1 \cap E_2 = \{x\}$, let B_ϵ denote the open ball centered at x and of diameter ϵ, and let $E^\epsilon = E \cap B_\epsilon^c$. Show that

$$m_\alpha^*(E^\epsilon) \geq \mathcal{H}_\alpha^\epsilon(E) \geq m_\alpha^*(E) - \mu(\epsilon) - \epsilon^\alpha,$$

where $\mu(\epsilon) \to 0$. Hence $m_\alpha^*(E^\epsilon) \to m_\alpha^*(E)$.]

3. Prove that if $f : [0, 1] \to \mathbb{R}$ satisfies a Lipschitz condition of exponent $\gamma > 1$, then f is a constant.

4. Suppose $f : [0, 1] \to [0, 1] \times [0, 1]$ is surjective and satisfies a Lipschitz condition

$$|f(x) - f(y)| \leq C|x - y|^\gamma.$$

Prove that $\gamma \leq 1/2$ directly, without using Theorem 2.2.

[Hint: Divide $[0, 1]$ into N intervals of equal length. The image of each sub-interval is contained in a ball of volume $O(N^{-2\gamma})$, and the union of all these balls must cover the square.]

5. Let $f(x) = x^k$ be defined on $\mathbb{R}$, where k is a positive integer and let E be a Borel subset of $\mathbb{R}$.

   (a) Show that if $m_\alpha(E) = 0$ for some α, then $m_\alpha(f(E)) = 0$.

   (b) Prove that $\dim(E) = \dim f(E)$.

6. Let $\{E_k\}$ be a sequence of Borel sets in $\mathbb{R}^d$. Show that if $\dim E_k \leq \alpha$ for some α and all k, then

$$\dim \bigcup_k E_k \leq \alpha.$$

7. Prove that the $(\log 2/\log 3)$-Hausdorff measure of the Cantor set is precisely equal to 1.

[Hint: Suppose we have a covering of $\mathcal{C}$ by finitely many closed intervals $\{I_j\}$. Then there exists another covering of $\mathcal{C}$ by intervals $\{I_\ell'\}$ each of length 3^{-k} for some k, such that $\sum_j |I_j|^\alpha \geq \sum_\ell |I_\ell'|^\alpha \geq 1$, where $\alpha = \log 2/\log 3$.]

8. Show that the Cantor set of constant dissection, $\mathcal{C}_\xi$, in Exercise 3 of Chapter 1 has strict Hausdorff dimension $\log 2/\log(2/(1-\xi))$.

9. Consider the set $\mathcal{C}_{\xi_1} \times \mathcal{C}_{\xi_2}$ in $\mathbb{R}^2$, with $\mathcal{C}_\xi$ as in the previous exercise. Show that $\mathcal{C}_{\xi_1} \times \mathcal{C}_{\xi_2}$ has strict Hausdorff dimension $\dim(\mathcal{C}_{\xi_1}) + \dim(\mathcal{C}_{\xi_2})$.

10. Construct a Cantor-like set (as in Exercise 4, Chapter 1) that has Lebesgue measure zero, yet Hausdorff dimension 1.

[Hint: Choose $\ell_1, \ell_2, \ldots, \ell_k, \ldots$ so that $1 - \sum_{j=1}^k 2^{j-1} \ell_j$ tends to zero sufficiently slowly as $k \to \infty$.]

11. Let $\mathcal{D} = \mathcal{D}_\mu$ be the Cantor dust in $\mathbb{R}^2$ given as the product $\mathcal{C}_\xi \times \mathcal{C}_\xi$, with $\mu = (1 - \xi)/2$.

(a) Show that for any real number λ, the set $\mathcal{C}_\xi + \lambda \mathcal{C}_\xi$ is similar to the projection of $\mathcal{D}$ on the line in $\mathbb{R}^2$ with slope $\lambda = \tan\theta$.

(b) Note that among the Cantor sets $\mathcal{C}_\xi$, the value $\xi = 1/2$ is critical in the construction of the Besicovitch set in Section 4.4. In fact, prove that with $\xi > 1/2$, then $\mathcal{C}_\xi + \lambda \mathcal{C}_\xi$ has Lebesgue measure zero for *every* λ. See also Problem 10 below.

[Hint: $m_\alpha(\mathcal{C}_\xi + \lambda \mathcal{C}_\xi) < \infty$ for $\alpha = \dim \mathcal{D}_\mu$.]

12. Define a primitive one-dimensional "measure" $\tilde{m}_1$ as

$$\tilde{m}_1 = \inf \sum_{k=1}^\infty \operatorname{diam} F_k, \qquad E \subset \bigcup_{k=1}^\infty F_k.$$

This is akin to the one-dimensional exterior measure m_α^*, $\alpha = 1$, except that no restriction is placed on the size of the diameters F_k.

Suppose I_1 and I_2 are two *disjoint* unit segments in $\mathbb{R}^d$, $d \geq 2$, with $I_1 = I_2 + h$, and $|h| < \epsilon$. Then observe that $\tilde{m}_1(I_1) = \tilde{m}_1(I_2) = 1$, while $\tilde{m}_1(I_1 \cup I_2) \leq 1 + \epsilon$. Thus

$$\tilde{m}_1(I_1 \cup I_2) < \tilde{m}_1(I_1) + \tilde{m}_1(I_2) \qquad \text{when } \epsilon < 1;$$

hence $\tilde{m}_1$ fails to be additive.

13. Consider the von Koch curve $\mathcal{K}^\ell$, $1/4 < \ell < 1/2$, as defined in Section 2.1. Prove for it the analogue of Theorem 2.7: the function $t \mapsto \mathcal{K}^\ell(t)$ satisfies a Lipschitz condition of exponent $\gamma = \log(1/\ell)/\log 4$. Moreover, show that the set $\mathcal{K}^\ell$ has strict Hausdorff dimension $\alpha = 1/\gamma$.

[Hint: Show that if $\mathcal{O}$ is the shaded open triangle indicated in Figure 14, then $\mathcal{O} \supset S_0(\mathcal{O}) \cup S_1(\mathcal{O}) \cup S_2(\mathcal{O}) \cup S_3(\mathcal{O})$, where $S_0(x) = \ell x$, $S_1(x) = \rho_\theta(\ell x) + a$, $S_2(x) = \rho_\theta^{-1}(\ell x) + c$, and $S_3(x) = \ell x + b$, with ρ_θ the rotation of angle θ. Note that the sets $S_j(\mathcal{O})$ are disjoint.]

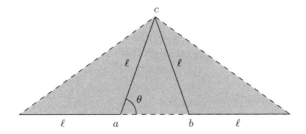

Figure 14. The open set $\mathcal{O}$ in Exercise 13

14. Show that if $\ell < 1/2$, the von Koch curve $t \mapsto \mathcal{K}^\ell(t)$ in Exercise 13 is a simple curve.

[Hint: Observe that if $t = \sum_{j=1}^{\infty} a_j/4^j$, with $a_j = 0, 1, 2,$ or 3, then

$$\{\mathcal{K}(t)\} = \bigcap_{j=1}^{\infty} S_{a_j}\left(\cdots S_{a_2}\left(S_{a_1}(\overline{\mathcal{O}})\right)\right).]$$

15. Note that if we take $\ell = 1/2$ in the definition of the von Koch curve in Exercise 13 we get a "space-filling" curve, one that fills the right triangle whose vertices are $(0,0)$, $(1,0)$, and $(1/2, 1/2)$. The first three steps of the construction are as in Figure 15, with the intervals traced out in the indicated order.

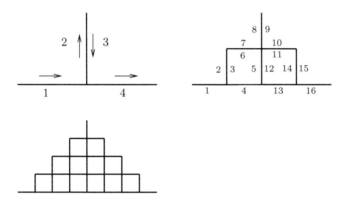

Figure 15. The first three steps of the von Koch curve when $\ell = 1/2$

16. Prove that the von Koch curve $t \mapsto \mathcal{K}^{\ell}(t)$, $1/4 < \ell \leq 1/2$ is continuous but nowhere differentiable.

[Hint: If $\mathcal{K}'(t)$ exists for some t, then

$$\lim_{n \to \infty} \frac{\mathcal{K}(u_n) - \mathcal{K}(v_n)}{u_n - v_n}$$

must exist, where $u_n \leq t \leq v_n$, and $u_n - v_n \to 0$. Choose $u_n = k/4^n$ and $v_n = (k+1)/4^n$.]

17. For a compact set E in $\mathbb{R}^d$, define $\#(\epsilon)$ to be the least number of balls of radius ϵ that cover E. Note that we always have $\#(\epsilon) = O(\epsilon^{-d})$ as $\epsilon \to 0$, and $\#(\epsilon) = O(1)$ if E is finite.

One defines the **covering dimension** of E, denoted by $\dim_C(E)$, as $\inf \beta$ such that $\#(\epsilon) = O(\epsilon^{-\beta})$, as $\epsilon \to 0$. Show that $\dim_C(E) = \dim_M(E)$, where $\dim_M$ is the Minkowski dimension discussed in Section 2.1, by proving the following inequalities for all $\delta > 0$:

(i) $m(E^\delta) \le c \#(\delta) \delta^d$.

(ii) $\#(\delta) \le c' m(E^\delta) \delta^{-d}$.

[Hint: To prove (ii), use Lemma 1.2 in Chapter 3 to find a collection of disjoint balls $B_1, B_2, \ldots, B_N$ of radius $\delta/3$, each centered at E, such that their "triples" $\tilde{B}_1, \tilde{B}_2, \ldots, \tilde{B}_N$ (of radius δ) cover E. Then $\#(\delta) \le N$, while $Nm(B_j) = cN\delta^d \le m(E^\delta)$, since the balls B_j are disjoint and are contained in E^δ.]

18. Let E be a compact set in $\mathbb{R}^d$.

(a) Prove that $\dim(E) \le \dim_M(E)$, where dim and $\dim_M$ are the Hausdorff and Minkowski dimensions, respectively.

(b) However, prove that if $E = \{0, 1/\log 2, 1/\log 3, \ldots, 1/\log n, \ldots\}$, then $\dim_M E = 1$, yet $\dim E = 0$.

19. Show that there is a constant c_d, dependent only on the dimension d, such that whenever E is a compact set,

$$m(E^{2\delta}) \le c_d m(E^\delta).$$

[Hint: Consider the maximal function f^*, with $f = \chi_{E^\delta}$, and take $c_d = 6^d$.]

20. Show that if F is the self-similar set considered in Theorem 2.12, then it has the same Minkowski dimension as Hausdorff dimension.

[Hint: Each F_k is the union of m^k balls of radius cr^k. In the converse direction one sees by Lemma 2.13 that if $\epsilon = r^k$, then each ball of radius ϵ can contain at most c' vertices of the kth generation. So it takes at least m^k/c' such balls to cover F.]

21. From the unit interval, remove the second and fourth quarters (open intervals). Repeat this process in the remaining two closed intervals, and so on. Let F be the limiting set, so that

$$F = \{x : x = \sum_{k=1}^{\infty} a_k/4^k \quad a_k = 0 \text{ or } 2\}.$$

Prove that $0 < m_{1/2}(F) < \infty$.

22. Suppose F is the self-similar set arising in Theorem 2.9.

(a) Show that if $m \le 1/r^d$, then $m_d(F_i \cap F_j) = 0$ if $i \ne j$.

(b) However, if $m \ge 1/r^d$, prove that $F_i \cap F_j$ is not empty for some $i \ne j$.

(c) Prove that under the hypothesis of Theorem 2.12

$$m_\alpha(F_i \cap F_j) = 0, \quad \text{with } \alpha = \log m/\log(1/r), \text{ whenever } i \ne j.$$

23. Suppose $S_1, \ldots, S_m$ are similarities with ratio r, $0 < r < 1$. For each set E, let

$$\tilde{S}(E) = S_1(E) \cup \cdots \cup S_m(E),$$

and suppose F denotes the unique non-empty compact set with $\tilde{S}(F) = F$.

(a) If $\bar{x} \in F$, show that the set of points $\{\tilde{S}^n(\bar{x})\}_{n=1}^\infty$ is dense in F.

(b) Show that F is **homogeneous** in the following sense: if $x_0 \in F$ and B is any open ball centered at x_0, then $F \cap B$ contains a set similar to F.

24. Suppose E is a Borel subset of $\mathbb{R}^d$ with $\dim E < 1$. Prove that E is totally disconnected, that is, any two distinct points in E belong to different connected components.

[Hint: Fix $x, y \in E$, and show that $f(t) = |t - x|$ is Lipschitz of order 1, and hence $\dim f(E) < 1$. Conclude that $f(E)$ has a dense complement in $\mathbb{R}$. Pick r in the complement of $f(E)$ so that $0 < r < f(y)$, and use the fact that $E = \{t \in E : |t - x| < r\} \cup \{t \in E : |t - x| > r\}$.]

25. Let $F(t)$ be an arbitrary non-negative measurable function on $\mathbb{R}$, and $\gamma \in S^{d-1}$. Then there exists a measurable set E in $\mathbb{R}^d$, such that $F(t) = m_{d-1}(E \cap \mathcal{P}_{t,\gamma})$.

26. Theorem 4.1 can be refined for $d \geq 4$ as follows.
   Define $C^{k,\alpha}$ to be the class of functions $F(t)$ on $\mathbb{R}$ that are C^k and for which $F^{(k)}(t)$ satisfies a Lipschitz condition of exponent α.
   If E has finite measure, then for a.e. $\gamma \in S^{d-1}$ the function $m(E \cap \mathcal{P}_{t,\gamma})$ is in $C^{k,\alpha}$ for $k = (d-3)/2$, $\alpha < 1/2$, if d is odd, $d \geq 3$; and for, $k = (d-4)/2$, $\alpha < 1$, if d is even, $d \geq 4$.

27. Show that the modification of the inequality (2) of Theorem 4.5 fails if we drop $\|f\|_{L^2(\mathbb{R}^d)}$ from the right-hand side.

[Hint: Consider $\mathcal{R}^*(f_\epsilon)$, with f_ϵ defined by $f_\epsilon(x) = (|x| + \epsilon)^{-d+\delta}$, for $|x| \leq 1$, with δ fixed, $0 < \delta < 1$, and $\epsilon \to 0$.]

28. Construct a compact set $E \subset \mathbb{R}^d$, $d \geq 3$, such that $m_d(E) = 0$, yet E contains translates of any segment of unit length in $\mathbb{R}^d$. (While particular examples of such sets can be easily obtained from the case $d = 2$, the determination of the least Hausdorff dimension among all such sets is an open problem.)

6 Problems

1. Carry out the construction below of two sets U and V so that

$$\dim U = \dim V = 0 \quad \text{but} \quad \dim(U \times V) \geq 1.$$

Let $I_1, \ldots, I_n, \ldots$ be given as follows:

- Each I_j is a finite sequence of consecutive positive integers; that is, for all j

$$I_j = \{n \in \mathbb{N} : A_j \le n \le B_j\} \quad \text{for some given } A_j \text{ and } B_j.$$

- For each j, I_{j+1} is to the right of I_j; that is, $A_{j+1} > B_j$.

Let $U \subset [0, 1]$ consist of all x which when written dyadically $x = .a_1 a_2 \cdots a_n \cdots$ have the property that $a_n = 0$ whenever $n \in \bigcup_j I_j$. Assume also that A_j and B_j tend to infinity (as $j \to \infty$) rapidly enough, say $B_j/A_j \to \infty$ and $A_{j+1}/B_j \to \infty$. Also, let J_j be the complementary blocks of integers, that is,

$$J_j = \{n \in \mathbb{N} : B_j < n < A_{j+1}\}.$$

Let $V \subset [0, 1]$ consist of those $x = .a_1 a_2 \cdots a_n \cdots$ with $a_n = 0$ if $n \in \bigcup_j J_j$.
  Prove that U and V have the desired property.

2.* The iso-diametric inequality states the following: If E is a bounded subset of $\mathbb{R}^d$ and diam $E = \sup\{|x - y| : x, y \in E\}$, then

$$m(E) \le v_d \left(\frac{\text{diam } E}{2} \right)^d,$$

where v_d denotes the volume of the unit ball in $\mathbb{R}^d$. In other words, among sets of a given diameter, the ball has maximum volume. Clearly, it suffices to prove the inequality for $\overline{E}$ instead of E, so we can assume that E is compact.

(a) Prove the inequality in the special case when E is symmetric, that is, $-x \in E$ whenever $x \in E$.

In general, one reduces to the symmetric case by using a technique called Steiner symmetrization. If e is a unit vector in $\mathbb{R}^d$, and $\mathcal{P}$ is a plane perpendicular to e, the Steiner symmetrization of E with respect to E is defined by

$$S(E, e) = \{x + te : x \in \mathcal{P}, \ |t| \le \frac{1}{2} L(E; e; x)\},$$

where $L(E; e; x) = m(\{t \in \mathbb{R} : \ x + t \cdot e \in E\})$, and m denotes the Lebesgue measure. Note that $x + te \in S(E, e)$ if and only if $x - te \in S(E, e)$.

(b) Prove that $S(E, e)$ is a bounded measurable subset of $\mathbb{R}^d$ that satisfies $m(S(E, e)) = m(E)$.

  [Hint: Use Fubini's theorem.]

(c) Show that diam $S(E, e) \le$ diam E.

(d) If ρ is a rotation that leaves E and $\mathcal{P}$ invariant, show that $\rho S(E, e) = S(E, e)$.

(e) Finally, consider the standard basis $\{e_1, \ldots, e_d\}$ of $\mathbb{R}^d$. Let $E_0 = E$, $E_1 = S(E_0, e_1)$, $E_2 = S(E_1, e_2)$, and so on. Use the fact that E_d is symmetric to prove the iso-diametric inequality.

(f) Use the iso-diametric inequality to show that $m(E) = \frac{v_d}{2^d} m_d(E)$ for any Borel set E in $\mathbb{R}^d$.

3. Suppose S is a similarity.

(a) Show that S maps a line segment to a line segment.

(b) Show that if L_1 and L_2 are two segments that make an angle α, then $S(L_1)$ and $S(L_2)$ make an angle α or $-\alpha$.

(c) Show that every similarity is a composition of a translation, a rotation (possibly improper), and a dilation.

4.* The following gives a generalization of the construction of the Cantor-Lebesgue function.

Let F be the compact set in Theorem 2.9 defined in terms of m similarities $S_1, S_2, \ldots, S_m$ with ratio $0 < r < 1$. There exists a unique Borel measure μ supported on F such that $\mu(F) = 1$ and

$$\mu(E) = \frac{1}{m} \sum_{j=1}^{m} \mu(S_j^{-1}(E)) \qquad \text{for any Borel set } E.$$

In the case when F is the Cantor set, the Cantor-Lebesgue function is $\mu([0, x])$.

5. Prove a theorem of Hausdorff: Any compact subset K of $\mathbb{R}^d$ is a continuous image of the Cantor set $\mathcal{C}$.

[Hint: Cover K by 2^{n_1} (some n_1) open balls of radius 1, say $B_1, \ldots, B_\ell$ (with possible repetitions). Let $K_{j_1} = K \cap \overline{B_{j_1}}$ and cover each K_{j_1} with 2^{n_2} balls of radius $1/2$ to obtain compact sets K_{j_1, j_2}, and so on. Express $t \in \mathcal{C}$ as a ternary expansion, and assign to t a unique point in K defined by the intersection $K_{j_1} \cap K_{j_1, j_2} \cap \cdots$ for appropriate $j_1, j_2, \ldots$. To prove continuity, observe that if two points in the Cantor set are close, then their ternary expansions agree to high order.]

6. A compact subset K of $\mathbb{R}^d$ is **uniformly locally connected** if given $\epsilon > 0$ there exists $\delta > 0$ so that whenever $x, y \in K$ and $|x - y| < \delta$, there is a continuous curve γ in K joining x to y, such that $\gamma \subset B_\epsilon(x)$ and $\gamma \subset B_\epsilon(y)$.

Using the previous problem, one can show that a compact subset K of $\mathbb{R}^d$ is the continuous image of the unit interval $[0, 1]$ if and only if K is uniformly locally connected.

7. Formulate and prove a generalization of Theorem 3.5 to the effect that once appropriate sets of measure zero are removed, there is a measure-preserving isomorphism of the unit interval in $\mathbb{R}$ and the unit cube in $\mathbb{R}^d$.

8.* There exists a *simple* continuous curve in the plane of positive two-dimensional measure.

9. Let E be a compact set in $\mathbb{R}^{d-1}$. Show that $\dim(E \times I) = \dim(E) + 1$, where I is the unit interval in $\mathbb{R}$.

10.* Let $\mathcal{C}_\xi$ be the Cantor set considered in Exercises 8 and 11. If $\xi < 1/2$, then $\mathcal{C}_\xi + \lambda \mathcal{C}_\xi$ has positive Lebesgue measure for almost every λ.

Notes and References

There are several excellent books that cover many of the subjects treated here. Among these texts are Riesz and Nagy [27], Wheeden and Zygmund [33], Folland [13], and Bruckner *et al.* [4].

Introduction
The citation is a translation of a passage in a letter from Hermite to Stieltjes [18].

Chapter 1
The citation is a translation from the French of a passage in [3].

We refer to Devlin [7] for more details about the axiom of choice, Hausdorff maximal principle, and well-ordering principle.

See the expository paper of Gardner [14] for a survey of results regarding the Brunn-Minkowski inequality.

Chapter 2
The citation is a passage from the preface to the first edition of Lebesgue's book on integration [20].

Devlin [7] contains a discussion of the continuum hypothesis.

Chapter 3
The citation is from Hardy and Littlewood's paper [15].

Hardy and Littlewood proved Theorem 1.1 in the one-dimensional case by using the idea of rearrangements. The present form is due to Wiener.

Our treatment of the isoperimetric inequality is based on Federer [11]. This work also contains significant generalizations and much additional material on geometric measure theory.

A proof of the Besicovitch covering in the lemma in Problem 3* is in Mattila [22].

For an account of functions of bounded variations in $\mathbb{R}^d$, see Evans and Gariepy [8].

An outline of the proof of Problem 7 (b)* can be found at the end of Chapter 5 in Book I.

The result in part (b) of Problem 8* is a theorem of S. Saks, and its proof as a consequence of part (a) can be found in Stein [31].

Chapter 4
The citation is translated from the introduction of Plancherel's article [25].

An account of the theory of almost periodic functions which is touched upon in Problem 2* can be found in Bohr [2].

The results in Problems 4* and 5* are in Zygmund [35], in Chapters V and VII, respectively.

Consult Birkhoff and Rota [1] for more on Sturm-Liouville systems, Legendre polynomials, and Hermite functions.

Chapter 5

See Courant [6] for an account of the Dirichlet principle and some of its applications. The solution of the Dirichlet problem for general domains in $\mathbb{R}^2$ and the related notion of logarithmic capacity of sets are treated in Ransford [26]. Folland [12] contains another solution to the Dirichlet problem (valid in $\mathbb{R}^d$, $d \geq 2$) by methods which do not use the Dirichlet principle.

The result regarding the existence of the conformal mapping stated in Problem 3* is in Chapter VII of Zygmund [35].

Chapter 6

The citation is a translation from the German of a passage in C. Carathéodory [5].

Petersen [24] gives a systematic presentation of ergodic theory, including a proof of the theorem in Problem 7*.

The facts about spherical harmonics needed in Problem 4* can be found in Chapter 4 in Stein and Weiss [32].

We refer to Hardy and Wright [16] for an introduction to continued fractions. Their connection to ergodic theory is discussed in Ryll-Nardzewski [28].

Chapter 7

The citation is a translation from the German of a passage in Hausdorff's article [17], while Mandelbrot's citation is from his book [21].

Mandelbrot's book also contains many interesting examples of fractals arising in a variety of different settings, including a discussion of Richardson's work on the length of coastlines. (See in particular Chapter 5.)

Falconer [10] gives a systematic treatment of fractals and Hausdorff dimension.

We refer to Sagan [29] for further details on space-filling curves, including the construction of a curve arising in Problem 8*.

The monograph of Falconer [10] also contains an alternate construction of the Besicovitch set, as well as the fact that such sets must necessarily have dimension two. The particular Besicovitch set described in the text appears in Kahane [19], but the fact that it has measure zero required further ideas which are contained, for instance, in Peres *et al.* [30].

Regularity of sets in $\mathbb{R}^d$, $d \geq 3$, and the estimates for the maximal function associated to the Radon transform are in Falconer [9], and Oberlin and Stein [23].

The theory of Besicovitch sets in higher dimensions, as well as a number of interesting related topics can be found in the survey of Wolff [34].

Bibliography

[1] G. Birkhoff and G. C. Rota. *Ordinary differential equations*. Wiley, New York, 1989.

[2] H. A. Bohr. *Almost periodic functions*. Chelsea Publishing Company, New York, 1947.

[3] E. Borel. *Leçons sur la théorie des fonctions*. Gauthiers-Villars, Paris, 1898.

[4] J. B. Bruckner, A. M. Bruckner, and B. S. Thomson. *Real Analysis*. Prentice Hall, Upper Saddle River, NJ, 1997.

[5] C. Carathéodory. *Vorlesungen über reelle Funktionen*. Leipzig, Berlin, B. G. Teubner, Leipzig and Berlin, 1918.

[6] R. Courant. *Dirichlet's principle, conformal mappings, and minimal surfaces*. Interscience Publishers, New York, 1950.

[7] K. J. Devlin. *The joy of sets: fundamentals of contemporary set theory*. Springer-Verlag, New York, 1997.

[8] L. C. Evans and R. F. Gariepy. *Measure theory and fine properties of functions*. CRC Press, Boca Raton, 1992.

[9] K. J. Falconer. Continuity properties of k-plane integrals and Besicovitch sets. *Math. Proc. Cambridge Philos. Soc*, 87:221–226, 1980.

[10] K. J. Falconer. *The geometry of fractal sets*. Cambridge University Press, 1985.

[11] H. Federer. *Geometric measure theory*. Springer, Berlin and New York, 1996.

[12] G. B. Folland. *Introduction to partial differential equations*. Princeton University Press, Princeton, NJ, second edition, 1995.

[13] G. B. Folland. *Real Analysis: modern techniques and their applications*. Wiley, New York, second edition, 1999.

[14] R. J. Gardner. The Brunn-Minkowski inequality. *Bull. Amer. Math. Soc*, 39:355–405, 2002.

[15] G. H. Hardy and J. E. Littlewood. A maximal theorem with function theoretic applications. *Acta. Math*, 54:81–116, 1930.

[16] G. H. Hardy and E. M. Wright. *An introduction to the Theory of Numbers*. Oxford University Press, London, fifth edition, 1979.

[17] F. Hausdorff. Dimension und äusseres Mass. *Math. Annalen*, 79:157–179, 1919.

[18] C. Hermite. *Correspondance d'Hermite et de Stieltjes*. Gauthier-Villars, Paris, 1905. Edited by B. Baillaud and H. Bourget.

[19] J. P. Kahane. Trois notes sur les ensembles parfaits linéaires. *Enseignement Math.*, 15:185–192, 1969.

[20] H. Lebesgue. *Leçons sur l'integration et la recherche des fonctions primitives*. Gauthier-Villars, Paris, 1904. Preface to the first edition.

[21] B. B. Mandelbrot. *The fractal geometry of nature*. W. H. Freeman, San Francisco, 1982.

[22] P. Mattila. *Geometry of sets and measures in Euclidean spaces*. Cambridge University Press, Cambridge, 1995.

[23] D. M. Oberlin and E. M. Stein. Mapping properties of the Radon transform. *Indiana Univ. Math. J*, 31:641–650, 1982.

[24] K. E. Petersen. *Ergodic theory*. Cambridge University Press, Cambridge, 1983.

[25] M. Plancherel. La théorie des équations intégrales. *L'Enseignement math.*, 14e Année:89–107, 1912.

[26] T. J. Ransford. *Potential theory in the complex plane*. London Mathematical Society student texts, 28. Cambridge, New York: Press Syndicate of the University of Cambridge, 1995.

[27] F. Riesz and B. Sz.-Nagy. *Functional Analysis*. New York, Ungar, 1955.

[28] C. Ryll-Nardzewski. On the ergodic theorem. ii. Ergodic theory of continued fractions. *Studia Math.*, 12:74–79, 1951.

[29] H. Sagan. *Space-filling curves*. Universitext. Springer-Verlag, New York, 1994.

[30] Y. Peres, K. Simon, and B. Solomyak. Fractals with positive length and zero Buffon needle probability. *Amer. Math. Monthly*, 110:314–325, 2003.

[31] E. M. Stein. *Harmonic analysis: real-variable methods, orthogonality, and oscillatory integrals.* Princeton University Press, Princeton, NJ, 1993.

[32] E. M. Stein and G. Weiss. *Introduction to Fourier Analysis on Euclidean Spaces.* Princeton University Press, Princeton, NJ, 1971.

[33] R. L. Wheeden and A. Zygmund. *Measure and integral: an introduction to real analysis.* Marcel Dekker, New York, 1977.

[34] T. Wolff. Recent work connected with the Kakeya problem. *Prospects in Mathematics, Princeton, NJ*, 31:129–162, 1996. Amer. Math. Soc., Providence, RI, 1999.

[35] A. Zygmund. *Trigonometric Series*, volume I and II. Cambridge University Press, Cambridge, second edition, 1959. Reprinted 1993.

Symbol Glossary

The page numbers on the right indicate the first time the symbol or notation is defined or used. As usual, $\mathbb{Z}$, $\mathbb{Q}$, $\mathbb{R}$, and $\mathbb{C}$ denote the integers, the rationals, the reals, and the complex numbers respectively.

Index

Relevant items that also arose in Book I or Book II are listed in this index, preceded by the numerals I or II, respectively.